Lecture Notes in Civil Engineering

Volume 623

Lecture Notes in Civil Engineering (LNCE) publishes the latest developments in Civil Engineering—quickly, informally, and in top quality. Though original research reported in proceedings and post-proceedings represents the core of LNCE, edited volumes of exceptionally high quality and interest may also be considered for publication. Volumes published in LNCE embrace all aspects and subfields of, as well as new challenges in, Civil Engineering. Topics in the series include:

- Construction and Structural Mechanics
- Building Materials
- Concrete, Steel and Timber Structures
- Geotechnical Engineering
- Earthquake Engineering
- Coastal Engineering
- Ocean and Offshore Engineering; Ships and Floating Structures
- Hydraulics, Hydrology and Water Resources Engineering
- Environmental Engineering and Sustainability
- Structural Health and Monitoring
- Surveying and Geographical Information Systems
- Indoor Environments
- Transportation and Traffic
- Risk Analysis
- Safety and Security

To submit a proposal or request further information, please contact the appropriate Springer Editor:

- Pierpaolo Riva at pierpaolo.riva@springer.com (Europe and Americas);
- Swati Meherishi at swati.meherishi@springer.com (Asia—except China, Australia, and New Zealand);
- Wayne Hu at wayne.hu@springer.com (China).

All books in the series now indexed by Scopus and EI Compendex database!

Indrajit Pal · Sangam Shrestha · Sarawut Ninsawat ·
Sreevalsa Kolathayar
Editors

Proceedings of the 3rd International Symposium on Disaster Resilience and Sustainable Development—Volume 1

Disaster and Climate Resilience

Springer

Editors
Indrajit Pal
Disaster Preparedness, Mitigation and Management (DPMM)
Asian Institute of Technology
Khlong Luang, Pathum Thani, Thailand

Sarawut Ninsawat
Remote Sensing and Geographic Information Systems
Asian Institute of Technology
Khlong Luang, Pathum Thani, Thailand

Sangam Shrestha
School of Engineering and Technology
Water Engineering and Management (WEM)
Asian Institute of Technology
Khlong Luang, Pathum Thani, Thailand

Sreevalsa Kolathayar
Department of Civil Engineering
National Institute of Technology Karnataka
Mangaluru, Karnataka, India

ISSN 2366-2557 ISSN 2366-2565 (electronic)
Lecture Notes in Civil Engineering
ISBN 978-981-96-2138-5 ISBN 978-981-96-2139-2 (eBook)
https://doi.org/10.1007/978-981-96-2139-2

This Springer imprint is published by the registered company Springer Nature Singapore Pte Ltd.
The registered company address is: 152 Beach Road, #21-01/04 Gateway East, Singapore 189721, Singapore

Preface

The increasing frequency and severity of natural hazards necessitate nuanced approaches and actions to manage future risks. This proceeding volume explores the multifaceted relationship between disaster risk management, climate change adaptation, resilience building, and sustainable development across various countries in the Asia Pacific. Through 18 chapters, the book delves into diverse sectors, including heritage conservation, land-use planning, urban development, sustainable transportation, business enterprises, legislation, and private sector involvement.

The conclusion underscores the interconnectedness of disaster risk management, public policy, community empowerment, and scientific knowledge. Locally-led actions, community-driven policies, and capacity building emerge as key recommendations. The book advocates for multi-stakeholder collaboration and communication to document and share learnings across the region, enhancing the effectiveness of actions and guiding policymakers and development practitioners in understanding the multiple aspects of disaster and climate resilience.

The proceeding volume is an outcome of the "3rd International Symposium on Disaster Resilience and Sustainable Development" organized by Disaster Preparedness, Mitigation and Management, Asian Institute of Technology, Thailand, and brings together research works from some of the critical issues of application of science, technology and public policy for disaster and climate resilience across the region. The proceeding volume consolidates the research findings, lessons, and learnings from implementation of various resilience and sustainability initiatives, providing valuable ground realities and insights for future actions.

Khlong Nueng, Thailand — Indrajit Pal
Khlong Nueng, Thailand — Sangam Shrestha
Khlong Nueng, Thailand — Sarawut Ninsawat
Suratkal, India — Sreevalsa Kolathayar

Contents

Disaster and Climate Resilience for Sustainable Development: Initiatives and Lessons Learned 1
Indrajit Pal, Satya Venkata Sai Aditya Bharadwaz Ganni, Ayush Baskota, and Jeeten Kumar

Analyzing Governments' Policies on Disaster Risk Management in the Borobudur Cultural Heritage Region, Central Java, Indonesia 15
Pusparini Dharma Putri and Lathifa Nursyamsu

Challenges of Sustainable Business Organization: Case Study of Young Craft Business Owner in Nepal 29
Richan Shrestha, Yurisha Upadhyaya, Prakriti Dangol, Sima Tamang, Fura Gelzen Sherpa, and Prakash C. Bhattarai

Constitutional Law Provisions Towards Disaster Management in Thailand 43
Wachiraporn Poungjinda and Shubham Pathak

Systematic Review of Community Resilience Framework Related to Disasters 59
Deepak Marahatta, Hari Devkota, and Shankar Tiwari

Drought Adaptation in Thailand: A National Governmental Perspective 73
Puvadol Doydee and Jessada Techamahasaranont

The Impact on PM2.5 in Thailand Society: An Investigation into Causes, Challenges, and Policy Gap in Implementations 87
Vasuthida Pinrattana and Oulavanh Sinsamphanh

Labor Law Implications on Disaster Susceptible Small and Medium Enterprises in Thailand 99
Weerayut Lasongyang and Shubham Pathak

Male Labor Migration and Women Empowerment: A Review Paper ... 113
Hari Devkota, Deepak Marahatta, Shankar Tiwari, and Bhim Raj Suwal

Compensation for Losses and Damages Caused by Climate Change: Challenges and Progress ... 129
Pratik Poudel and Mokbul Moshed Ahmad

Evaluating the Effectiveness of Sustainable Transportation Policies in Mitigating Climate Change in Metro Manila, Philippines: The Public Utility Vehicle Modernization Program (PUVMP) ... 139
Tabassam Raza, Hannah Barbara Gonzalez, Maria Belen Vasquez Cresp, Carmelita R. E. U. Liwag, Zuhura Mtoi Kanyawanah, Dayanara Bermudez, Bassam Mohammed Zainudeen, Mark Anthony M. Morales, Shaker Mahmood Mayo, and Dina C. Magnaye

Indigenous Knowledge of Afro-Asian Tribal Communities: An Alternative Science of Natural Disaster Management ... 165
Tapas Pal

Resilience Starts Locally: Enhancing Sustainability Through Localization in Humanitarian Response ... 193
Aniruddha Dey, Sheikh Tawhidul Islam, Swarnabha Bandyopadhaya, Nandini Sanyal, Al Jubaer, and Dipanjan Mishra

Disaster and Climate Resilience in Quezon City: Evaluating Local Development Planning for Sustainable Development Goals ... 205
Tabassam Raza, Indrajit Pal, Carmelita R. E. U. Liwag, Jade S. Agapinan, Isabella M. Prodigalidad, Danica Marie V. Arriesgado, Abegail Rose L. Valenzuela, Bassam Mohammed Zainudeen, Julia Angela Mae E. Collado, and Mark Anthony M. Morales

Identifying Challenges to the Optimal Utilization of Early Warning Systems for Flood Management in Metro Manila: Advancing SDGs 11 and 13 ... 221
Tabassam Raza, Krishna Buenaventura, Maryam Mazouz, Carmelita R. E. U. Liwag, Emmanuel Yasin, Jean Baptist Muhoza, Darling Ximena Velasco Caicedo, Mark Anthony M. Morales, and Dina C. Magnaye

Private Sector Engagement for Community Resilience and Sustainable Development ... 245
Kilian T. Murphy and Edwin M. Salonga

Land Readjustment Approach as an Urban Development Tool Towards Disaster Resilience: A Study of Japan's Experience and Its Applicability in the Philippines 259
Arlene D. R. Santiago and Tabassam Raza

Role of Micro-Finance in Poverty Reduction: A Case Study from Neelkantha Municipality, Nepal 289
Shiva Raj Khanal, Romi Ghimire, and Pradeep Shrestha

Disaster and Climate Resilience for Sustainable Development: Initiatives and Lessons Learned

Indrajit Pal, Satya Venkata Sai Aditya Bharadwaz Ganni, Ayush Baskota, and Jeeten Kumar

Abstract This chapter investigates the essential function of resilience in disaster risk reduction and climate adaptation within the built environments of the Asia–Pacific region. The escalating impacts of climate change are increasing the frequency and severity of extreme events, posing significant risks to ecological systems and infrastructure stability. Establishing resilient infrastructure is critical in mitigating vulnerabilities, allowing systems to adapt, endure shocks, and recover effectively. It is paramount to adopt a socio-technical perspective, which integrates scientific knowledge with local expertise to strengthen climate resilience. A series of case studies included highlight adaptive strategies to address specific regional challenges such as urban heat stress, water security, and infrastructure resilience against seismic and extreme weather events. The authors advocate for inclusive, governance-led approaches and emphasize innovative practices, including green infrastructure, climate-smart agriculture, and smart urban planning, to align resilience objectives with sustainable development goals. This work offers a comprehensive framework for bolstering climate resilience across urban and rural settings, providing valuable insights for policymakers, practitioners, and communities.

Keywords Disaster risk reduction · Sustainable development · Resilient infrastructure · Built environment · Vulnerability assessment

1 Introduction

The world is experiencing an increasing frequency and intensity of disasters, including floods, hurricanes, wildfires, and droughts, which are exacerbated by climate change. This introductory chapter examines the intricate relationship between disasters and climate resilience, examining the multifaceted strategies necessary to

I. Pal (✉) · S. V. S. A. B. Ganni · A. Baskota · J. Kumar
Disaster Preparedness, Mitigation and Management, Asian Institute of Technology, Pathum Thani, Thailand
e-mail: Indrajit-pal@ait.asia

I. Pal et al. (eds.), *Proceedings of the 3rd International Symposium on Disaster Resilience and Sustainable Development—Volume 1*, Lecture Notes in Civil Engineering 623, https://doi.org/10.1007/978-981-96-2139-2_1

improve community-level resilience and the critical role of governments, nature-based solutions, and innovative strategies in disaster risk management. These events not only disrupt lives and livelihoods but also challenge the very fabric of communities, pushing them to the boundaries of vulnerability. Exacerbated by climate change, disasters have evolved into a persistent and escalating threat rather than being uncommon or isolated events.

Disaster resilience incorporates various strategies, capabilities, and systems that enable communities and organizations to endure, recover, and enhance their resilience post-disaster [3]. It embodies a holistic perspective beyond survival, emphasizing preserving and restoring essential functions and structures. Communities are acknowledged to be at significant risk due to the ageing infrastructure and the escalating risks posed by climate change. Therefore, it is imperative that governments at all levels take action to respond to hazards, mitigate risks, prepare for impacts and apply the lessons of past experiences to cultivate a higher level of resilience.

2 Disasters and Climate Change

The extent of climate change as a driver of disaster risk is becoming more widely acknowledged. Climate change and natural hazards are intricately interlinked, as climate change influences the frequency, intensity, and distribution of extreme weather events that can result in disasters. Several studies have highlighted that the occurrence and intensity of weather and climate extreme events have been altered as a result of climate change, which has significant implications for disaster risk [23]. IPCC's Six Assessment Report provides additional detail regarding the evolving characteristics of climate extremes and weather, noting that every region on Earth has experienced a degree of changes in climate and weather patterns, which include heatwaves, excessive precipitation, droughts, and tropical cyclones, among others. These modifications are believed to be the result of human-induced climate change and are expected to persist in the future. Moreover, although climate change projections provide a macroscopic understanding of what lies ahead, the uncertainties in the global emission pathways and limited micro-level data lead to ambiguity in potential risks and mitigation planning.

Therefore, a multifaceted approach is necessary to comprehend and mitigate disasters. It entails not only monitoring and attribution of changes in extreme weather events but also improving our understanding and assessment of communities' vulnerability and exposure to these events. It advocates for the integration of catastrophe risk reduction strategies to fortify resilience in the face of the escalating risks associated with a changing climate.

3 Defining Climate Resilience

Climate resilience is the ability of social, economic, physical, and environmental systems to withstand, respond or adjust to climate change-induced events, trends, or disruptions. This involves understanding and managing not just the risks that climate variability and change carry but also the opportunities that may arise [24]. It encompasses a dual focus: reducing vulnerability to potential harm and improving the ability to recover from climate-related shocks [2]. At its core, climate resilience is the process of "*improving a system's capacity to absorb stress, maintain function in the presence of external stresses imposed by climate change, and adapt, reorganize, and evolve into more desirable configurations that enhance the sustainability of the system, thereby better preparing it for future climate change impacts*" [18]. This encompasses the physical, socio-economic and environmental dimensions that contribute to the community's overall resilience [31].

Building climate resilience typically entails incorporating climate change adaptation measures into development planning, operations, and decision-making processes of organizations and governments in the policy and planning domains [28]. This may encompass initiatives such as the promotion of sustainable land use practices, implementation of early warning systems, and the climate-proofing of infrastructure [9]. Moreover, the process is dynamic and necessitates continuous learning and adaptation [6].

4 Climate Resilience and Development

Climate resilience has emerged as a critical concept in the face of global climate change, which poses significant risks to natural and human systems worldwide. The importance of climate resilience can be understood by examining its global and local impacts. To protect the interconnected systems that are essential for human existence, climate resilience is imperative at the global level. The IPCC has consistently underscored the significance of improving climate resilience as a strategy for decreasing the susceptibility of societies to the adverse consequences of climate change [24]. This is expected to have a cascading effect on global agricultural security, water availability, and health outcomes [25].

Additionally, the economic implications are substantial. According to the Global Commission on Adaptation [1], the absence of investment in climate resilience could lead to a substantial economic downturn, with the potential for a substantial loss of trillions of dollars in global economic activity over the next decade. Additionally, resource scarcity can result in conflict and displacement, which can exacerbate geopolitical tensions [35].

International cooperation is necessary to reduce greenhouse gas emissions, share knowledge and resources, and implement adaptation strategies that can withstand the effects of climate change to build climate resilience at the global level. The

Paris Agreement adopted in 2015 is a global initiative that aims to fortify the global response to the threat of climate change by improving adaptive capacity and resilience [40]. Deeply interconnected are the global and local consequences of climate change. Local actions can have global consequences and vice versa. In contrast, the Green Climate Fund and other international funding mechanisms can facilitate the implementation of adaptation initiatives by local communities in developing countries [40]. Similarly, the integration of climate change and resilience in several targets of the UN SDGs also portrays the importance of mainstreaming climate adaptation in development, and jots out a suitable pathway, actions and targets for governments and donor organizations to invest in such endeavour.

5 Climate Change Adaptation

Research indicates that a comprehensive understanding of the linkages between climate change and disasters is essential for effective disaster risk management and adaptation strategies in the face of a changing climate. Climate change is one of the most pressing issues of our era, necessitating a dual strategy: adaptation and mitigation. Mitigation is the process of taking measures to reduce or prevent the emission of greenhouse gases, whereas adaptation is the process of adapting to the current and anticipated effects of climate change. Mitigation is intended to stabilise greenhouse gas levels within a timeframe that is conducive to the natural adaptation of ecosystems, the protection of food production, and the promotion of sustainable economic development. The Paris Agreement of 2015 set global targets for mitigation, with nearly every country agreeing to zero out their greenhouse gas emissions to limit global warming to well below 2 °C, and ideally to 1.5 °C. However, current mitigation efforts are not on track to meet these goals, and emissions continue to rise.

Adaptation is the process of modifying our lifestyles to mitigate the negative consequences of climate change. This encompasses constructing or retrofitting infrastructure to withstand adverse weather events, restoring natural barriers such as wetlands to serve as hurricane buffers, and implementing policy and behavioural changes. Proactive adaptation is essential, as it is designed to construct systems that can withstand both current and future climate impacts. The strategies for adaptation will differ based on the local conditions and requirements, as the effects of climate change are experienced differently in different regions.

Therefore, a harmonious combination of both is necessary for effective climate action. Mitigation has historically received the majority of funding for climate change, while adaptation has received a lesser portion. The necessity for adaptation is increasing at an accelerated pace, and many low- and middle-income countries may encounter significant challenges in adapting swiftly enough to prevent significant hardships in the absence of adequate financial and technological support.

It is necessary to establish a multi-level approach that connects global and local endeavours that encompasses physical infrastructure and social and institutional systems that can adjust to evolving circumstances. Education, community engagement, and awareness-raising are essential components of adaptation, as they enable individuals and communities to engage in resilience-building activities [16].

6 Frameworks and Policies

International agreements and climate change legislation are the foundation of global endeavours to confront the obstacles presented by a warming planet. These frameworks and policies are intended to serve as a guide for climate change action, including managing calamity risk, adaptation, and mitigation in various sectors. The UNFCCC establishes the fundamental legal framework and principles for international climate change cooperation. Its objective is to prevent hazardous anthropogenic interference with the climate system by stabilizing atmospheric concentrations of greenhouse gases. In 1997, the Kyoto Protocol was implemented, which required industrialized nations to establish quantified emissions reduction objectives for six greenhouse gases.

The United Nations Framework Convention on Climate Change (UNFCCC) is an international treaty that is designed to mitigate the danger of catastrophic climate change and slow the pace of climate change. 154 countries have signed the treaty since its inception, and it currently has 197 member countries. The UNFCCC, which is headquartered in Bonn, Germany, convenes annually to establish policies to combat global warming. The AOSIS (The Association of Small Island Developing States) initiated the concept of receiving compensation from developed countries for the destruction caused by climatic hazards. However, national and local laws have often been found to be inadequate in addressing and incorporating disaster and climate-induced risks and damages.

Decision-makers are challenged to go beyond conventional approaches to prepare for and adapt to climate change. Various frameworks can be used to address the complexities of climate change, with an iterative risk management framework being one of the most effective for climate-related decisions. This strategy encompasses both adaptation and mitigation, considering the impacts of climate change, co-benefits, sustainability, equity, and attitudes towards risk. It is an ongoing process that identifies, assesses, prioritizes, manages, and reevaluates climate change consequences and policies in response to new information and experiences.

7 Outline of the Book

The book consists of eighteen chapters covering a wide range of topics in disaster risk management, climate change adaptation and sustainable development, highlighting diverse global perspectives, success stories, challenges and lessons learned.

The second chapter, "*Analyzing Governments' Policies on Disaster Risk Management in the Borobudur Cultural Heritage Region, Central Java, Indonesia*" presents a case study of Borobudur Cultural Heritage Region in Indonesia, which experienced rapid urban development after its designation as a National Strategic Tourism Area in 2011. Considering its strategic and economic importance and exposure to multiple hazards including volcanic eruption, earthquakes, floods and tropical storms, authors Pusparini Dharma Putri and Lathifa Nursyamsu have identified gaps in disaster risk management, and highlighted the need for better inter-ministerial coordination and comprehensive disaster preparedness planning for the region.

The sustainability of businesses is considered pivotal in economic resilience. Human capital, resource management and institutional policies all contribute to sustainability [36, 7, 38]. In the third chapter, "*Challenges of Sustainable Business Organization: Case Study of Young Craft Business Owner in Nepal*", Richan Shrestha et al., use qualitative research to present the case of five young craft producers and business enterprises in Nepal who face significant challenges due to lack of skilled youth, dependency on raw materials, market acceptance and policy barriers. Their findings highlight the need to increase awareness and education among businesses and enterprises and targeted policies and strategies to enhance resilience.

The role of strong legislation for the sustainability of business enterprises is further elaborated in the fourth chapter, "*Constitutional Law Provisions Towards Disaster Management in Thailand*" by Wachiraporn Poungjinda and Shubham Pathak which explores the implications of legal frameworks in Thailand on the business sector, amid exposure to disasters. SMEs play a significant role in most of economies across the world, including in Thailand, but face significant challenges such as limited access to financing, lack of government support and regulatory requirements. Poungjinda and Pathak argue that gaps in the promulgation and implementation of constitutional laws and provisions are one of the major reasons for ineffectiveness in coping with disaster impacts, particularly in rural areas. They recommend a comprehensive review of existing policies and implementation mechanisms to enhance sustainability of business sector amid disasters.

The role of risk and resilience assessment processes is widely acknowledged in guiding effective disaster risk management policies and actions and promoting sustainability [5]. However, given the broad spectrum and multi-disciplinary characteristic of resilience research, various frameworks and indicators have emerged and are currently in use [11]. Deepak Marahatta, Hari Devkota and Shankar Tiwari conducted a review of thirty-two resilience assessment frameworks in their chapter, "*Systematic Review of Community Resilience Framework Related to Disasters*" to identify key themes and sectors where resilience assessment studies have focussed on. Their study highlights that social and economic dimensions are highly prioritized in resilience assessment frameworks and that researchers most frequently use quantitative methods and indicator-based approaches. The study therefore highlights the importance of comprehensive and contextually relevant approach to assess resilience and emphasizes the need for balanced consideration of physical, environmental and cultural factors.

The sixth chapter, "*Drought Adaptation in Thailand: A National Governmental Perspective*" analyses the Government of Thailand's capacity for drought risk adaptation, highlighting the national laws and regulations and detailed roles of various ministries and departments in managing water resources. This chapter continues the elaboration on the role of strong government institutions and policies to mitigate disaster risks, such as the Disaster Prevention and Mitigation Act 2007 of Thailand which delineates and empowers government entities to undertake various disaster management actions [38]. With the increase in frequency, severity and impacts of droughts in the country over the past two decades, the authors Puvadol Doydee and Jessada Techamahasaranont emphasize the importance of inter-agency coordination and communication, and multi-sectoral approach across national strategies and action plans for mitigation of drought impacts.

In recent periods, environmental hazards and risks such as water and air pollution and biodiversity loss have become a significant concern. Several studies have shown that similar to other conventional natural hazards, vulnerable groups, including the elderly, disabled, women and children, are at high risk for air pollution [19, 28]. In this backdrop, Chapter 7 "*The Impact on PM2.5 in Thailand Society: An Investigation into Causes, Challenges, and Policy Gap in Implementations*" investigates the scenario of PM 2.5 pollution in Thailand from 2019 to 2022, highlighting the impacts on health, tourism and agriculture. In the chapter, Vasuthida Pinrattana and Oulavanh Sinsamphanh analyzed extant policies in Thailand with regard to air pollution control and their effectiveness in addressing the issues. With several identified gaps, the author emphasized the need for stricter policies, better coordination and increasing public awareness as important measures to combat air pollution and reduce its negative impacts. Moreover, the authors stress the need for Thailand to leverage lessons from global frameworks and case studies to enhance policy implementation.

In the eighth chapter, "*Labor Law Implications on Disaster Susceptible Small and and Medium Enterprises in Thailand*", Weerayut Lasongyang and Shubham Pathak examine the extent to which labor laws in Thailand affect the small and medium enterprises during disasters. Disasters, particularly the COVID-19 crisis had a severe impact on labor and migration [32], which impacted their well-being and rights as well as hampered productivity among SMEs. Using qualitative interviews and secondary data, the chapter highlights issues such as reduced labor rights, inadequate implementation of minimum wages and unequal employment opportunities as key issues surrounding SMEs and workers and argues for robust laws, better governance and effective policy implementation. Furthermore, the authors also highlight the role of collaboration among government, broker agencies and migrant workers to improve labor conditions amidst disasters.

Sustainable development and resilience are also closely linked with complex socio-dynamics, including aspects of migration. This has been discussed in Chapter 9 "*Male Labor Migration and Women Empowerment: A Review Paper*", where authors Hari Devkota et al. who examined the impact of male labor migration on women's empowerment in South Asia, with a focus on Nepal. The chapter argues that migration trends among male labor in Nepal have impacts on women's personal, economic, social and political empowerment. However, these impacts are found to be both positive, such as increased decision-making capacity, financial freedom and social engagement [22, 13, 33] and negative, such as increased workload, psychological issues and social stigma [8, 14]. Hence, the authors recommend a deeper synthesis of the role of political empowerment and social support mechanisms in these outcomes to have a better understanding.

A key component of disaster and climate risk management involves supporting affected entities, including individuals, communities and businesses, through damage and loss compensation. However, loss assessment and compensation are complex, involving social, economic and political dimensions. Pratik Poudel and Mokbul Morshed Ahmad in their chapter "*Compensation for Losses and Damages Caused by Climate Change: Challenges and Progress*" reviewed the evolution of the concept of loss and damage due to climate change, highlighting challenges in implementation of these practices due to gaps in determining accountability [15], distinguishing climate change related events [34] and impacts and addressing non-economic and intangible losses [26]. They discuss various financial mechanisms adopted in addressing the issue and emphasize the need for comprehensive approaches, including distributive and compensatory justice frameworks.

Similarly, Chapter 11 "*Evaluating the Effectiveness of Sustainable Transportation Policies in Mitigating Climate Change in Metro Manila, Philippines: The Public Utility Vehicle Modernization Program (PUVMP)*" authored by Tabassam Raza et al. discusses the role of public transportation systems and modernized vehicles to enhance safety, efficiency and environmental sustainability, including improvement of air quality and mitigating climate change through emission reduction. The chapter elaborates on the challenges faced in adoption of such programs in Manilla, and provides recommendations including enhancing policy effectiveness, stakeholder communication and infrastructure development.

The role of indigenous knowledge and practices in adaptation and risk reduction has been noted in Chapter 12 "*Indigenous Knowledge of Afro-Asian Tribal Communities: An Alternative Science of Natural Disaster Management*" authored by Tapas Pal. The study explores how Afro-Asian tribes use traditional methods for preparedness and adaptation, including forecasting of natural hazards [37], including droughts and floods based on environmental cues. However, documentation of indigenous knowledge lags behind, with the potential risk of fragmentation. The chapter emphasizes the importance of documentation and preservation of indigenous knowledge and integration with modern science [30] and technology for effective disaster management strategies [4].

Similarly, localization approaches are increasingly acknowledged as vital components in resilience building, by empowering communities through knowledge, skills and strategic mindset to implement adaptation and risk reduction actions [17, 27]. In Chapter 13 "*Resilience Starts Locally: Enhancing Sustainability Through Localization in Humanitarian Response*", Aniruddha Dey et al., explore the role of localization in empowering individuals to address local challenges using indigenous resources and knowledge in the Bengal Basin in Bangladesh. The authors highlight the interconnectedness between localization and UN SDGs, and advocate for capacity building and integration of localization strategies to build resilience.

Stakeholders at local levels, both governments and communities face gaps in institutional, information and resource capacities to mainstream DRR and CCA approaches in development [10]. Local level planning and actions are further highlighted in Chapter 14, "*Disaster and Climate Resilience in Quezon City: Evaluating Local Development Planning for Sustainable Development Goals*" where Tabassam Raza et al. analyze City Development Plans of Quezon City, Philippines and explore the state of integration of climate change adaptation, disaster risk reduction and sustainable development goals. The authors highlight the necessity of integrating comprehensive risk assessments into existing land use, socio-economic development, and disaster risk management plans to enhance resilience. Moreover, the paper also emphasizes the need for public awareness, stakeholder collaboration and communication and increased funding to strengthen local government capacity to achieve urban resilience.

Early warnings are key components of disaster risk management, particularly in large urban settlements. However, as explored in Chapter 15, "*Identifying Challenges to the Optimal Utilization of Early Warning Systems for Flood Management in Metro Manila: Advancing SDGs 11 and 13*" by Krishna Buenaventura et al., inadequate infrastructure, financial constraints, poor coordination and limited community access to technology hinders the utilization of EWS in many developing nations. The chapter further emphasizes the importance of comprehensive public awareness campaigns, community engagement, policy reforms, improved stakeholder coordination and increased funding and resources to enhance EWS effectiveness in urban cities.

Several strategies and initiatives have been employed to address the resource-gaps in disaster risk management. The involvement of private sector entities can be particularly helpful to alleviate the burden on public sector and communities, and have especially become increasingly involved in disaster preparations and relief through cash donations, in-kind contributions, infrastructure development and insurance [20]. In Chapter 16 "*Private Sector Engagement for Community Resilience and Sustainable Development*", Kilian T. Murphy and Edwin M. Salonga have explored the role of private sector in disaster risk management and sustainable development. The paper discusses the expanding role of private sector in enhancing community resilience and provide evidences of how these partnerships can leverage mutual benefits and enable businesses to safeguard their operations while contributing to community development.

Chapter 17, "*Land Readjustment Approach as an Urban Development Tool Towards Disaster Resilience: A Study of Japan's Experience and Its Applicability in the Philippines*", discusses Japan's land readjustment practices including land pooling and redistribution, integrated infrastructure development as potential tools for urban disaster resilience and economic growth. Across the world, land readjustment approaches have been widely used and effective in managing urban growth [22], high-density urban population and building back better in post-disaster recovery [41]. The authors, Arlene D.R. Santiago and Tabassam Raza emphasize the successful implementation of these approaches in the post-earthquake reconstruction processes in Japan and highlight the necessity of policy adaptations to improve the practice of land readjustment in Philippines to overcome extant legal barriers and resource constraints.

Rehabilitation of local economy is a vital component of post-disaster recovery process, to mitigate the potential exacerbation of vulnerabilities and inequities arising from disaster impacts. In such cases, micro-finances have played a significant role in rural communities in Nepal, as demonstrated in Chapter 18 "*Role of Micro-Finance in Poverty Reduction: A Case Study from from Neelkantha Municipality, Nepal*" authored by Shiva Raj Khanal, Romi Ghimire and Pradeep Shrestha. The chapter discusses the role and analyses the impact of microfinance institutions in supporting poverty reduction in Nepal in the aftermath of the 2015 Gorkha Earthquake. Findings from their study showed that micro-credits helped rural communities shift from a traditional agriculture to diversified income generating activities, enhancing economic stability. Moreover, findings of the chapter reveal positive improvements on household income, savings, and access to health, education and housing through microfinance institutions, emphasizing the role of financial capital in enhancing community resilience.

8 Conclusion

The past two decades have seen a significant advancement in how disasters are perceived and managed, with notable strides in risk mitigation and preparedness. However, the increasing frequency and severity of natural hazards due to climate change demands nuanced approaches and actions to manage future risks. Undeniably, disaster and climate resilience will require a multi-sectoral, multi-disciplinary and multi-pronged approach across all levels of governance.

This book explores the multifaceted relationship between disaster risk management, climate change adaptation, resilience building and sustainable development across various countries in Asia Pacific. The case studies presented here offer diverse perspectives on challenges, solutions and lessons in these areas. Across the eighteen chapters, the book has delved into several sectors, ranging from heritage conservation, land-use planning, urban development, sustainable transportation, businesses and enterprises, legislations and policies, localization and the involvement of private sector. Similarly, the book also discusses migration and damage compensation issues, providing unique insights into the complexities of implementation of risk management and adaptation actions across the region.

Collectively, these chapters underscore the interconnectedness of disaster risk management, public policy, community empowerment and scientific knowledge. Across all case studies, locally-led actions, community-driven policies action plans and capacity building emerge as key recommendations. Furthermore, given the immense shortages of resources, it is also imperative to document and share learnings across the region to adopt innovative ideas, avoid pitfalls and ultimately enhance the effectiveness of actions. It, therefore, highlights the immense need for multi-stakeholder collaboration and communication. The diverse case studies provide a holistic view of the current landscape and future directions for enhancing resilience in Asia Pacific. The book will serve as a guidance note for policymakers and development practitioners in comprehending the multiple aspects of disaster and climate resilience, the policies and practices adopted across the region and the lessons learned.

References

1. Adapt now: a global call for leadership on climate resilience (2019) Adapt now: a global call for leadership on climate resilience. https://doi.org/10.1596/32362
2. Adger WN (2000) Social and ecological resilience: are they related? Prog Hum Geogr 24(3):347–364. https://doi.org/10.1191/030913200701540465
3. Aksha SK, Emrich CT (2020) Benchmarking community disaster resilience in Nepal. Int J Environ Res Public Health 17(6). https://doi.org/10.3390/IJERPH17061985
4. Asghar S, Alahakoon A, Churilov L. (2004) A hybrid decision support system model for disaster management. In: Fourth international conference on hybrid intelligent systems (HIS'04). IEEE. https://doi.org/10.1109/ICHIS.2004.5. ISBN:0-7695-2291-2
5. Béné C, Wood RG, Newsham A, Davies M (2012) Resilience: new Utopia or New Tyranny? Reflection about the potentials and limits of the concept of resilience in relation to vulnerability reduction programmes. IDS Work Pap 2012(405). https://doi.org/10.1111/j.2040-0209.2012.00405.x
6. Berkes F, Ross H (2013) Community resilience: toward an integrated approach. Soc Nat Resour 26(1):5–20. https://doi.org/10.1080/08941920.2012.736605
7. Biermann R, Harsch M (2017) Resource dependence theory. Palgrave handbook of inter-organizational relations in world politics, pp 135–155
8. Binzel C, Assaad R (2011) Egyptian men working abroad: labour supply responses by the women left behind. Labour Econ 18:S98–S114. https://doi.org/10.1016/j.labeco.2011.03.002
9. Birkmann J, von Teichman K (2010) Integrating disaster risk reduction and climate change adaptation: key challenges-scales, knowledge, and norms. Sustain Sci 5(2):171–184. https://doi.org/10.1007/s11625-010-0108-y
10. Cuevas SC (2015) Challenges in mainstreaming climate change adaptation into local land use planning: evidence from Albay, Philippines. Int J Clim Chang Impacts Responses 7(3):45–65. https://doi.org/10.18848/1835-7156/CGP/v07i03/37246
11. Cutter SL, Ash KD, Emrich CT (2014) The geographies of community disaster resilience. Glob Environ Chang 29:65–77. https://doi.org/10.1016/j.gloenvcha.2014.08.005
12. Cutter SL, Barnes L, Berry M, Burton C, Evans E, Tate E, Webb J (2008) A place-based model for understanding community resilience to natural disasters. Glob Environ Chang 18(4):598–606. https://doi.org/10.1016/j.gloenvcha.2008.07.013
13. Datta A, Mishra SK (2011) Glimpses of women's lives in rural Bihar: impact of male migration. Indian J Labour Econ 54(3):457–477
14. De Haas H (2010) The internal dynamics of migration processes: a theoretical inquiry. J Ethn Migr Stud 36(10):1587–1617. https://doi.org/10.1080/1369183X.2010.489361
15. Dow K, (2014) Climate change, limits to adaptation and the 'loss and damage'debate. E-international relations
16. Few R, Brown K, Tompkins EL (2007) Climate change and coastal management decisions: insights from Christchurch Bay, UK. Coast Manag 35(2–3). https://doi.org/10.1080/08920750601042328
17. Fisher A, Fukuda-Parr S (2019) Introduction–data, knowledge, politics and localizing the SDGs. J Hum Dev Capab 20(4):375–385
18. Folke C, Carpenter SR, Walker B, Scheffer M, Chapin T, Rockström J (2010) Resilience thinking: integrating resilience, adaptability and transformability. Ecol Soc 15(4). https://doi.org/10.5751/ES-03610-150420

19. Fu X, Li L, Lei Y, Wu S, Yan D, Luo X, Luo H (2020) The economic loss of health effect damages from PM 2.5 pollution in the central plains urban agglomeration. Environ Sci Pollut Res 27:25434–25449
20. Fuchs A, Öhler H (2021) Does private aid follow the flag? An empirical analysis of humanitarian assistance. The World Economy, 44(3):671–705
21. Gartaula HN (2009) International migration and local development in Nepal. Contrib Nepal Stud 36(1):37–65
22. Home R (2007) Land readjustment as a method of development land assembly: a comparative overview. Town Plan Rev 78(4):459–483. http://www.jstor.org/
23. Intergovernmental Panel on Climate Change (IPCC) (2021) Weather and climate extreme events in a changing climate. In: Climate change 2021—the physical science basis. https://doi.org/10.1017/9781009157896.013
24. IPCC (2014) Climate change 2014 part A: global and sectoral aspects. In: Climate change 2014: impacts, adaptation, and vulnerability. Part A: global and sectoral aspects. Contribution of working group ii to the fifth assessment report of the intergovernmental panel on climate change. papers2://publication/uuid/B8BF5043-C873-4AFD-97F9-A630782E590D
25. IPCC (2018) Global warming of 1.5°C: an IPCC special report on impacts of global warming of 1.5°C above pre-industrial levels in context of strengthening response to climate change, sustainable development, and efforts to eradicate pover. Global warming of 1.5°C, pp 1–24. https://www.cambridge.org/core/product/identifier/9781009157940%23prf2/type/book_part
26. James RA (2019) Attribution: how is it relevant for loss and damage policy and practice? In: Reinhard Mechler L M-B, Loss and damage from climate change. Springer, Cham, pp. 113–154
27. Khan AK, Kontinen T (2022) Impediments to localization agenda: humanitarian space in the Rohingya response in Bangladesh. In: Khan AK, Kontinen T (2022) Impediments to localization agenda: humanitarian space in the Rohingya response in Bangladesh. J Hum Dev Capab 7(14):1–15
28. Leichenko R (2011) Climate change and urban resilience. Curr Opin Environ Sustain 3(3):164–168. https://doi.org/10.1016/j.cosust.2010.12.014
29. Maung TZ, Bishop JE, Holt E, Turner AM, Pfrang C (2022) Indoor air pollution and the health of vulnerable groups: a systematic review focused on particulate matter (PM), volatile organic compounds (VOCs) and their effects on children and people with pre-existing lung disease. Int J Environ Res Public Health 19(14):8752
30. Krichen M, Abdalzaher MS, Elwekeil M, Fouda MM (2024) Managing natural disasters: an analysis of technological advancements, opportunities, and challenges. Internet Things Cyber-Phys Syst. 4:99–109
31. Patel SS, Rogers MB, Amlôt R, Rubin GJ (2017) What do we mean by "community resilience"? A systematic literature review of how it is defined in the literature. PLoS Curr 9. https://doi.org/10.1371/currents.dis.db775aff25efc5ac4f0660ad9c9f7db2
32. Pathak S, Laikram S (2022) Cooperative education during Covid-19 pandemic: enhancing legal rights and professional development of interns in Thailand. Int J Disaster Resil Built Environ 13(2):133–149
33. Rajkarnikar PJ (2017) The impacts of foreign labor migration of men on women's empowerment in Nepal. University of Massachusetts Amherst
34. Roberts E (2018) Climate change-related loss and damage: translating the global policy agenda for national policy processes. Climate Dev 10(1):4–17
35. Smith D, Vivekananda J (2007) Aclimate of Conflict. the links between climate change, peace and war. Negotiation: From theory to practice, november. https://www.preventionweb.net/files/7948_AClimateOfConflict1.pdf
36. Strober MH (1990) Human capital theory: Implications for HR managers. Ind Relat: J Econ Soc 29(2):214–239
37. Stigter CJ, Zheng Dawei LOZ, Onyewotu, Xurong M (2005) Using traditional methods and indigenous technologies for coping with climate variabilit. Springer Nature 70:255–271
38. Thanyalakmetha S (2020) Thailand country report 2020. Visiting researcher program (FY 2020B). Asian disaster reduction center (ADRC). Department of disaster prevention and mitigation (DDMP), p 9

39. Tina Dacin M, Goodstein J, Richard Scott W (2002) Institutional theory and institutional change: introduction to the special research forum. Acad Manag J 45(1):45–56
40. UNFCCC (2015) UNFCCC 2015 Paris agreement. United Nations framework convention on cliamte change, December
41. Wismansyah AR, Dewi MS, Rahmi, Syahzaeni T, Hastut W (2019) Dynamic system development: an approach for land re-adjustment in managing slum areas. https://doi.org/10.18502/kss.v3i21.4960

Analyzing Governments' Policies on Disaster Risk Management in the Borobudur Cultural Heritage Region, Central Java, Indonesia

Pusparini Dharma Putri and Lathifa Nursyamsu

Abstract The Borobudur region is rich in cultural heritage, and its assets are more than the three Buddhist temples (Borobudur, Pawon, and Mendhut Temples). There are also various traditional house settlements, and how the people interact with nature is marvellous. That is why many tourists and researchers come and go to this region, and in 2019, it was assigned as a National Tourism Strategic Area of Indonesia. As a National Tourism Strategic Area, the governments prioritize this area, and all development will be implemented here as soon as possible. Their point of view is to emphasize the tourism aspect, if good management is established, it will have a positive impact. As predicted, the negative impacts were not considered to happen because of the unwell-prepared development. This is an urban development disaster; other disasters, such as volcanic eruptions, earthquakes, floods, and hurricanes, negatively impacted the Borobudur region. By analyzing various official documents, field observation, and conducting Focus Group Discussions with the community, governments, experts, and researchers to find the loopholes in government action, why that can happen, and how they formulate regulations, act, and manage it. As a result, it shows that the government already has a strategic action plan for each ministry to develop the Borobudur Region. However, they need coordination, which leads to ambiguity in regulations. Somehow, it also found that the government tried to change the original regulation to fight off their work. From another perspective, the government already has a good scheme for tackling such natural disasters. With all findings, it is acknowledged that the government should establish good coordination between each ministry involved in Borobudur Region to fulfil the needs of good Disaster Risk Management in this cultural heritage region of Borobudur.

P. D. Putri (✉)
PhD Program on Architecture, Faculty of Architecture, Chiang Mai University, Chiang Mai, Thailand
e-mail: pusparinidharmaput_p@cmu.ac.th

L. Nursyamsu
Architecture Department, Faculty of Civil Engineering and Planning, Universitas Internasional Batam, Batam, Indonesia

I. Pal et al. (eds.), *Proceedings of the 3rd International Symposium on Disaster Resilience and Sustainable Development—Volume 1*, Lecture Notes in Civil Engineering 623, https://doi.org/10.1007/978-981-96-2139-2_2

Keywords Disaster risk management · Sustainable development · Heritage regulation · Borobudur cultural heritage

1 Introduction

Borobudur region is a cultural landscape around Borobudur Temple in Magelang, Central Java, Indonesia. According to [11], a cultural landscape is an imprint of human history and a window to our past and culture. Therefore, preserving the area is a crucial thing to do, both for individuals and the government itself. Borobudur region is designated as a world heritage site by UNESCO in 1991 and has three Buddhist temples at the heart of this region: Borobudur Temple, Pawon Temple, and Mendut Temple [3]. Borobudur region has both tangible and intangible cultural heritage value. The heritage value of the Borobudur region comes not only from the Buddhist temples around the area but also from the various traditional house settlements. The community still holds This local culture closely and how the people interact with nature. That is why a lot of tourists and researchers come and go to this region.

Many natural elements surround the Borobudur region, such as hills, rivers, mountains, and other natural forms. This area is surrounded by several volcanoes (Merapi, Sumbing, and Merbabu Mountain), mountains (Tidar, Andong, and Telomoyo Mountain), and a mountain hill (Menoreh). Borobudur region also passes by rivers, such as the Elo and Progo Rivers, and other small rivers that also become water sources and support the area's agriculture [9]. Although these natural resources benefit the area, they can also be a source of disaster. Geologist researcher [7] revealed that the Borobudur region is the former Ancient Lake area. As a former ancient lake area, the Borobudur region has the potential to experience geological phenomena that have occurred in the past, such as earthquakes and volcanic eruptions [8]. Natural disasters related to geology and other natural disasters often occur in the Borobudur area. Some of these disasters include floods, landslides, drought, and windstorms.

Every year, there are at least 60 cases of landslide that occur surrounds Borobudur region. A windstorm, last February, damaged buildings in 7 villages in the Borobudur region. They also destroyed one of the communal buildings-space in that manage by Wanurejo villagers. This disaster occurred because of the high rainfall intensity in the Borobudur region, and its existence surrounded by hills makes this area prone to windstorms. Based on hazard data that occurred in surrounding the Borobudur region (data is accessed from lusaka.magelangkab.go.id), in 2022 there were 664 disasters. The most frequent hazard were landslides, with 223 incidents (40.47%), and windstorms, 238 incidents (43.19%). The number of disasters that have occurred in the Borobudur region over the last 3 years has increased: 324 incidents in 2020, 567 incidents in 2021, and 551 incidents in 2022. This shows that the Borobudur and surrounding are vulnerable to hazard. For more specific, data from 21 existing subdistricts shows that Borobudur District is the second highest number of hazards, with 54 incidents. Obviously, it will impact the sustainability of the Borobudur Temple

Table 1 Magelang regency disaster data

No	Type of disaster	Number of events	Percentage (%)
1	Landslide	223	40.47
2	Windstorm	238	43.19
3	Fire	48	8.71
4	Others	32	5.81
	Flood	9	1.63
	Volcano eruption	1	0.18
Total		664	

Source [6]

and other cultural heritage objects in there. The following are details of disaster events in the Borobudur area during 2022 (see Table 1).

Borobudur region also vulnerable to hazard caused by human activities, known as human-induced hazards. Based on Disaster Risk Management of Cultural Heritage in Urban Areas: A Training Guide by UNESCO [5], human-induced hazards are disasters caused by human activities, such as arson, armed conflicts, infrastructure failures, planning issues, and long-term effects of climate change. One of the human-induced phenomena that happened occur in 2019, when Indonesian government assigned Borobudur Region as a National Tourism Strategic Area of Indonesia government. Consequently, this is a major development in various areas in the Borobudur region. The negative impact of this program is that there is a lot of interference in development, which causes the loss of local values, socio-cultural changes in society, and disruption of conservation and preservation efforts in the Borobudur region.

In Borobudur District, the Disaster Risk Management Organization is tasked with dealing with and anticipating disasters in the area. However, this organization only plays a role after the disaster and when losses arise. Moreover, this organization only operates in the Borobudur District Area. On a national and regional scale, there are several institutions or organizations formed by the government that are related to disaster management, such as The National Agency for Disaster Countermeasure (*Badan Nasional Penanggulangan Bencana/BNPB*) from the central government, the Regional Disaster Management Agency (*Badan Penanggulangan Bencana Daerah/BPBD*) of Magelang Regency, the Center for Volcanology and Geological Disaster Mitigation (*Pusat Vulkanologi dan Mitigasi Bencana Geologi/PVMBG*) from the central government, and the Magelang Regency Tourism and Culture Office (*Dinas Pariwisata dan Kebudayaan*). These institutions and organizations are tasked with disaster management from the pre-disaster stage, during a disaster/emergency response, to post-disaster. Furthermore, again, this institution only plays a role when a disaster has occurred or after a disaster so that the losses cannot be avoided. From this explanation, many institutions or organizations have been formed to deal with disasters in the Borobudur region. However, precise action has yet to be taken to prevent disasters in this area, which has many historical and cultural values.

In this article, we meticulously explore the government's disaster risk management policies in the Borobudur Cultural Heritage Region. We aim to unravel the intricate web of strategies and measures implemented by local, regional, and national authorities to safeguard this cultural heritage region from the ever-looming spectre of natural and human-induced disasters. By delving into the nuances of these policies, we aim to shed light on the effectiveness of current approaches, identify potential gaps, and propose considerations for future improvements in disaster preparedness and mitigation efforts. As we navigate the landscape of governmental initiatives, this analysis strives to contribute valuable insights to the ongoing discourse surrounding the intersection of cultural heritage preservation and disaster resilience in one of Indonesia's most revered regions.

2 Literatures

This section will explain two main topics that the writer used as references to create the same understanding of the result and discussion section. First, talk about Disasters in Heritage Context, where this section will explain the categorization of hazards by UNESCO[1] as the highest referee on heritage context and the Indonesia Government as the state party that owns the objects. The second part will emphasize sustainable development growth regarding the urban heritage context. The last topic is trying to capture the same understanding of what is important in the effort to have sustainable development in a heritage context.

2.1 Hazard on Heritage Context

Heritage buildings are very vulnerable to damage. 3 words are commonly used to identify threats that are susceptible to being experienced in a heritage context: disaster, hazard, and vulnerability. The UNISDR[2] [14] defines a disaster as a significant disturbance to the operations of a community or society, characterized by extensive human, material, or environmental damages and consequences that surpass the affected community's capacity to manage solely with its resources. Disasters are often associated with the hazards that cause them to occur. A hazard is defined as a dangerous phenomenon, substance, human action, or situation that poses the risk of causing fatalities, injuries, health issues, property destruction, disruption of livelihoods and services, socio-economic disturbances, or harm to the environment [14]. Hazards may arise individually, in a series, or combination, each with distinct origins and outcomes. Furthermore, they are vulnerability that have a direct connection with

---

[1] United Nations Educational, Scientific and Cultural Organization.

[2] United Nations International Strategy for Disaster Reduction, now United Nations Office for Disaster Risk Reduction (UNDRR).

hazards. Vulnerability refers to the traits and conditions of a community, system, or asset that render it prone to the detrimental consequences of a hazard. It encompasses a range of existing or resultant conditions stemming from diverse physical, social, economic, and environmental factors, heightening a community's susceptibility to hazard impacts [14].

In the past, the term "disaster" was generally used for accidents that resulted in losses in nature or the human environment (for example natural disasters), but nowadays the term "hazard" is used more often. The change from the use of the word "disaster" to "hazard" also has an impact on the understanding of the heritage context. As mentioned by UNESCO (2010),[3] there are three groups of hazards on heritage sites: natural, human-induced, and indirect (see Fig. 1). Two of them, human-induced and indirect, can be prevented, even though they still occur no matter what we have tried, where, in the end, we tried to minimize the impacts. But UNESCO, in the book Managing Disaster Risk for World Heritage, just directly mentions those hazards, even though various indirect hazards happened in the Heritage Context by development or, we can say, urbanization. The indirect hazards that impact heritage objects are such as urban heat islands, floods caused by poor drainage system, or mass tourism. In those cases, ICOMOS[4] 2011 launched Guidance on Heritage Impact Assessment for Cultural World Heritage Properties,[5] followed by Heritage in Urban Contexts: Impacts of Development Projects on World Heritage Properties in Cities,[6] and Guidance and Toolkit for Impact Assessment.[7]

2.2 *Sustainable Development: Awareness of Disaster Risk Management on Heritage Site*

Effective disaster risk management strategies tailored to cultural heritage must be developed according to the unique traits of cultural assets and the types of hazards prevalent in a particular region. This necessitates the involvement of competent professionals, administrators, and policymakers capable of considering diverse factors to craft risk management plans specifically addressing the preservation of cultural heritage. The most important thing about this plan is that the management is not only implemented after a disaster occurs but also long before a disaster occurs. Disaster risk management must start from the risk identification phase. It may be undertaken at individual heritage buildings and heritage sites at the urban level. The risk identification process can be done in 3 ways; site observation, in-depth to the people's experience, and analysis of the statistical data [5] (Fig. 2).

---

3 [12]

4 International Council of Monument and Sites.

5 [4]

6 [11]

7 [1]

	Natural	Human-induced	Indirect/Secondary
Meteorological	Hurricane Lightning Heavy precipitation		Flooding (coastal/rivers) Fire Mass movement
Hydrological (caused by high rainfall)	Flash flood Landslide / volcanic ash / lava / ice damming of a river Tsunami	Hydrological infrastructure failure (dams, levees, reservoirs, drainage system) Coastal protection failure (sea walls)	Disease epidemic Pollution
Volcanic	Lava flows Pyroclastic flows Ash and block falls Gases	Mining-induced (e.g. mud volcano)	Lahars (mudflows) Landslides Tsunami Fire
Seismic	Faulting Transient shaking Permanent deformation (e.g. folds) Induced movement (liquefaction and mass movement)	Dam- and reservoir-induced mass movement Mining-induced Explosion/nuclear induced	Mass movement Fire Flood
Mass movement (of snow, ice, rock, soil mud, etc.) (induced by slow-acting erosion or one of the above)	Falls Slumps Slides Flows	Unstable mining/construction waste spoil heaps	

Fig. 1 Relationship of natural and human-induced hazards. *Source* [12]

Historic Urban Landscapes (HUL) also become interesting to be seen as a holistic point of view in managing urban areas, not just for current heritage but to ensure future heritage. ICOMOS 2011 published guidance regarding HUL implementation, followed by a UNESCO book in 2016. This approach looks not just at all layers that shape an urban area but also deeply into all layers of the urban area, including layers of history, economy, social culture, and each urban area's uniqueness. Nevertheless, the truth is that HUL approaches are familiar in managing urban contexts. HUL's core is to know what the previous-now-future is and what the urban element itself is; somehow, this point of view is just like seeing the urban area with typology and morphology and getting added social and cultural features. HUL may help to solve the key governance challenges affecting heritage in an urban context [2]. The HUL approach also assumes that promoting a more integrated urban management system, including a broader range of stakeholders, residents, investors, experts, and consequently, heritage, is to broaden and interlink diverse categories and attributes. A good implementation can help stakeholders make an objective movement by understanding the layer of the urban area and can predict the impact. HUL may help to solve the key governance challenges affecting heritage in an urban context [2]. ICOMOS 2011 published guidance regarding HUL implementation, followed by a UNESCO book in 2016. The HUL approach also assumes that promoting a more integrated urban

Fig. 2 Disaster risk management cycle for cultural heritage sites. *Source* (Jigyasu and Aurora 2011)

management system, including a broader range of stakeholders, residents, investors, experts, and heritage, will broaden and interlink diverse categories and attributes.

3 Research Methodology

This research critically analyzes several cases of heritage management by the Indonesian government by reflecting on how the world is doing. The output will focus on the result of Indonesia's government regulation in handling heritage sites and managing disaster risk. The data used in this paper is obtained from cases in 2019 and advanced after the Presidential of Indonesia Regulation regarding the National Super Priority Tourism Area was issued. The data was obtained by literature review, and some events happened in the Borobudur area, Magelang District. The analysis was brought by comparing variables from disaster risk guidelines and what has been done by the

Indonesian government. It should be acknowledged that there is still a need for further research and discussion with all sectors involved in heritage management in Indonesia, especially those undertaking disaster risk management.

4 Findings and Discussion

This paper grouped the findings into three main topics (see Fig. 3), which influence chronological events. The first finding is that hazards nowadays have unclear triggers, and discussing the correlation between them is mandatory to know how to prevent or minimize the impact. The second finding is the bias of regulation on managing hazards on heritage sites, especially when it is related to urban development. That happened because of the inability to state the cause of the hazard, where it cannot be denied. The ambiguity of who should take responsibility sometimes impacts the action, either for prevention or mitigation. Lastly, it discusses the actions and limitations of Indonesia's National Disaster Management Agency as an Indonesian government body that handles post-disaster mitigation, especially with the heritage context or site (Fig. 4).

4.1 Human Induced or Natural Hazard: Prevention and Management of Hazard

Disasters in the context of heritage sites can be grouped into two categories: natural disasters and human induced [5]. It is already known that Java Island has a high frequency of earthquakes. Besides, in Magelang, many typhoons happen due to the changing seasons. One of the massive damages to Heritage objects in Magelang Regency was a volcanic eruption in 2010. Before that, an earthquake in 2006 also caused an impact not just on the object but also on social-economy surroundings. When this occurred, there still needed to be strict regulation or guidance in managing the disaster at heritage sites, which made the act just immediate action without specific guidance for heritage sites. After this occasion, the government became aware that having regulations or guidance in managing post-disaster heritage sites is necessary.

With the rapidness of urban development, and because of the issues about disasters led by government regulation, which started even before the presidential regulation regarding labelling the Borobudur area as a National Super Priority, more unpredicted disasters have happened. UNESCO, in 2011, already established guidance about Heritage Impact Assessment (HIA), which was led by urban development's surrounding heritage (tangible or intangible). To prevent a big loss like what already happened, they encourage all state parties owning a world heritage to conduct HIA before implementing the project. The successful regulation can also lead to natural

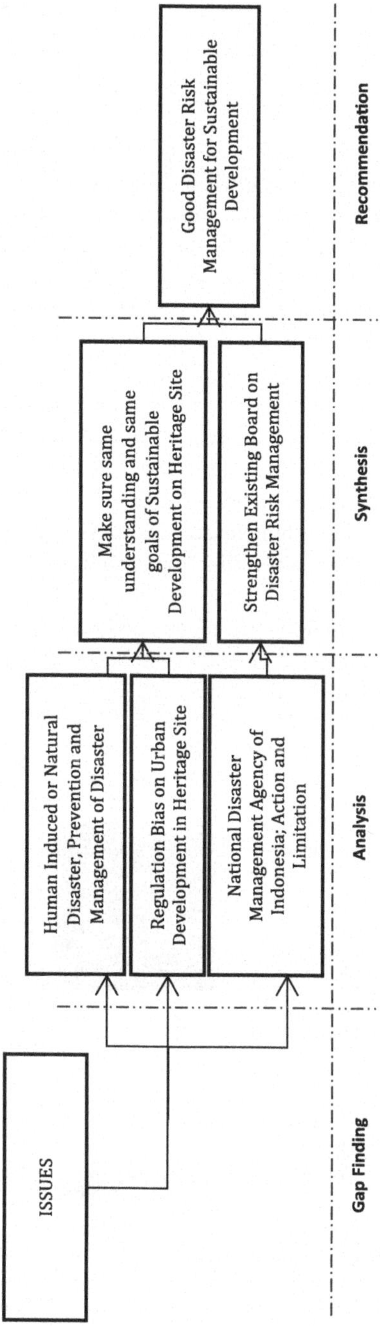

Fig. 3 Flow of research

Fig. 4 Influences on findings

disasters, such as landslides. This happened in Magelang even before the government's proposed project got approval. The site's condition is alert; there is already a landslide on some edge of the site. With the proposed project functioning as a tourism attraction with a high-intensity load, it can be sure there will be a landslide. Why so? Because the proposed project did not provide the action to prevent such an event, it just focused on the aesthetic side of the sites.

Over the last few years, whether it is natural or unnatural, the hazard has become clear. Floods and landslides, for instance, are common natural disasters driven by humans. In the heritage context, those can be judged after looking back at what has occurred. Separation blunders are occurring in the Borobudur area, as government projects that lack proper planning have many consequences. So, regardless of whether the disaster is natural or caused by humans, all disasters have an impact (direct or indirect). Then, disaster prevention and mitigation cannot be separated; they must be discussed together (Table 2).

We cannot deny that disaster will never happen and impact the heritage, whether prevention or not. UNESCO, in 2011, already established guidance about Heritage

Table 2 Example of natural hazard and human induced that happend in borobudur area

Natural disaster		Human induced	
	Volcanic eruptions directly impact heritage objects (in 2010). Conservers need to cover the stupa to protect it from the acid reaction caused by the ash. Just after that, the government made protection guidelines regarding volcanic eruptions		Because of massive intervention in unnecessary design surrounding the Borobudur area, pedestrian ways have become an economic space. The brightness of the light attracts people to come and occupy the space
	In 2023, some village buildings collapsed due to the typhoon. All structures need to be reconstructed. This brings economic impact where some events should be rescheduled, and the place closed for a week		As part of presidential regulation, a huge incentive is given to people who live around to "make up" their houses to look as in the photos. This creates a bias of the truth of architectural expression

Impact Assessment (HIA), which was led by urban development's surrounding heritage (tangible or intangible). To prevent a big loss like what already happened, they encourage all state parties owning a world heritage to conduct HIA before implementing the project. This studies the impact of a planned phenomenon/event related to heritage. This is one of the Disaster Risk Management actions for heritage sites.

4.2 Regulation Bias on Urban Development in Heritage Site

The ambiguity in government regulation is not just because there is more than one sentence that should be held, but more importantly is because there is a layer for who should act regarding there is a disaster that occurs on a heritage site and damages the heritage, the responsibility will be ambiguity in some way. As an example, what happens when many tourists come at some time, making the suburb chaotic? The villagers complain to the local government. As the first layer that faces what happened in the field, they tend to take immediate action, followed by reporting to the higher level. At a higher level, these things did not catch their attention. Moreover, when the national government visits the site, sees what is going on, and gets a direct report, they act like locals should report to the central government first, and this kind of cycle repeatedly happens in the Borobudur area. When the bias is examined more closely, it becomes clear that all regulation is attempting to support the exploitation of heritage property. To make it wiser, the government makes regulations become layers and confusing where some advantage can be exploited, as happened in the Borobudur Temple Compound. All actors should have the same goals to protect heritage property with all means to achieve sustainable development, not as a constraint.

Strengthen the loyalty between each actor who is involved in the Borobudur area is the key to unite the regulation in this heritage site. Do not just establish a new board because it proves it cannot work effectively. Like what *Badan Otorita* first initiated, this board hoped to help organize and coordinate everything in Borobudur. However, this board cannot fulfil its jobs fully and is now focusing on tourism. It is a good achievement that Borobudur as a heritage is managed not just by one board but various boards and ministries. It should be a good move to prevent and manage disaster risks that will highly happen in the Borobudur area (Fig. 5).

5 Bridging Institutional, Community, and Heritage Management

In this modern, we agree that the understanding of cultural heritage is fluid and dynamic [10]. This all includes tangible and intangible aspects of heritage itself. These aspects cannot be viewed as two different groups, but they should be seen as a part of each other, which means even though they are two different forms,

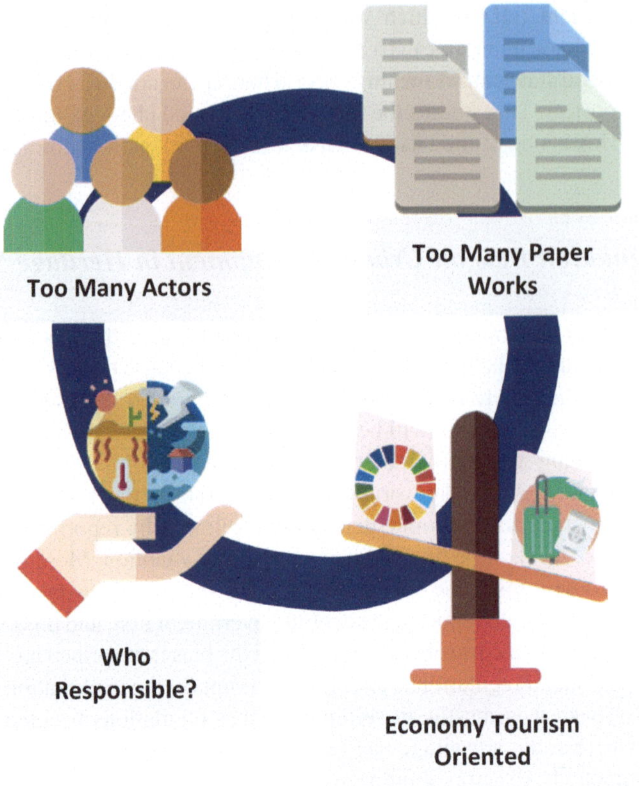

Fig. 5 Looping problems in handling disaster management

they stand as one entity, as cultural heritage. With a good understanding of cultural heritage, UNESCO adapts what UNDP[8] mentions in its agenda. In the Agenda 2030, they recognized that heritage conservation is a step to sustainable development by ensuring that development efforts are contextualized and closely attuned to the place and local social culture.

5.1 National Disaster Management Agency of Indonesia: Action and Limitation

It should be appreciated that the Indonesia government established The National Agency for Disaster Countermeasure (*Badan Nasional Penanggulangan Bencana/ BNPB*) in 2008. This board has existed since 1945 and is focused on handling the

[8] United Nations Development Programme.

issues regarding the families of war survivors after Indonesia became legally independent. The BNPB was legally established after Presidential Regulation 83 of 2005 concerning the National Disaster Management Coordinating Agency. After issuing Law Number 24 of 2007 concerning Disaster Management, the government issued Presidential Regulation Number 8 of 2008 concerning the BNPB. This board consists of a head, a directing element for disaster management, and an implementing element for disaster management.

This board will formulate regulations regarding disaster risk management (prevention and handling post-disaster) and coordinate disaster. The function is to coordinate the implementation of disaster management activities in a planned, integrated, and comprehensive manner. However, although in the first place, this board also helps with the impact of social-urban disasters, nowadays, they focus on natural disasters and refugee matters. Also, this board is not involved in some regulatory processes that will have an impact (as a disaster) following it.

6 Conclusion

In conclusion, Indonesia's disaster risk management regulations must be revised to ideal criteria. Existing organizations do not play the role they should do and do not have authority in development in their area, which can cause negative effects (human-induced hazards). Regulations regarding disaster risk management still need to integrate prevention and mitigation comprehensively. Moreover, the excessive number of interested parties and existing regulations in this area make it difficult to take action to protect cultural heritage. Government of Indonesia, national, regional, and district should have the same understanding of what sustainable development on heritage sites is. Then, with the support from other outsider of heritage object, strong local community, and tourist awareness, government can become stronger in managing disaster on heritage sites. Therefore, the author provides several recommendations:

1. Comprehensive Disaster Risk Management plans and clear regulations: All regulations should integrate and have a clear statement that mentions who and how obligated the system is. All regulations should follow the owner and who is the actor in handling the hazard.
2. Active Involvement of Disaster Risk Management Organizations (DRMO) as focused one organization: DRMO must be a caretaker in the pre-hazard stage. It should be done because they know the impact and how to handle it in the short- and long-term.
3. Community Involvement and Awareness: The government and the community must work together to raise awareness of pre- and post- hazards, with the aim to minimize the impact and sub-impact.

Acknowledgements The authors wish to acknowledge all researchers and assistant researchers under the Center for Heritage Conservation, Faculty of Engineering, Universitas Gadjah Mada for their work and dedication over the past two years of research in Borobudur Temple Compound Area. This droves the authors to publish the current topic to make an awareness of the importance of Managing Disasters on Heritage Sites.

References

1. Court S, Jo E, Mackay R, Murai M, Therivel R (2022) Guidance and toolkit for impact assessment. In: Wickenden J (ed) UNESCO
2. Cunha Ferreira T, Rey-Pérez J, Pereira Roders A, Tarrafa Silva A, Coimbra I, Breda Vazquez I (2023) The historic urban landscape approach and the governance of world heritage in urban contexts: reflections from three European cities. Land 12(5):1–15. https://doi.org/10.3390/land12051020
3. ICOMOS (1990) Advisory body evaluation: Borobudur temple compound. https://whc.unesco.org/document/153832
4. ICOMOS and World Heritage Centre (2011) Guidance on heritage impact assessments for cultural world heritage properties, vol 1. https://www.iccrom.org/sites/default/files/2018-07/icomos_guidance_on_heritage_impact_assessments_for_cultural_world_heritage_properties.pdf
5. Jigyasu R, Arora V (2011) A training guide: disaster risk management of cultural heritage in urban areas. Ritsumeikan University (RitsDMUCH), Kyoto, Japan, Research center for disaster mitigation of urban cultural heritage
6. Magelang Regency Disaster Data (2022) Regional Disaster Management Agency.
7. Murwanto H, Purwoarminta A (2019) The ancient Borobudur lake, history, and its evidences to develop geo-archeotourism in Indonesia. Indones J Geosci 6(1):103–113. https://doi.org/10.17014/ijog.6.1.103-113
8. Permana NR, Rahman F, Zani FR, Shafa D, Fadrian DF (2023) Identification of the ancient lakes based on gravity methods in Borobudur temple. JGE (Jurnal Geofisika Eksplorasi) 9(1):32–42. https://doi.org/10.23960/jge.v9i1.237
9. Rahmi DH, Tamimi S (2023) Traditional domestic architecture in the rural cultural landscape of Borobudur, Indonesia. ISVS E-J 10(03)
10. Ripp M, Rodwell D (2017) Governance in UNESCO world heritage sites: reframing the role of management plans as a tool to improve community engagement. In: Aspects of management planning for cultural world heritage sites: principles, approaches and practices. Springer International Publishing, pp 241–253. https://doi.org/10.1007/978-3-319-69856-4_18
11. Taylor K (2003) Cultural landscape as open air museum. Humlties Res 10(2):51–62
12. UNESCO, ICCROM, ICOMOS, and IUCN (2010) Managing disaster risks for world heritage. UNESCO
13. UNESCO World Heritage Centre; ICOMOS; ICCROM (2020) Heritage in urban contexts: impacts of development projects on world heritage properties in cities, pp 1–20
14. UNISDR (2009) UNISDR terminology on disaster risk reduction. United Nation, vol 1. www.preventionweb.net

Challenges of Sustainable Business Organization: Case Study of Young Craft Business Owner in Nepal

Richan Shrestha, Yurisha Upadhyaya, Prakriti Dangol, Sima Tamang, Fura Gelzen Sherpa, and Prakash C. Bhattarai

Abstract Sustainable business enterprises are an essential part of economics, environment, and society. However, there are challenges to sustain such enterprises. This study explores the challenges faced by young craft business owners in Nepal to sustain their businesses. The study used qualitative research design using a case study approach. Based on the literature review and consultation with agencies, five young craft businesses were identified for the detail case study. After reviewing establishment and growth process, in-depth interviews were conducted with those five young craft business owners in Kathmandu. The study has pinpointed that a lack of skilled youth, dependency on raw materials, market acceptance, and policy barriers as major challenges faced by such organizations. This study suggests that education and awareness about craft businesses are necessary for making craft businesses resilient. The findings of this study are useful for youth, policymakers, and craft business owners.

Keywords Sustainable development enterprises · Youth · Craft business · Resiliency · Nepal

1 Introduction

The concept of sustainability and sustainable development is growing worldwide with the aim of meeting the needs of the future without compromising the ability of future generations to meet their own needs. The term sustainable development was first reported by the United Nation (UN) in the 1987 Brundtland Report and since then has been gaining a lot of popularity [2]. According to this concept, a sustainable enterprise should aim to leave the world in a better place than it found, or at least no worse off [9]. According to McDonough and Braungart [24], the concept of sustainable development incorporates the three aspects generally considered to be

R. Shrestha (✉) · Y. Upadhyaya · P. Dangol · S. Tamang · F. G. Sherpa · P. C. Bhattarai
School of Education, Kathmandu University, Lalitpur, Nepal
e-mail: richan.shrestha@kusoed.edu.np

I. Pal et al. (eds.), *Proceedings of the 3rd International Symposium on Disaster Resilience and Sustainable Development—Volume 1*, Lecture Notes in Civil Engineering 623, https://doi.org/10.1007/978-981-96-2139-2_3

3P, or Triple Bottom Line (TBL), which are people, planet, and profit. In addition to that, in the concept of sustainable enterprises or businesses, there are additional 3 aspects of Ethics, Effectiveness, and Efficiency (3E) or Triple Top Line (TTL). Eskildsen et al. [12] explain that these 3Es—Ethics, which means 'values of life that become standards are based on moral judgment', Effective, which means 'choosing the correct method', and Efficient, which means 'doing things right—constitute a group of governance in a sustainable enterprise and are essential variables that must exist in the sustainable enterprise. Based on this theory, it can be discussed that Sustainable Development Enterprises (SDEs) should have six dimensions to look after: Self-capital, Ecology, Economy, Environment, Society, and profit. According to the International Labour Organization (ILO), Sustainable Business takes into account the impact on the environment as well as the economic and social aspirations of those who depend on it both inside and outside the organization. It operates a business to be viable, grow, and earn a profit.

With growing demand, increasing competition, and changing business strategies, companies have begun to realize the importance of a balance between their economic, environmental, and social priorities. In order to maintain successful development, they implement the principles of sustainable development into their business processes, including management, cooperation with stakeholders (suppliers, customers, local communities), planning of production, marketing, and sale of products [11]. These businesses aim to satisfy the requirements of both current and future generations of customers and stakeholders by creating and innovating new business strategies and initiatives that promote social change, protect and preserve environmental integrity, and enhance business performance.

The opportunities still lie in expanding the SDE with increasing consumer demand for sustainable products and services. Consumers are increasingly interested in purchasing products and services that align with their values around sustainability. SDEs can tap into this demand and create new markets for themselves [10]. Additionally, access to specialized financing with a wide range of financing options available specifically for SDEs in numerous countries has made it possible for impact investing and green bonds, thereby helping SDEs overcome the barrier of limited access to funding and grow in the market [16]. Similarly, according to Ngo et al. [26], numerous countries have started adopting a supportive policy environment where governments and international organizations recognize the importance of sustainable development and implement policies and programs to support SDEs. Thus, there are extensive barriers to implementing sustainability practices in SMEs, which involve not only the entire government but also the entire business sector, community issues, and individual limitations. However, if the identified drivers could be neutralized with the facilitators of sustainability implementation and the identification of potential areas for growth, then there could be a way to change the mindset and improve the desired behavior.

However, SDEs also face challenges. Despite the growing need and market for sustainable enterprises, there are still different barriers and challenges that hold companies back in their sustainability transformation. According to Ngo et al. [26], the major challenge for developing nations like Vietnam and others is a traditional organizational culture where companies are not open to new ideas and innovation due to the low popularity and recognition of the SDE concept. Another major issue is the lack of resources and the organization's limited access to financing. According to Hockerts and Wustenhagen [16], small sustainable enterprises often struggle to secure funding as investors may not be familiar with the unique risks and opportunities associated with sustainable development. Another challenge is the lack of skilled human resources in organizations. SDEs often require specialized knowledge and skills, which are usually lacking in companies, and due to the limited resources and infrastructure, there is difficulty scaling up beyond a certain size [10]. In addition to this, from the employer's perspective, it is identified that it limits employee freedom to be autonomous. Furthermore, high staff turnover and cynicism about whether the organization really wants to change are also mentioned in an article by RMIT [28]. Another organizational internal barrier is that senior leadership groups are not leading the change or are not committed to it. Additionally, due to legislative and institutional challenges such as bureaucracy, corruption, and political instability, employees are not motivated enough to acquire new skills for sustainable practices [5]. Also, existing regulations that are not designed to support SDEs are making it difficult for enterprises to succeed in certain markets [10, 26].

From the consumer's perspective, it can be identified that some consumers may not be willing to pay the premium prices that SDEs need to be profitable, which can limit the potential market for these enterprises [23]. At the individual level, personal barriers such as time crises, lack of connection with the natural environment, materialism, overconsumption, a lack of a common frame of reference regarding where to start, a lack of conservation ethics, an attitude-behavior gap, resistance to change, etc. are among the few identified barriers to sustainable practice adoption [37, 39]. Therefore, the four primary reasons for individuals not adopting sustainable practices are a lack of understanding, resigned or surrendered lifestyles, selfishness, and associated costs and taxes, which were also underlined by a global survey on sustainable consumption by WBCSD, as cited in Partidario et al. [27].

In the context of Nepal as well, even though small and medium-sized enterprises are growing and making their mark in the market [18], these businesses are facing challenges that impede their sustainable growth and development. As stated in the news article by Chhetri [8], businesses in Nepal are struggling with economic and political instability, restricted access to financing, infrastructural constraints, regulatory barriers, and a lack of skilled human capital. With consumers perceiving these sustainable products from small sustainable businesses to be expensive and substandard, it is evident that sustainable enterprises have a long way to go to secure a stable market position. Furthermore, it is stated that Nepal has adopted the Enabling Environment for Sustainable Enterprises (EESE) process, which was developed by ILO following the 96th International Labor Conference in Geneva in 2007 [7]. This process divides the enabling environment for businesses into four major categories: political, economic, social, and environmental. However, if we look at the ground level, Nepal falls far behind in meeting these categories. The increasing challenges from limited human capital, financial services, socio-economic divisions, and unequal linkage to communication and transportation are highly identified as another major bottleneck to the growth and development of such small and medium-sized enterprises. The study of dimensions of challenge of SDEs needs more rigor. However, challenging dimensions of SDEs in Nepal have hardly been studied. Thus, the main purpose of this study is to identify the challenges of SDEs (craft business) in Nepal and understand their resiliency.

In order to meet the objective, this study has used the theoretical lens of human capital theory, resource dependency theory, and institutional theory to integrate diverse challenges faced by the sustainable development organization. As discussed above one of the challenges is lack of skilled youth for organization. As per human capital theory, skilled human resources are the important factors for the growth of the organization [33]. The lack of skilled manpower will hinder the growth of the organization [32]. Thus, lack of skilled human resource is challenge for the organization. Likewise, another challenge discussed above is dependency on raw materials. Resource dependency theory suggests that for the survival of the organization, they need critical resources (Johnson Jr 1995). The lack of resources poses severe challenge for the organization to survive [4]. In line to this, above discussion suggests market acceptance and policy barrier as challenges for the organization. This can be explained with institutional theory which suggests rules, routines, and norms significantly influence organization on their development [36].

2 Methodology

To meet the objective, this study used qualitative research design utilizing the case study approach. Case study helps in analyzing the single phenomenon of a single individual, incidents, or events to test the theoretical issues [34]. Philosophical foundation of the case study can be either post-positivist or non-positivist. With the use of a post-positivist philosophical foundation, objective reality can be assumed and measured independently [29]. Many researchers like Yin (2014) utilized a positivist foundation to undertake research using a case study approach. However, researchers with a non-positivist foundation perceive reality of their own and they also believe in multiple realities [14]. Merriam [25] and Stake [31] have implemented a non-positivist foundation while conducting their research integrating the case study approach in their case study research. This study used a non-positivist foundation. Likewise, as suggested by Yin and Hollweck [38], information rich cases were selected. Participants were selected from the organizations registered with Fair Trade Group, Kathmandu (umbrella organization which ensures organizational sustainable practices). Only those organizations are included in the study that are members and follows the 10 principles of Fair Trade Group [13].

Table 1 Profile of the participants

S.no	Name of the participants	Gender	Organization	Position	Year of experiences
1	Ram	Male	Organization 1	Managing director	8
2	Shyam	Male	Organization 2	Owner	10
3	Gita	Female	Organization 3	Managing director	8
4	Sita	Female	Organization 4	Owner	7
5	Riya	Female	Organization 5	Chair-person	12

2.1 Data Collection

The participants for this study were craft business owners (two males and three females). Male participants are Ram and Shyam, whereas female participants are Gita, Sita, and Riya. The profiles of the participants are shown in Table 1.

The participants were contacted through telephone and asked for their availability. Once they agreed, the interview date was scheduled. To take interviews, this study implemented semi-structured interviews. This method of taking interviews helped researchers to focus on a few defined themes. Each interview lasted for 30–35 min. During the interview, researchers were patient and listened carefully to the thoughts of the participants. They were also given a chance to express their additional experience at the end of the interview which we failed to capture. The consent of every participant was taken for recording their interview. Researchers focused on minute details of their behavior during the interview. At the end of the interview, we asked them to add up their experience which we failed to capture. After the field work, every aspect of the interview was noted in notebooks. Also, pseudo names were used for participants.

2.2 Data Analysis

The recorded interviews were transcribed properly to analyze the data. Then, transcribed data was checked thoroughly for meaning-making and other language and grammar checks. After that, the transcribed data were coded. Following that, the coded data were categorized, and a proper theme was developed. Developed themes were checked with previous literature and connected it with relation to institutional theory. This theory suggests that organization is affected by external environments like pre-existing norms, rules, and beliefs Lammers and Garcia, [22]. These external challenges hamper the sustainability of the organization. This study also use this lens to see what are the challenges faced by organization and hampering their sustainability. According to Hanna et al [15], changes in societal value, legislation, and technology affects the sustainable activities and management of organization.

Similarly, various documents from the Fair Trade Organization and those five organizations were checked to verify the participants and organizational sustainability practices. Likewise, researchers observed the working process and sustainable practices of those organizations. For maintaining the credibility for quality standards, researchers did not utilize their preset thinking and minds during the whole process. As said earlier, the backgrounds of participants were thoroughly checked. A good rapport was developed with them. Final themes and their opinion were shared with them. Researchers maintained their ethics in their critical and reflective roles during the entire research process.

3 Findings of the Study

This study has found diversified challenges in craft business in Nepal. They are a lack of skilled youth, limited innovation, dependency on raw materials, market acceptance, and policy barriers. Participants also shared that there is no concrete plan for resiliency. Finally, they believe in education and awareness as a hope for the sustainability of the craft business. The findings are presented in the key themes as generated from codes.

3.1 Lack of Skilled Youth

During the interview, all the participants agreed on the challenge of the lack of skilled workers in the craft business. They believe youngsters today are not willing to join the craft business. In this regard, Participant Ram shared.

The major challenges for us are to attract youth to this business. Young generations are more willing to pursue their career in technology based careers or white-collar professions like office jobs. They understand the value of craft business.

This sharing shows that youngsters are not attracted to traditional craft business. Although there are good demands for these products, it is a big challenge for them to fulfill those demands due to a shortage of skilled labor. Participant Riya shared similar feelings.

There aren't enough employees in the workplace. We are therefore worried that we won't be able to fill every purchase if we receive a lot of them. The people working in the weaving area are mostly older. None of the people in the present generation have stated that they want to work in the handloom department. Since the quantity is so large and it will take more time to create them, we are unable to fulfill the order when it is received. We are only able to handle so much, even when we have an order for ceramics. This has also had an impact on the industry's overall growth. For this reason, the younger generation must develop their skills in order to continue this sector.

Thus, the lack of youth in this field is one of the major challenges for the craft business. Many participants also shared that many youths are willing to go abroad, creating a shortage of human resources in Nepal. In this regard, participants Gita shared.

There are fewer workers on the job every day. Everyone wants to go abroad either for further studies or for foreign jobs. Here are ample opportunities but they are fascinated with the lifestyles of abroad. Youth migration to other countries is an issue for our industry.

These sharing of the participants suggest that youth are not willing to pursue a career in craft business as they are more inclined towards technology based jobs, white collar jobs or willing to go abroad. And this trend has created the shortage of human resources in Nepal.

3.2 Dependency on Raw Materials

One of the main challenges faced by craft business is dependency on the raw materials to other countries. In Nepal there is no sufficient production of raw materials. Thus, craft businesses import such materials from other countries like India, China, and New Zealand. This has increased the production cost. In this regard, participant Riya shared.

Being a landlocked nation, we rely on another nation for transit for importing. The outcome is that the product is extremely expensive. Frequently, we face two types of challenges. To begin with, the cost of the fundamental components is increasing because it is a natural product. For example, copper is getting more expensive. In a similar vein, cotton was originally bought from Nepalgunj. Because Nepal lacks raw materials suitable for industrial usage, we are dependent on wool from India and the Netherlands.

Thus, due to dependency on raw materials, Nepali craft business is losing its competitiveness in terms of price. Dependency of raw materials has also added to the management complexities. According to participant Shyam,

In terms of raw materials for clay manufacture, red clay is purchased from Dang, while white clay is purchased from India. We assisted a technical person in the clay department in establishing a place in his house, and he then refines and supplies us with the refined material made from raw material supplied from India. This process has added extra burden in our business.

Nepal is rich in biodiversity. But, craft business heavily depends on other countries for raw materials. This has increased the production cost. Thus, it is difficult for Nepalese companies to decrease their production cost which is hampering their competitiveness.

3.3 Market Acceptance

Nepalese craft products are expensive as they are handmade as well as their production cost is high because of various reasons like dependency of raw materials as stated above. Thus, consumers are not willing to buy these products. On this participant Sita shared.

It is very hard for us to convince the importance of our product. Our products are handmade. Also, the production cost is high. It can be identified that some consumers may not be willing to pay the premium prices that SDEs need in order to be profitable, which can limit the potential market for these enterprises.

People are also less aware about these sustainable products. Consumers care about the price but not the benefits of sustainable products. Participant Ram shared.

They must be aware of the environmental initiatives we have done in terms of raw materials, such as packaging in the form of paper-based goods and water treatment. Our equipment is powered by solar. We also have a system for collecting rainwater. For the reservoir needed for rainwater collection, we have spent two to three lacks. We consider our obligation to protect the environment as a result. However, customers do not see these efforts. They ignored our products because of the price.

These shares suggest that consumers in Nepali market are not aware about the importance of sustainable products. Since the production costs of craft business are high, their prices are higher compared to other products. This has led to the non-acceptance of craft products in Nepalese markets.

3.4 Policy Barriers

Nepal's policy is also not helping the craft business. The policy barrier is another major challenge for the craft business. The policy does not address the concern of the craft business. In this regard, participant Riya shared.

A big impediment is the policy barrier. The government has done a lot of lobbying, and we've had several discussions about it. There are many things that are required of us. We bring a lot of foreign cash, but there are no appealing benefits to exporting.

Likewise, the dependency of the raw materials is high for craft business in Nepal. But the government is doing nothing. Regarding this, participant Ram shared.

In our country, there are no raw materials. There was once cotton from Nepalgunj. But now, they do not supply us. We rely heavily on wool from India and the Netherlands since we use it so much. Farming is not conducted at the industrial level. The government has various intervention possibilities, yet it is doing nothing.

Thus, the policy of Nepal is not helping craft business. Although there are opportunities, the Nepalese government is not able to bring policies that support craft business.

3.5 *Lack of Resilient Strategy: Education and Awareness as a Hope*

Another challenge is not having a concrete strategy to cope with the challenges among craft businesses. They are just running their business without any planning. On this, participant Riya shared.

As of now, I am running my business as per the situation. I have no concrete planning or any long-term strategy. Whatever problems come, I try to deal with them accordingly. Sometimes it works and sometimes, it does not. Somehow, I am surviving. However, with growing challenges, I fear I may have to shut down my business.

There is no plan for sustainability of the business. However, they hope education and awareness about craft business among youth will help to sustain craft business. In this regard, Participant Sita shared.

The only hope is to educate and motivate youth to consider craftsmanship and its business as their career.

With educated and aware customers, they will understand the importance of craft products for sustainability. Thus helping the business grow. Participant Shyam has a similar opinion.

For the sustainability of the craft business, there must be good education and awareness among consumers and youth. If consumers are aware, they demand craft products irrespective of price. Likewise, if youth are aware, they understand the importance of craft products. It will motivate them to engage in the craft business.

This sharing shows that there is no concrete planning for the resiliency of the craft business. For resiliency of the craft business, the education and awareness regarding craft business are important.

4 Discussion

Thus, from the sharing of participants in the previous section, synthesize the following challenges: lack of skilled youth, dependency on raw materials, market acceptance, and policy barriers. These challenges of craft business are consistent with human capital theory, resource dependency theory, and institutional theory respectively.

One of the major challenges found by this study is lack of skilled youth in the craft business. Due to the lack of skilled youth, it is difficult for the craft business to sustain their business. This challenge is also mentioned by Ngo et al. [26], where poor management skills and limited quality human resources are considered to play a decisive role in the survival and success of enterprises. Although the demand is good for craft products, due to less human resources, craft businesses are not able to fulfill the demand. In line to this, the study of Khanal and Maharjan [20] also claimed that lack of human resource is limiting them to fulfill the demand and hampering their business. This is consistent to human capital theory. As suggested by human capital theory, lack of skilled human resource hinders the growth of organization [32].

Another challenge for craft business is dependency on raw materials in other countries. Numerous studies, including Bhandari et al. [3], Mangal [23], and Ngo et al. [26], have also noted the lack of raw materials, increased dependency on raw materials, and undersupply of sustainable raw materials as major organizational external barriers faced globally. Due to the scarcity of locally accessible raw materials, the organization is increasingly relying on imported raw materials from foreign countries. This has increased their production cost and hence the prices of the product are higher. This has reduced the competitiveness of the craft business. In line to this, the study of Mangal [23] claimed that due to dependency on raw materials, their production cost is high. This has led to the higher price of the product. Thus, as per Mangal [23], consumers are no longer willing to pay the premium price, which is limiting the market potential of these enterprises in Nepal. This is consistent to resource dependency theory. According to resource dependency theory, lack of critical resources derails the growth of the organization [4].

Likewise another challenge faced by the craft business is market acceptance. Nepalese craft products are expensive as they are handmade as well as their production cost is high because of various reasons like dependency of raw materials as stated above. Thus, consumers are not willing to buy these products. Mangal [23] also claimed that consumers are no longer willing to pay the premium prices, which is limiting the market potential of these enterprises in Nepal. The consumers in Nepal are not aware of the importance of sustainable products. Thus, they are not willing to pay a higher price. The findings also coincide with that of Alayon et al. [1], highlighting that non-acceptance of the product impacts an organization's development.

Finally, the other challenge faced by craft businesses is the policy barrier. The policy barrier has also emerged as a significant obstacle for sustainable enterprises. Other studies also claimed policy barriers as a major hurdle. For example, despite lobbying efforts and discussions with the government, there is a lack of appealing benefits and support for exporting [21]. The availability of raw materials is limited, and the enterprise heavily relies on wool imports from countries like India and the Netherlands. In addition to this, the increased cost of taxes has created a barrier to the enterprise's growth. A study by Mangal [23] also indicated a lack of guidance from the government, which prevented them from improving the scope of their sustainability programs. Another finding by Timsina [35] also mentioned that exporters are not getting pre-export loans and that the frequent changes in government are changing the policy and creating an administrative hurdle in export procedures. This is consistent with institutional theory. The rules or norms influence the growth of the organization [36].

Another challenge for the organization is lack of concrete plan for the resiliency of the organization. There is no plan for sustainability of the business. They are just operating on the mentality of as it goes. However, there must be alternative plans like use of technology to sustain the organization as suggested by Hussain [17]. Similarly, education and awareness are required for saving craft business. Numbers of studies [6,

30] have also considered education and awareness as decisive components. Educating youth and interested adults on craftsmanship and technology is a must in order to attract consumer engagement as well as grow the market for businesses, which is constantly facing downfall.

5 Conclusion

Businesses aim to satisfy the requirements of both current and future generations of customers and stakeholders by creating and innovating new business strategies and initiatives that promote social change, protect and preserve environmental integrity, and enhance business performance. However, there are extensive barriers to the implementation of sustainability practices in SMEs, which involve not only the entire government but also the entire business sector, community issues, and individual limitations. Similar is the case with craft business in Nepal. With the interview among young craft businesses in Kathmandu, this study identified lack of skilled youth, dependency on raw materials, market acceptance, and policy barriers as major challenges faced by such organizations. This study also suggests that education and awareness about craft businesses are necessary to make craft businesses sustainable.

The strength of this paper is the contemporary issue it raised. In Nepal, various organizations are facing challenges for their sustainability. This study provides the insights on the issue. The findings from this rigorous study provide knowledge on challenges and highlighted the importance of education for resilient strategy among organizations. However, there are certain limitations on this study. First of all, this study implements the qualitative case study research design. Secondly, this paper is focused on sustainable development enterprises only. Finally, the study area of this research is Kathmandu only. These limitations hinder the generalizability of the findings.

The findings of this study are helpful for youth as they can understand the importance of craft business as their career. Likewise, this study is important for policymakers. They can create a policy regarding education and awareness of the craft business. Finally, this study is important to craft business owners to understand the challenges that exist in the business.

References

1. Alayon CL, Safsten K, Johansson G (2022) Barriers and enablers for the adoption of sustainable manufacturing by manufacturing SMEs. Sustainability 14(4):2364
2. Bawono A, Rahmana F (2021) Sustainable development toward sustainable enterprise. Manag Sci Lett 11(2):657–668
3. Bhandari N, Garza-Reyes JA, Rocha-Lona L, Kumar A, Naz F, Joshi R (2022) Barriers to sustainable sourcing in the apparel and fashion luxury industry. Sustain Prod Consum 31:220–235

4. Biermann R, Harsch M (2017) Resource dependence theory. Palgrave handbook of inter-organizational relations in world politics, pp 135–155
5. Bocken NM, Geradts TH (2020) Barriers and drivers to sustainable business model innovation: organization design and dynamic capabilities. Long Range Plan 53(4):101950
6. Cagno E, Trianni A, Spallina G, Marchesani F (2017) Drivers for energy efficiency and their effect on barriers: empirical evidence from Italian manufacturing enterprises. Energ Effi 10:855–869
7. Chalise A (2021) Sustainable businesses in Nepal: in between sustaining and surviving. The Farsight
8. Chhetri B (2023) Overcoming business challenges in Nepal: strategies for sustainable growth. Online Khabar English news
9. Collings T (2023) What is a sustainable enterprise? GreenMBA. What is a sustainable enterprise?—The green MBA
10. Costache C, Dumitrascu DD, Maniu I (2021) Facilitators of and barriers to sustainable development in small and medium-sized enterprises: a descriptive exploratory study in Romania. Sustainability 13(6):3213
11. Dvorakova L, Zborkova J (2014) Integration of sustainable development at enterprise level. Procedia Eng 69:686–695
12. Eskildsen JK, Edgeman R, Mønster D, Olesen JV (2013) Response based segmentation in satisfaction surveys. In: Performance management association Australasia conference
13. Fair Trade Group (2024) 10 principles of fair trade. Fair trade group. https://www.fairtradegroupnepal.org/principles
14. Greener S (2008) Business research methods. Ventus Publishing ApS, Frederiksberg, Denmark
15. Hanna B, Xu G, Wang X, Hossain J (2023) Blockchain-enabled humanitarian supply chain management: sustainability and responsibility. In: Blockchain in a volatile-uncertain-complex-ambiguous world. Elsevier, pp 251–276
16. Hockerts K, Wustenhagen R (2010) Greening goliaths versus emerging Davids—theorizing about the role of incumbents and new entrants in sustainable entrepreneurship. J Bus Ventur 25(5):481–492
17. Hussain Z (2021) Paradigm of technological convergence and digital transformation: the challenges of CH sectors in the global COVID-19 pandemic and commencing resilience-based structure for the post-COVID-19 era. Digit Appl Archaeol Cult Herit 21:e00182
18. ILO (2018) Enabling environment for sustainable enterprises in Nepal. International labor organization
19. Johnson Jr BL (1995) Resource dependence theory: a political economy model of organizations
20. Khanal NP, Maharjan KL (2010) Sustainability of community-based seed production enterprises in Nepal: institutional issues. Nepal Agric Res J 10:34–41
21. Kharel P, Dahal K (2020) Small and medium-sized enterprises in Nepal: examining constraints on exporting, no 1166. ADBI Working Paper Series
22. Lammers JC, Garcia MA (2017) Institutional theory. The SAGE handbook of organizational communication: advances in theory, research, and methods, pp 195–216
23. Mangal B (2022) Assessment of barriers in sustainable business management of small and medium sized enterprises: a systematic literature review. SSRN 4294111
24. McDonough W, Braungart M (2002) Design for the triple top line: new tools for sustainable commerce. Corp Environ Strateg 9(3):251–258
25. Merriam SB (1998) Qualitative research and case study applications in education. Revised and expanded from case study research in education. Jossey-Bass Publishers, 350 Sansome St, San Francisco, CA 94104
26. Ngo TL, Ngo QD, Tran MD (2019) Sustainable development of enterprises: barriers and policy recommendations. Sustain Dev 10(16)
27. Partidario MR, Vicente G, Belchior C (2010) Can new perspectives on sustainability drive lifestyles? Sustainability 2(9):2849–2872
28. RMIT (2021) Barriers to change. RMIT University, FutureLearn, Barriers to change

29. Rashid Y, Rashid A, Warraich MA, Sabir SS, Waseem A (2019) Case study method: a step-by-step guide for business researchers. Int J Qual Methods 18:1–13. https://doi.org/10.1177/1609406919862424
30. Shi H, Peng SZ, Liu Y, Zhong P (2008) Barriers to the implementation of cleaner production in Chinese SMEs: government, industry and expert stakeholders perspectives. J Clean Prod 16(7):842–852
31. Stake RE (2005) Qualitative case studies. In: Denzin NK, Lincoln YS (eds) The Sage handbook of qualitative research, 3rd edn. Sage, Thousand Oaks, CA, pp 443–466
32. Strober MH (1990) Human capital theory: Implications for HR managers. Ind Relat: J Econ Soc 29(2):214–239
33. Sweetland SR (1996) Human capital theory: foundations of a field of inquiry. Rev Educ Res 66(3):341–359
34. Thomas G (2019) Case study foundation entries. London. CA: Sage. https://doi.org/10.4135/9781526421036
35. Timsina P (2019) A study on prospects and problems of Nepalese handicraft export. [Master's Thesis, Tribhuvan University]. https://elibrary.tucl.edu.np/bitstream/123456789/8534/1/All%20theshis.pdf
36. Tina Dacin M, Goodstein J, Richard Scott W (2002) Institutional theory and institutional change: introduction to the special research forum. Acad Manag J 45(1):45–56
37. Ukaga O, Maser C, Reichenbach M (eds) (2010) Sustainable development: principles, frameworks, and case studies. CRC Press
38. Yin RK, Hollweck T (2014) Case study research design and methods. Can J Program Eval 30(1):108–110. https://doi.org/10.3138/cjpe.30.1.108
39. Young W, Hwang K, McDonald S, Oates CJ (2010) Sustainable consumption: green consumer behavior when purchasing products. Sustain Dev 18(1):20–31

Constitutional Law Provisions Towards Disaster Management in Thailand

Wachiraporn Poungjinda and Shubham Pathak

Abstract Democracy involves fundamental rights, minimizing social, political, and economic risks governed by legal philosophies. The Thai Constitutional law is epitome of the law system. It governs all sectors including public and private and aims at ensuring fairness, justices and implying with laws. This research explores, understands and analyses the various legal implications upon the business sector in Thailand. The methodology adopted for this research aims at exploring, understanding and analysing the existing legal framework which has a direct effect upon business sector amid disasters. The data collection involved various legal Acts, government reports and publications. To support the government data, several research and online publications have been also analyzed to reach a holistic and comprehensive theoretical model to achieve sustainable future. The findings include the ineffective implementation of Thai laws during disasters in the rural and remote areas, mismanagement of the business sector resulting in several closures. The research provides for a comprehensive discussion and recommendations to enhance sustainability in Thailand amid disasters such as COVID-19 pandemic, floods, droughts, earthquakes and Tsunami. This research could be replicated or adopted in the other developing countries in the Association of Southeast Asian Nations (ASEAN) region. The limitations of this research including the language barrier have been tackled with official English language translations of the Thai Constitution and other legal documentations. The research paves way for future development and growth of the existing legal, business and disaster management scenarios in the developing world.

Keywords Constitutional law · Business management · Disasters · Thailand · ASEAN · Sustainable development

---

W. Poungjinda
School of Law, Walailak University, Nakhon Si Thammarat, Thailand

S. Pathak (✉)
School of Accounting and Finance, Center of Excellence in Sustainable Disaster Management (CESDM), Walailak University, Nakhon Si Thammarat, Thailand
e-mail: shubhampathak@gmail.com

I. Pal et al. (eds.), *Proceedings of the 3rd International Symposium on Disaster Resilience and Sustainable Development—Volume 1*, Lecture Notes in Civil Engineering 623, https://doi.org/10.1007/978-981-96-2139-2_4

1 Introduction

Constitution is the basis of democracies. Democracy involves fundamental rights, minimizing social, political, and economic risks, governance governed by legal philosophies, justice for all, principle of equality and fairness. Thailand has a mixed system of Monarchy led constitutional (Constitutional Monarchy) system of governance [2]. Current version of The Constitution of Thailand B.E. 2560 (2017) (a.k.a. People's Constitution) provides articles related with disaster management. Thailand boasts a unique and admirable system of governance, as it operates under a mixed system of constitutional monarchy.

It is crucial that we ensure our internal domestic land laws are in line with international SDG (Sustainable Development Goals) norms. By doing so, we can pave the way for a more sustainable and equitable future. These norms encompass a wide range of issues, from poverty eradication to responsible consumption and production, and they provide a comprehensive framework for sustainable development. By aligning our laws with these norms, we can create a more just and sustainable society that benefits everyone, from individuals to entire nations.

The monarchy holds a ceremonial and symbolic role in the administration, while the government is led by a written constitution. This system has proven to be effective in promoting stability and peace in the country, while also allowing for a fair and democratic process of governance. It is a system that has been existing and continues to be a source of pride for the Thai people. The constitution of Thailand has been revised several times due to political instability and coups in the past. However, the legal framework has been strong during the uneven governance system in Thailand.

2 Development of Constitutional Law and Disaster Management in Thailand

Due to the presence of Thai Constitution, the Thai people have a positive perception towards the judicial system. The Thai Constitutional law is the embodiment of the law system. It governs all sectors including public and private and aims at ensuring fairness, justice and implying with laws. Apart from various existing rights, the 2017 constitution version added the right to a healthy environment as well.

The Constitution of Thailand protects and formulates the legal rights of the Thai people and the governance throughout the country. The rights include basic human rights, comprehensive protection of people of Thailand including social, cultural and economic preservations. The Thai legal and political framework adopts the international standards and legal norms to ensure the amalgamated flow of legal rights protection. The European Union law including the British legal framework is the guiding format that the Thai Legal system follows to be adopted through the Thai Constitution. However, the divided focus and guiding forces results in inadequate adaptation in Thailand. Therefore, Thailand requires to revisit the existing laws and ensure the adequacies of the legal framework.

The existing governance framework, under the leadership of Thai Monarchy governs social, cultural, political and economic environments. Sovereignty of Thailand has been an essential aspect of Thai legal and economic systems. The parliament and department governance is directly related with each ministry and department in Thailand.

Thailand has been struggling with rights, liberties, and changes in the legal and social setup. The protection of basic rights is protected; however, the implementation is limited due to several different governance and implementations channels in Thailand, especially at the local levels. The amalgamated complex framework between the judiciary, economic and democratic rights become difficult to implement and be successful.

Thailand has a long history of political instability and then drastic changes in the frameworks at all levels of the government. The discrepancies in the legal and related political framework leads to the inadequate protection of rights of the Thai people. These implementations become more restricted in the times of disaster events and disruptive events. The legal protection of rights in Thai Constitution lacks in terms of apt implementation and ensure sustainable development in Thailand.

The research focuses upon various articles of the Thai constitution and following Acts, bylaws, ordinances have which leads to the understanding and favourable recommendations to enhance the existing framework. Several rights and democratic protection have been included in the People's Constitution, however, the implementation of these is found to be lacking and require revisiting and amendments at various levels.

There are several commissions which define the rights and responsibilities for the Thai nationals, for instance the following Section 247, defining the protection levels for the Thai people (Constitution B.E. 2560 (2017), 2021). The recent pandemic of COVID-19 led to the exposure of the inadequate legal and political framework. The youth movement 2020 and the feeling of lack of rights protection among the Thai people during the pandemic depicts the lack of the comprehensive protection of democratic rights of the Thai people [13, 18]. The essentialities of the holistic legal framework have been lacking in the existing framework and requires to be explored, understood, and amended in future to ensure resilient and sustainable environment in Thailand [12].

Thailand is a party to the Berne Convention for the Protection of Literary and Artistic Works since July 17, 1931, and ratified for the international court of justice in 1980 (Berne Convention, 2021). This provides for the protection of copyrights in Thailand and is incorporated into the constitution of Thailand. This has been continued with the collaborative Human rights declaration among the Association of Southeast Asian nations (ASEAN) nations [4].

European Union (EU) has been epitome of preserving justice in terms of constitutional, human rights and copyright. The successful implementation of preserving the rights of the population throughout the continent with ensuring sustainable development in Europe could be adopted by the developing nations. Thailand has much to gain by adopting the exemplary business laws of the European Union. These laws have proven to be effective and successful, providing a solid foundation for businesses to thrive and grow. By following the EU's lead, Thailand can improve its business environment and attract more investment, leading to greater economic growth and prosperity for all [13].

The European Union laws has been propagated in several developing economies including Thailand [5]. There has been jurisprudence towards acceptance, understanding and regulatory implementation of the rights towards national population in EU as well as in Thailand. However, there is a profound gap between the adequate implementation of the rights at the local and national levels.

European Union constitutional law has been developed over a period and provides for a comprehensive tool to govern the rights of the citizens of its member nations. The Treaty on Functioning of the European Union (TFEU) has been an exemplary tool since 2007, on how to preserve the justice and rights of the people of various member states [32]. The human rights have been protected under the Article 218, 6(a) (ii). Likewise, the Article 118 protects the intellectual property rights which is applicable throughout the European Union.

The rights are adequately protected along with the inclusion of general principles in the constitutional law in European Union [39, 23]. These inclusion and legal framework of protection of rights of people in European Union are to be adopted in Thai legal context. The various similar approach towards justice and equality for all has been the center of the Thai laws and policies. Hence, it is crucial to incorporate the diverse legal frameworks that have been proven successful in the European Union.

The legal courts in European Union have been divided into the two major courts: Court of Justice of the European Union (CJEU) and European Court of Human Rights (ECHR) [26]. The security of human rights and copyrights have been an increasing concern in the European Union in the past decades. The tried and tested policies and adequate implementation measures are ready to be adapted by the developing countries. Thailand is member state to several international legal framework and would be benefitted by the adoption of European Union laws and legal approach towards protect of rights of the people.

Thailand needs to ratify several agreements and declaration to achieve transparency and accountability among its legal structure [8]. For instance, Thailand implemented the Copyright Act (1994) and thereof went on to ratify the Free Trade Agreement (FTA), however, Thailand still need to adopt the World Intellectual Property Organization (WIPO) Performances and Phonograms Treaty 1996 (WPPT). Similarly, Thailand requires to strengthen the regional growth among ASEAN member nations in the lines of the policies adopted and implement in the European Union, such as Common Foreign and Security Policy (CFSP) in the European Union member states [37].

Thailand lacks the adoption of the human rights for its citizens. The civil disobedience is a direct result of the absence of the policies and infrastructure towards ensuring the protection of the rights of the Thai nationals. One of the appropriate examples from the European Union is the setting up of the national human rights institution

(NHRI) which protects the human rights through the litigation-based support at the European Court of Human Rights [17].

Thailand's persistent failure to adopt and enforce human rights protections is having a devastating impact on its ability to develop and grow as a nation. The absence of these fundamental safeguards leaves the country's citizens vulnerable to exploitation, discrimination, and other forms of abuse, which in turn creates an environment that is highly unstable and unpredictable for businesses and investors. It is imperative that Thailand takes decisive action to address this issue and provide its people with the basic dignity and respect they deserve, to build a more equitable and prosperous society for all.

Thailand's blatant disregard for human rights is severely impeding its potential for business growth and overall development. The lack of protection for citizens leaves them vulnerable to exploitation and abuse, which creates a highly unstable and unreliable environment for investors. It is imperative that Thailand takes immediate action to prioritize the well-being of its people, and uphold basic human rights, to unlock its full potential for economic prosperity.

The disaster management act was created after the disastrous 2004 Tsunami, known as Disaster Prevention and Mitigation Act (DPM Act 2007). Furthermore, the National Disaster Risk Management Plan (2015) was implemented and revised in 2020.

Following the devastating Tsunami that hit Thailand in 2004, the country took an important step by creating the Disaster Management Act, also known as the Disaster Prevention and Mitigation Act (DPM Act 2007). This Act was designed to help protect the people of Thailand by providing guidelines for disaster prevention and mitigation. Additionally, the National Disaster Risk Management Plan (2015) was put into effect and recently updated in 2020. This plan helps to establish a clear framework for disaster response and recovery efforts.

This study aims to provide a comprehensive analysis of the various legal implications that the Disaster Management Act and the National Disaster Risk Management Plan have for Thailand's business sector. By examining these legal implications, businesses in Thailand can better understand how they can prepare for and respond to disasters. By taking proactive measures, businesses can help to mitigate potential risks and minimize the impact of disasters on their operations. This study provides an important resource for businesses in Thailand looking to stay ahead of the curve in disaster management. This research explores, understands and analyzes the various legal implications upon the business sector in Thailand.

Disaster management requires robust legal framework to ensure adequate disaster management and faster disaster recovery. Thailand has the legal framework with its constitution at the highest level of legal document and all the ordinances follow the provisions [33].

Thailand has a mixed system of Monarchy led democratic system of governance [2]. The current version of the Thai constitution is Thai constitution B.E. 2540 (1997) [15, 21]. Apart from various existing rights the 2017 constitution added the right to a healthy environment as well [25].

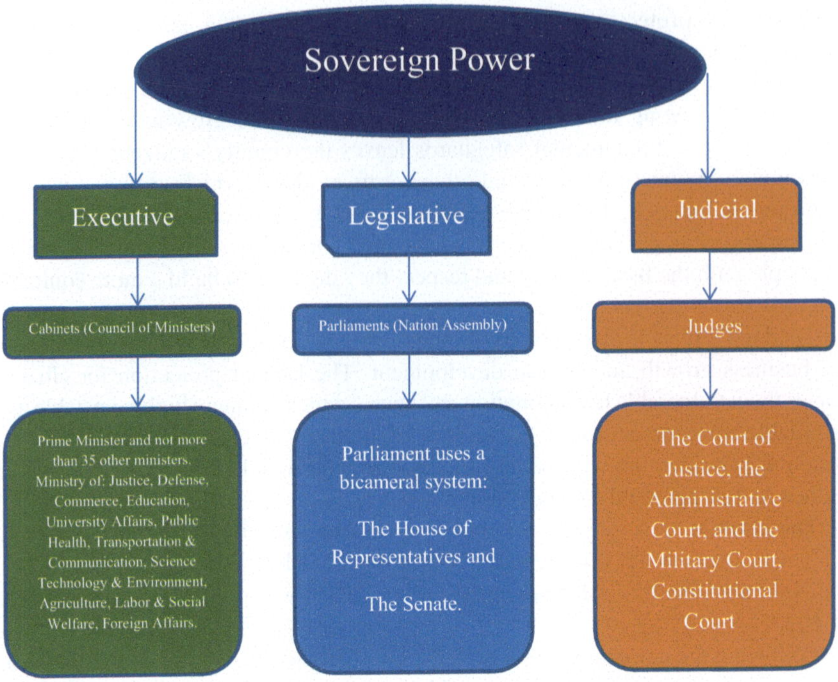

Disaster Risk Reduction 2015-2030 (Center, 2015; Marome & Shaw, 2021).

Fig. 1 Legal and political framework

The support for disaster management through the constitution is provided under the office of Disaster Prevention and Mitigation Office (DPMO) [36]. The department has been working in close contact with national as well as international framework such as the Sendai framework for Disaster Risk Reduction 2015–2030 [7, 22].

However, the Thai constitution has been criticized for being struggling with the pulls and pushes of the monarchy, government and political parties which limit its scope to provide the holistic protection towards Thai citizens [20] (Fig. 1).

3 Methodology

The methodology adopted for this research aims at exploring, understanding and analysing the existing legal framework which has a direct effect upon business sector amid disasters. The data has been collected from various legal Acts, government reports and publications. To support the government data, several research and online publications have been also analyzed to reach a holistic and comprehensive theoretical model to achieve sustainable future. The secondary data has been collected through several published sources in National Library of Thailand and online printed law and official translated documents. The untranslated Thai documents were translated with the co-author who is a Thai national and native Thai speaker.

The data analysis involves tools inclusive of content analysis and Strengths, Weaknesses, Opportunities and Threats (SWOT) analysis. The discussions along with quotations from the key informant respondents provides for the in-depth understanding and analysis of the current legal scenario in Thailand and comparative analysis with European Union laws.

Several researches have been comparing the Thai laws with the EU laws in terms of understanding and adoption of the legal framework [34, 35, 1, 3]. However, there is a lack of pertinent literature to support adoption of the EU exemplary in Thailand. This study supports the existing legal framework and constitutional implications towards human rights and copyright laws.

4 Findings and Discussions

The findings include the ineffective implementation of Thai laws during disasters. The Cabinet consists of the Prime Minister and not more than thirty-five other ministers. Parliaments (Nation Assembly); Parliament uses a bicameral system: the House of Representatives and the Senate. The House of Representatives consists of five hundred members of the House of Representatives, 350 from constituency elections and 150 from political party lists. The section on courts has three courts: the Court of Justice, the Administrative Court, and the Military Court, with the Constitutional Court separated into a separate section (Fig. 2).

Constitutional laws are related with protection of the rights of its citizen and maintaining the sovereignty of the state. The complexities of the constitutional reforms require focused expertise to minimize any loophole in the legal system in the state [29]. The success of these constitutional laws is to be understood with the response from the public participation [16]. Thailand lacks this approach of inclusion of the public as the core stakeholder among the legal infrastructure. The best international examples are from European Union where the Council of Europe assists during the careful reforms adopted at various levels of legal and governance system [11].

Small and medium-sized enterprises (SMEs) play a crucial role in the economic development of Thailand. They contribute significantly to job creation, innovation, and economic growth across the country. However, many of these businesses face significant challenges that hinder their success [30]. These challenges include limited access to financing, lack of government support, and regulatory burdens.

To address these issues, it is essential to amend the Constitution to provide more comprehensive support for SMEs. This can be achieved by introducing policies that facilitate access to credit, simplify regulatory procedures, and enhance government support. Such measures would help SMEs to operate more efficiently, reduce their costs, and strengthen their competitiveness in the market.

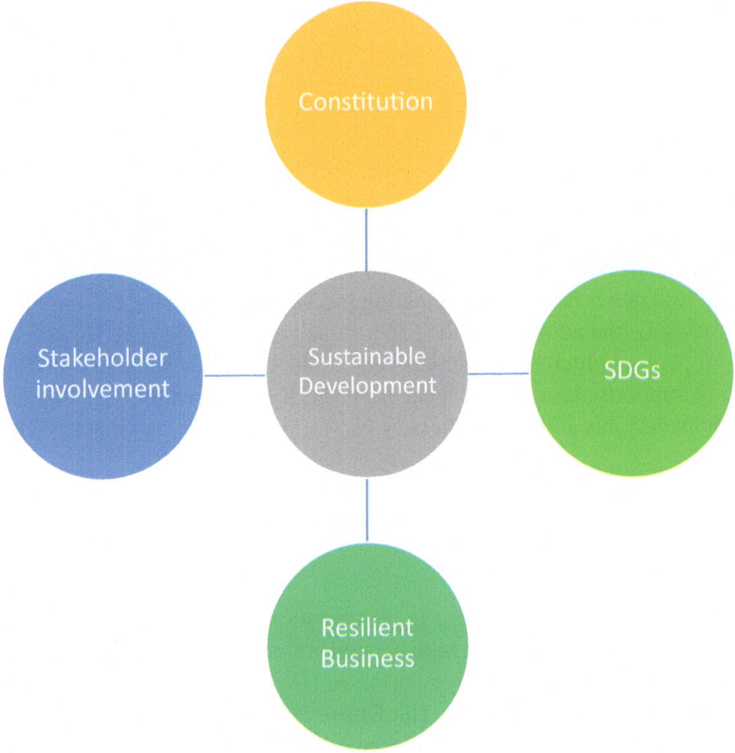

Fig. 2 Sustainable business development in Thailand

Additionally, the government could provide targeted support to specific sectors such as agriculture, tourism, and manufacturing, where SMEs are prevalent. This could include tax incentives, subsidies, and financial assistance to help these businesses grow and create more jobs.

By supporting SMEs, we can create a more inclusive and prosperous economy that benefits everyone. Therefore, amending the Constitution to provide enhanced support for SMEs is not only necessary but also crucial for the continued growth and success of Thailand's economy.

Democracy is where the people are the center of the legal reforms and economy [14]. Constitutional law in Thailand requires strong and citizen-oriented approach to strengthen the rights of the general population. European Union provides for robust exemplary laws and policies with upholding the rights of the people along with the sovereignty of each member state. Thailand requires to change the approach of constitutional reforms towards the people rather than an inadequate mashup of monarchy, democracy and strict top to bottom governance [27].

Human rights have been gaining considerable interest in every state and ensures the protection of the public despite of the social, economic and political scenario in the state. However, in Thailand, there has been several violent and non-violent incidences and events of public disobedience and civil unrest. European Union has been a successful example of maintaining the human rights at the regional as well as the national level of governance. The jest of the legal protection in European Union has been the human rights preservation [38]. The repulsive attitude and action by the government towards the non-violent civil protests has been a direct violation of the human rights in the recent 2020 civil unrest. The presence of the Covid-19 pandemic has enhanced the vulnerabilities of the people during such unfortunate events where the government actions were heavily criticized.

The research provides for a comprehensive discussion and recommendations to enhance sustainability in Thailand amid disasters such as COVID-19 pandemic, floods, droughts, earthquakes and Tsunami.

This research could be replicated or adopted in the other developing countries in the Association of Southeast Asian Nations (ASEAN) region.

The comprehensive planning and adequate revision of the disaster management policies and law in accordance with the updated constitution would enable sustainability and resilience in Thailand as well as in ASEAN [24].

Thailand has a comprehensive disaster management plan in place to effectively respond to any natural or man-made disasters. The plan includes a framework for preparedness, response, and recovery activities, and it involves collaboration between various government agencies, non-governmental organizations (NGOs), and the private sector [28].

The plan includes measures to mitigate the impact of disasters, such as early warning systems, evacuation plans, and the construction of flood barriers and other infrastructure. It also outlines procedures for search and rescue operations, medical assistance, and the provision of necessities such as food and shelter.

In addition, the disaster management plan emphasizes the importance of community participation and the involvement of local leaders and volunteers in disaster response and recovery efforts. This ensures that the needs of affected communities are met in a timely and effective manner.

Overall, Thailand's disaster management plan is designed to minimize the loss of life and property, and to support the country's efforts to build a more resilient and sustainable future [19] (Fig. 3).

Thailand is found to be in the growing state in terms of economic and legal system. The reforms in Thai laws are essential to provide justice to all [31]. The study found out that sustainable development is to be achieved through adequate planning within the scope of the law. Internal domestic law must guide by the international norms and guidelines which are provided through the SDGs. However, the actual local implementation needs to be done through the local level implementation through the stakeholders. These include the government departments, agencies, NGOs and local communities.

The key to achieving successful implementation of any policy or program lies in the active involvement and participation of various stakeholders. When it comes to implementing policies or programs at the local level, the process becomes even more complex and challenging. This is where the collaborative efforts of diverse

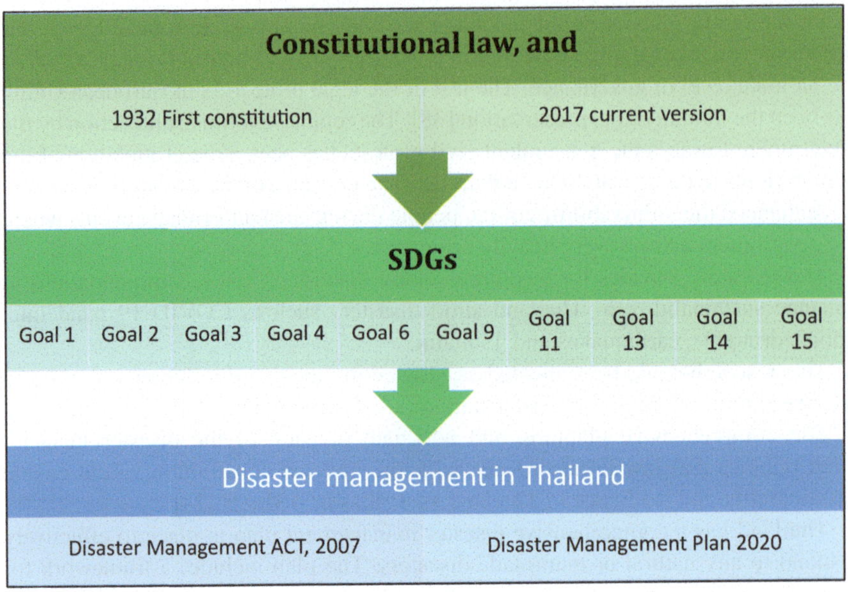

Fig. 3 SDG related with Thailand's legal framework

groups such as government departments, agencies, non-governmental organizations (NGOs), and local communities come into play.

Government departments and agencies play a crucial role in providing the necessary resources, funding, and infrastructure to support the implementation process. They developed policies and guidelines that will ensure effective implementation and monitoring of the program. NGOs, on the other hand, can be instrumental in providing valuable expertise, knowledge, and support to ensure that the policy or program meets the needs of the local community.

Local communities are also essential stakeholders in the implementation process. They are the ones who will be directly impacted by the policy or program, and their participation and feedback are crucial in ensuring that the implementation meets their needs and expectations. Local communities can provide valuable insights into the unique challenges and opportunities of the local context, and their feedback can help ensure that the policy or program is tailored to meet their specific needs.

In conclusion, successful implementation of policies or programs at the local level requires the active participation and collaboration of all relevant stakeholders. This includes government departments, agencies, NGOs, and local communities, who each have a unique role to play in ensuring effective and efficient implementation of the policy or program (Fig. 4).

Strengths: It was found that there is a vast and detailed legal and economic framework being setup at national and provincial levels in Thailand. The Ministries, departments work closely with the departments and institutions to provide for the preservation and protection of the rights of Thai People. The judiciary, governance

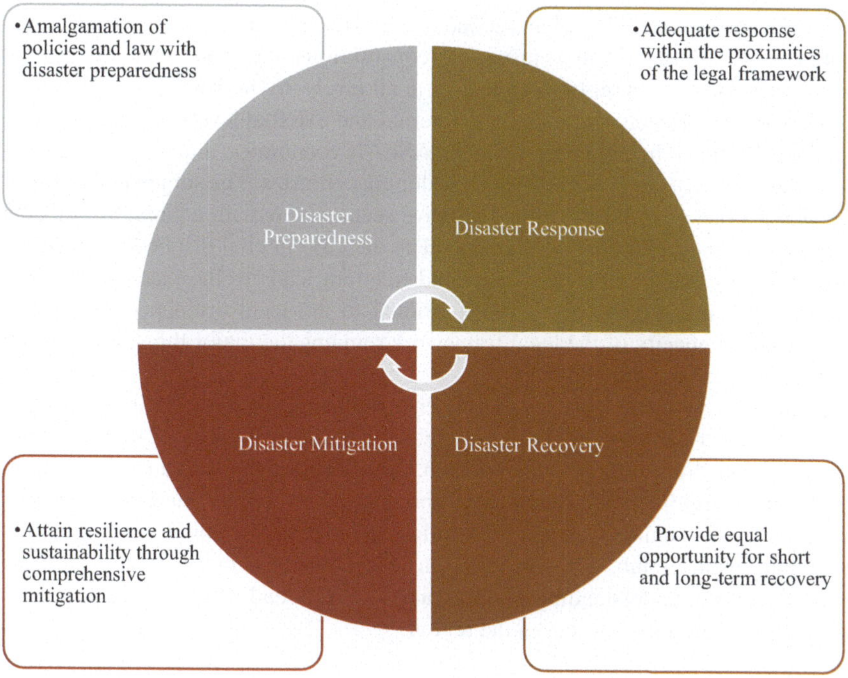

Fig. 4 SWOT analysis for the study

and implementation has been followed in strict top to bottom approach in Thailand. To ensure the reach and awareness to all levels of governance in Thailand the local ordinances work closely with the provincial and national streamlines. The adaptations from international norms and guidelines, for instance Sustainability Development Goals (SDGs), to ensure not only Thailand but also the progress of whole Association of Southeast Asian Nations (ASEAN) region.

Weaknesses: The political instability and the rigid framework to amend the law system remains the basic weakness of the governance and legal frameworks. The democratic government is under the Monarchy system, and it enhances the complexities of the existing system. The economic, social, and legal policies are strictly restricted to the national level institutions which in turn impact the local level needs and reunification of the legal protection. The presence of military, judiciary and department inefficient collaborations lead to the complex and inaccurate governance channels that was well-written in the Thai constitution, however, contradictions at the local governance limits the sustainability of the economy.

Opportunities: The hurdles of the weaknesses in the Thai legal system are omnipresent, however, there are several opportunities to improve and achieve sustainable development. There are several successful exemplary models from the international scenarios that could be adopted in Thailand. The democratic systems, protection of legal rights, dispute resolutions and adequate legal amendments could be adopted from the world. The global economies and legal protection for each sovereign State has been drafted in their respective constitutions, therefore, there are several exemplariness to be learnt from and adopt at all levels in Thailand.

Threats: The threats remain at both internal and external levels of governance in Thailand. Thailand being the leader in the ASEAN economies, require providing the exemplary scenarios to other regional developing countries. The subjugated rights of the Thai people, especially during disruptive scenarios, remain a threat towards the resilience and sustainability of Thailand. The increase in civil disobedience among Thai nationals due to restricted rights and freedom leads to the inadequacies and require legal amendments from Constitutional to the local ordinances. However, the restricted capacity of the legal framework remains the major threat towards the resilient and sustainable Thailand.

Thailand's business laws wield immense influence over the ASEAN region, making them critical for businesses operating within the country or those looking to expand their footprint in Southeast Asia. From corporate governance to taxation, employment, and intellectual property rights, these laws govern multiple aspects of business. Given Thailand's strategic location and vibrant economy, it has emerged as a vital business hub, attracting foreign investment and multinational corporations. Therefore, understanding the intricacies of Thailand's business laws is vital for enterprises seeking success in the region.

5 Conclusion and Recommendations

Thailand is a country with a unique legal system that combines both civil law and common law elements. Its constitutional law lays the foundation for the country's political and legal framework, while its business practices are shaped by cultural and economic factors. Understanding the interplay between constitutional law and business practices is essential for anyone seeking to navigate Thailand's legal and business landscape.

Disaster Preparedness, Response, Recovery and Mitigation depends directly upon the robustness of the legal framework of any country. This research provides for following recommendation for enhancing the adequateness of Thai disaster management framework.

- Set up committees at the national levels to ensure the local implementation of the law.
- Inclusiveness of all the concerned stakeholders, especially the local communities, into the disaster governance.
- Amending the constitution according to the global change scenarios.

This research focused and documented the lack of adequate legal acts, rules and procedures in the Thai Constitution as well as the local ordinances and norms. Therefore, the researchers provide for the focused recommendations at the legal and governance frameworks in Thailand as follows:

- Amendments under the Section 247 of the Thai Constitution to ensure protection of the basic human rights. The focused amendments must include the protection these rights during the disaster events.
- Governance framework to be strengthened, especially at the local levels in Thailand, following the top to bottom and inclusion of bottom to top approach to ensure the ground level implementations according to the local needs of the Thai People.
- The Thai people and institutions require to be aware of their enhanced and strengthened rights, therefore, amendments int eh Thai constitution and related laws must cater to awareness generation among local communities through public and private institutions.
- Stakeholder platform creation at the local level governance in line with the guidelines provided in Thai constitution and strict implementation with inclusion of all related and involved stakeholders.

The limitations of this research including the language barrier have been tackled with official English language translations of the Thai Constitution and other legal documentations. The official translations were utilized along with Thai co-author to cross check the Thai version of the constitution to minimize inadequate interpretation of the existing law system.

5.1 Way Ahead

The research paves way for future development and growth of the existing legal, business and disaster management scenarios in the developing world. The revision of the constitutional articles and sub sections would enhance the rights of the Thai people. The success stories from the amendments could be Adopted in other countries including ASEAN member nations and beyond.

Acknowledgements The uninterrupted support from Ammonrat Ammartsena, Wijittra Petchakit and other colleagues and family members is highly appreciated.

References

1. Ahmad N (2021) Protecting the rights of minorities under international law and implications of COVID-19: an overview of the Indian context. Laws 10(1):17
2. Anckar C (2021) Constitutional monarchies and semi-constitutional monarchies: a global historical study, 1800–2017. Contemp Polit 27(1):23–40
3. Androutsopoulos G (2021) The right of religious freedom in light of the Coronavirus pandemic: the Greek case. Laws 10(1):14
4. ASEAN (2021) ASEAN human rights declaration, November 19, 2012. [Online] Available at: https://asean.org/asean-human-rights-declaration/

5. Bennett KL (2013) Assessing the implementation of the European union guidelines on human rights defenders-the cases of Kyrgyzstan, Thailand and Tunisia., Belgium: European Union
6. Berne Convention (2021) Berne convention for the protection of literary and artistic works. https://www.wipo.int/export/sites/www/treaties/en/documents/pdf/berne.pdf
7. Center ADR (2015) Sendai framework for disaster risk reduction 2015–2030, Geneva, Switzerland: United Nations office for disaster risk reduction
8. Connors MK (1999) Political reform and the state in Thailand. J Contemp Asia 29(2):202–226
9. Constitution B.E. 2560 (2017) (2021) Constitution of the Kingdom of Thailand. https://www.constitutionalcourt.or.th/occ_en/download/article_20170410173022.pdf
10. Copyright Act B.E. 2537 (1994) (2021) Copyright act B.E. 2537 (1994). https://www.thailandlawonline.com/translations/copyright-law-thailand-intellectual-property
11. Dakolias M (2006) Are we there yet?: Measuring success of constitutional reform. Vanderbilt J Transnatl Law 39(4):1210–1211
12. Drobnik MJ, Bimbilovski I, Pathak S (2023) Space law and space mining, exploring new horizons amid COVID-19 pandemic. Emerg Sci J 7:114–127
13. Drobnik M, Pathak S (2021) Sustainable development and poverty-thai perspective amid COVID-19 pandemic. J Leg Ethical Regul Issues 24:1
14. Farrell DM, Suiter J, Harris C, Cunningham K (2020) The effects of mixed membership in a deliberative forum: the Irish constitutional convention of 2012–2014. Polit Stud 68(1):54–73
15. Ginsburg T (2017) Constitutional afterlife: the continuing impact of Thailand's postpolitical constitution. In: Public Law in East Asia. s.l.:Routledge, pp 289–311
16. Ginsburg T, Elkins Z, Blount J (2009) Does the process of constitution-making matter? Annu Rev Law Soc Sci 5:201–223
17. Haglund J, Welch RM (2021) From litigation to rights: the case of the European court of human rights. Int Stud Q 65(1):210–222
18. Human Rights Watch (2021). Thailand country chapter. https://www.hrw.org/world-report/2021/country-chapters/thailand
19. Inpin W, Dania M (2023) Mapping research on disaster risk reduction for sustainable development in Thailand: thematic analysis approach. Asia-Pac Soc Sci Rev 23(2)
20. Leelapatana R, Asanasak S (2022) Constitutional struggles and polarised identities in Thailand: the constitutional court and the gravitational pull of Thai-ness upon liberal constitutionalism. Fed Law Rev 50(2):156–173
21. Lhakard P (2022) Thailand's administration and policies in response to the COVID-19 pandemic. Interdiscip Res Rev 17(3):23–27
22. Marome W, Shaw R (2021) COVID-19 response in Thailand and its implications on future preparedness. Int J Environ Res Public Health 18(3):1089
23. Miller V, Clark E (2010) The European Union: a guide to terminology, procedures and sources. Briefing paper, pp 1–14
24. Mukherjee M et al (2023) Extent and evaluation of critical infrastructure, the status of resilience and its future dimensions in South Asia. Prog Disaster Sci 17:100275
25. Nugroho SS et al (2020) The right for a healthy environment in Indonesia: comparison of global laws. J Glob Pharma Technol 12(09):123–129
26. O'Meara N (2011) A more secure Europe of rights? The European court of human rights, the court of justice of the European Union and EU accession to the ECHR. Ger Law J 12(10):1813–1832
27. Pathak S, Ahmad MM (2018) Role of government in flood disaster recovery for SMEs in Pathumthani province Thailand. Nat Hazards 93(2):957–966
28. Poungjinda W, Pathak S (2022) Educational reforms amid COVID-19 in Thailand. Front Educ 7:905445
29. Renwick A (2014) After the referendum: options for a constitutional convention. The Constitution Society, London
30. Saengtabtim K, Tang J, Leelawat N, Egawa S (2023) Universal health coverage mitigated COVID-19 health-related consequences in Asia Oceania. Int J Disaster Risk Reduct 92:103725

31. Sangkaew T (2021) Political party finance reform and the rise of dominant party in emerging democracies: a case study of Thailand. Central European University, s.l.
32. Shaw J (2005) Mainstreaming equality and diversity in European Union law and policy. Curr Leg Probl 58(1):255
33. Sudhipongpracha T (2014) Local emergency management in decentralized Thailand: analysis of Thai municipal administrators' perceptions of democratic accountabilities in the post-decentralization era. J Asian Public Policy 7(3):259–274
34. Supasiripongchai N (2014) The protection of performer's rights under the copyright law in Thailand: the proposed reform in the light of the prospective free trade agreements with the United States and European Union. Intellect Prop Rights: Open Access J 2(4):1–19
35. Suwanbamrung C, Nukan N, Sripon S, Somrongthong R (2010) Community capacity for sustainable community-based dengue prevention and control: study of a sub–district in Southern Thailand. Asian Pac J Trop Med 3(3):215–219
36. Taekratok T, Phanthunane P, Taekratok T (2018) Disaster-based budgeting in new public health policy of Thailand. Procedia Eng 715–722. Bangkok, Thailand
37. Verellen T (2021) Constitutional law of the EU's common foreign and security policy: competence and institutions in external relations. Am J Comp Law 69(1):173–176
38. Von Bogdandy A (2000) The European Union as a human rights organization? Human rights and the core of the European Union. Common Mark Law Rev 38(6):1307–1338
39. Wagner E, Bech S, Martínez J (2014) Translating for the European Union institutions. Routledge, London

Systematic Review of Community Resilience Framework Related to Disasters

Deepak Marahatta, Hari Devkota, and Shankar Tiwari

Abstract Disaster resilience assessment is critical to disaster risk reduction strategies and sustainable development efforts. This study presents a desk review analysis of 32 existing disaster resilience assessment frameworks to identify common themes, indicators, and measurement methods utilized in the field. The frameworks were systematically selected based on relevance, publication year, hazard focus, and type. Thematic analysis classified the indicators into ten themes, emphasizing social capital (84.4%) and economic dimensions (62.5%). Quantitative methods (47%) were the most frequently employed for measuring indicators. This study contributes valuable insights to enhance understanding of disaster resilience assessment methodologies. It informs the development of comprehensive and contextually relevant approaches to promote community resilience and sustainable livelihoods in the face of disasters.

Keywords Disaster · Framework · Livelihood · Resilience · Sustainable

1 Introduction

Disasters of natural or anthropogenic origin persist as an enduring global challenge, exacting a toll that knows no geographical or temporal boundaries. Earthquakes, pandemics, and other catastrophic events continually reshape communities and societies, placing the fundamental pillars of human existence—security, stability, and sustainability—on the precipice of uncertainty [54]. The concept of resilience, rooted in the Latin term 'resalire,' meaning 'to leap back,' has become a pivotal notion in the discourse surrounding disasters and their aftermath [19]. Its genesis can be traced to ecology, where C.S. Holling introduced it in 1973, defining resilience as a system's ability to absorb perturbations and shift to alternative states when necessary [21].

D. Marahatta (✉) · H. Devkota · S. Tiwari
Tribhuvan University Central Department of Rural Development, Kathmandu, Nepal
e-mail: deepak.207730@cdrd.tu.edu.np

I. Pal et al. (eds.), *Proceedings of the 3rd International Symposium on Disaster Resilience and Sustainable Development—Volume 1*, Lecture Notes in Civil Engineering 623, https://doi.org/10.1007/978-981-96-2139-2_5

The concept transcended disciplinary boundaries, making substantial inroads into psychology, climate change, and disaster literature. In psychology, resilience characterizes an individual's capacity to adapt psychologically to stressful circumstances [27]. While resilience has long been discussed in academic circles, its prominence in the context of hazards and disasters gained momentum in recent years, particularly with the endorsement of the United Nations Hyogo Framework for Action in 2005, which elevated disaster risk reduction and resilience-building as top priorities for governments and local communities [52].

In line with the Hyogo Framework of Action, numerous scholars define resilience as a community's ability to respond positively to crises and potential hazards [24, 26, 34, 55]. Some liken resilience to 'bouncing back,' highlighting the need for swift recovery when disasters strike [20]. These definitions underscore the importance of resilience in fortifying a system's capacity to withstand and recover from disasters while fostering adaptability to reduce future risks.

Resilience's integration into disaster literature has led to the recognition that a comprehensive understanding requires considering the multifaceted aspects of disasters affecting individuals and communities [25]. In this context, resilience encompasses the ability of individuals and communities to access the necessary resources for mitigating and responding effectively to crises [42]. The increasing frequency and severity of natural and man-made disasters have elevated disaster resilience as a critical strategy for reducing vulnerability and enhancing community adaptability [14]. Disaster resilience encompasses a complex interplay of social, economic, environmental, and political factors contributing to a community's ability to cope with, recover from, and adapt to hazardous events [28, 33, 53].

Disaster resilience incorporates various strategies, capabilities, and systems that enable communities and organizations to endure, recover, and enhance their resilience post-disaster [3, 52]. It embodies a holistic perspective beyond survival, emphasizing preserving and restoring essential functions and structures [52]. Other scholars and organizations have expanded the scope further, defining disaster resilience as the capacity of a system, community, or society to withstand, assimilate, adapt to, and recover from hazards efficiently and within a reasonable timeframe [5, 16, 24, 33, 52]. This includes safeguarding and reinstating fundamental structures and operations.

Recent scholarship has also characterized resilience as rapidly reducing the disruption caused by a disaster and returning to pre-event conditions [3, 37]. These definitions collectively highlight the intricate nature of disaster resilience, emphasizing the capacity to endure, adapt, innovate, and evolve in the face of diverse threats. Moreover, some scholars and institutions view disaster resilience through sustainability, emphasizing the community's ability to manage change to sustain living standards without compromising long-term development goals after a disaster [10, 16]. This perspective underscores the importance of balancing livelihood preservation with broader developmental objectives. Conversely, others view disaster resilience as a process and condition through which a community strengthens its capacity to withstand and recover from disruptions with minimal or no external support [40], highlighting the significance of internal strengths and resources.

Additionally, scholars have debated whether disaster resilience is a continuous process, an outcome, or a combination of both. Some describe it as an ongoing process, while others consider it an end state or outcome. Alternatively, some suggest it is a dynamic interplay between process and outcome [14]. Moreover, disaster resilience is not solely about returning to pre-disaster conditions, it also aims to achieve a more feasible and sustainable equilibrium [7]. Some argue that resilience involves "bouncing back" and actively leaping forward and adapting [43]. These diverse perspectives on the nature of resilience underscore its dynamic and forward-looking characteristics, emphasizing its potential to facilitate constructive adaptation in the face of challenges and opportunities.

In the evolving landscape of disaster resilience, definitions have evolved dynamically, fostering ongoing discourse within academic and disaster management communities [19]. This evolution underscores an increasing awareness of resilience's multifaceted nature, highlighting its proactive dimensions and adaptive capacities. Such diverse perspectives contribute to a more comprehensive understanding of resilience, enriching disaster management and community development discussions.

The assessment and measurement of disaster resilience play pivotal roles in guiding effective disaster risk reduction (DRR) strategies and promoting sustainable livelihoods [6, 49]. Responding to this need, various resilience assessment frameworks and indicators have emerged, offering unique insights into resilience's multidimensional character [13]. These frameworks are indispensable tools for policymakers, researchers, and practitioners alike, enabling them to evaluate community capacities, pinpoint vulnerabilities, and prioritize interventions for enhancing resilience.

However, the abundance of existing resilience assessment frameworks poses a challenge, making it imperative to discern pertinent and universally applicable indicators [2, 6]. Understanding which factors are consistently emphasized and underrepresented across these frameworks becomes crucial for refining resilience measurement approaches. This study, therefore, endeavors to provide a comprehensive review and analysis of existing disaster resilience assessment frameworks. Through systematic examination and categorization of indicators employed in these frameworks, the objective is to identify common themes, areas of emphasis, and potential gaps. Additionally, this research aims to investigate the methods used to measure these indicators, providing insights into the various approaches taken in resilience assessment.

2 Study Methods

This article employs a desk review meths to comprehensively analyze existing disaster resilience assessment frameworks. The study is based on a systematic review of scholarly literature and relevant reports, encompassing frameworks published between 2005 and 2022. This review aims to identify, categorize, and analyze indicators used in the selected frameworks and understand the methods employed for measuring these indicators.

The data collection process involves rigorous searches of academic databases, such as Scopus, Web of Science, and Google Scholar. Additionally, grey literature from reputable disaster management organizations, governmental agencies, and international bodies is also included in the search. The search terms used include 'disaster resilience measuring indicators,' 'community resilience indicators,' and 'disaster resilience indicators.' The study is limited to frameworks to assess a community's resilience to natural disasters or risks induced by climate change.

2.1 Framework Selection Criteria

The selection of frameworks is carried out through a systematic approach to ensure the inclusion of relevant and significant sources. Each framework is critically evaluated based on the following criteria:

- Relevance: Frameworks directly focused on disaster resilience assessment or incorporating resilience indicators as a critical component are considered relevant.
- Publication Year: Only frameworks published between 2005 and 2022 are included to encompass the most recent developments in the field.
- Hazard Focus: Frameworks must be targeted towards assessing resilience to natural disasters or climate-induced risks.
- Type of Framework: The study includes indices, scorecards, and toolkits that measure and evaluate community disaster resilience.

After searching several frameworks, only 32 have been chosen on community resilience frameworks applied at the local level in different settings that meet the selection criteria, i.e., contain indicators of disaster resilience at a community level and are tools or indexes.

Once the relevant frameworks are identified, the study conducts a thematic analysis of the indicators used. The indicators are categorized into different themes: community development, disaster management, preparedness and recovery, and the seven capitals (economic, natural, human, physical, social-cultural, political, and others). The classification aims to understand the perspectives on which each framework is built, and the dimensions emphasized.

The data analysis also focuses on the methods used for measuring the identified indicators. The study categorizes the methods into quantitative, qualitative, and mixed methods, aiming to grasp the diversity of approaches taken in resilience assessment.

3 Results

3.1 Disaster Resilience Indicators and Measurement Framework

The disaster resilience indicators and measurement frameworks selected for review are listed in the Table in alphabetical order, along with the authors' names, types of frameworks, types of disasters that the framework studies, study methods, and study area. Among the 32 frameworks, six are related to the Community Disaster Resilience Index (CDRI), five are related to the Baseline Resilience Index for Communities (BRIC), five are related to the Community Resilience Index (CRI), and the remaining frameworks are associated with other indicators.

Notably, half of the frameworks (16 out of 32, 50%) adopted an all-hazards approach, encompassing multiple types of disasters. Additionally, specific frameworks were developed to study various types of hazards, including earthquake hazards (5 frameworks), tsunami and storm (4 frameworks), hurricane (2 frameworks), flood (2 frameworks), landslide (2 frameworks), drought (2 frameworks), and one framework focused on climate-induced hazards (Table 1).

Disaster resilience assessment frameworks are typically categorized into indices, scorecards, and toolkits (Cutter, 2016). Among the 32 selected frameworks, approximately 53% (17 frameworks) are indices, approximately 44% (14 frameworks) are toolkits, and the remaining one is a score sheet. An index summarizes observations and measures by aggregating multiple indicators into a single value, while toolkits guide resilience assessment using two or more approaches. Scorecards are used to evaluate performance or progress toward a specific goal and are often implemented as checklists.

Regarding the methods used for measuring the indicators, approximately 47% (15 frameworks) employed quantitative methods, approximately 25% (8 frameworks) used mixed methods, and approximately 22% (7 frameworks) utilized qualitative methods. The two frameworks did not explicitly define the method (approximately 6%).

In terms of the country of study, approximately 28% (9 frameworks) focused on the USA, approximately 19% (6 frameworks) were conducted in Asia (excluding South Asian countries), and two frameworks each were conducted in Asia (including South Asia), South Asia, Australia, and Europe. Also, four frameworks were conducted in Africa, while one each was conducted in Haiti and Canada. Three frameworks were not specific to any particular country.

Table 1 Selected disaster resilience measuring frameworks

Number of frameworks selected for study = 32					
Framework/Tool	Author	Type	Types of disaster	Method	Geography
Baseline Resilience Index for Communities (BRIC1)	[8]	Toolkit	Hurricane	Quantitative	USA
Baseline Resilience Index for Communities (BRIC2)	[15]	Index	Multi	Quantitative	USA
Baseline Resilience Index for Communities (BRIC3)	[13]	Index	Multi	Quantitative	USA
Baseline Resilience Index for Communities (BRIC4)	[47]	Index	Flood	Quantitative	Asia
Climate Disaster Resilience Index (CDRI)	[39]	Index	Climate	Mixed	South Asia
Coastal Resilience Index (CRI1)	[11]	Toolkit	Tsunami and Storm	Qualitative	Asia (South Asia Included)
Coastal Resilience Index (CRI2)	[44]	Scoresheet	Storm	Not Define	USA
Community Advancing Resilience Toolkit (CART)	[38]	Toolkit	Multi	Qualitative	Not Specific
Community-Based Resilience Analysis (CoBRA)	[50]	Toolkit	Drought	Qualitative	Africa
Community Disaster Resilience Index (CDRI1)	[17]	Toolkit	Earthquake	Mixed	Haiti
Community Disaster Resilience Index [CDRI2]	[30]	Index	Hurricane	Mixed	USA
Community Disaster Resilience Index (CDRI3)	[36]	Index	Multi	Mixed	Australia
Community Disaster Resilience Index (CDRI4)	[57]	Index	Multi	Quantitative	Asia

(continued)

Table 1 (continued)

Number of frameworks selected for study = 32					
Framework/Tool	Author	Type	Types of disaster	Method	Geography
Community Disaster Resilience Index (CDRI5)	[35]	Index	Tsunami and Storm	Quantitative	Asia
Community Disaster Resilience Index (CDRI6)	[45]	Toolkit	Tsunami and Storm	Qualitative	Asia (South Asia Included)
Community Disaster Scorecard and Toolkit (CDRST)	[4]	Toolkit	Multi	Qualitative	Australia
Conjoint Community Resilience Assessment Measurement (CCRAM)	[9]	Toolkit	Multi	Mixed	Asia
Composite Resilience Index (CRI3)	[29]	Index	Multi	Quantitative	Africa
Community Resilience Index (CRI4)	[24]	Index	Multi	Quantitative	Asia
Community Resilience Index (CRI5)	[46]	Index	Multi	Quantitative	USA
Community Resilience Index (CRI6)	[48]	Toolkit	Earthquake	Not Define	Europe
Community Resilience Index (CRI7)	[33]	Index	Multi	Quantitative	Not Specific
Community Resilience Index (CRI8)	[1]	Index	Earthquake	Quantitative	South Asia
Household Resilience to Flood (HRF)	[32]	Index	Floods	Qualitative	Asia
PEOPLES	[41]	Toolkit	Earthquake	Quantitative	USA
Oxfam Resilience Framework (ORF)	[22]	Index	Drought	Quantitative	Africa

(continued)

Table 1 (continued)

Number of frameworks selected for study = 32					
Framework/Tool	Author	Type	Types of disaster	Method	Geography
ResilSim	[23]	Toolkit	Multi	Mixed	Europe
ResilUS	[31]	Toolkit	Multi	Quantitative	USA
RIMA	[18]	Index	Multi	Quantitative	Africa
Rural Resilience Index (RRI)	[12]	Index	Multi	Qualitative	Canada
Seismic Resilience Index (SRI)	[56]	Toolkit	Earthquake	Mixed	USA
UNISDR	[51]	Toolkit	Multi	Mixed	Not Specific

3.2 *Classification of the Major Indicators Used in Frameworks*

The indicators used in the frameworks can be divided into ten different themes. These themes include community development, disaster management, preparedness and recovery, seven capital (economic capital, natural capital, human capital, physical capital, social-cultural capital, political capital), and others. These thematic indicators and their associated framework are presented in Table 2.

The thematic classification of indicators provides valuable insights into the perspectives on which various frameworks are built. Most frameworks (84.4 percent) strongly emphasize the significance of social capital and assets in reducing disaster vulnerability. Additionally, 62.5 percent of the frameworks include indicators related to the economic dimension, 59.4 percent include indicators related to human capital, and 53.1 percent include indicators related to physical capital. Only half of the frameworks (50%) consider environment-related indicators when evaluating disaster resilience, even though natural disasters cause the loss of natural resources. Political capital-related measures are emphasized in only 43.8 percent of the frameworks, and the measures of cultural capital are the least represented, with only 9.4 percent of the frameworks focusing on them. Additionally, 25 percent of the frameworks measure catastrophe resilience using additional metrics alongside the seven capitals.

4 Discussion

Disaster resilience, defined as the ability of a community to adapt, cope, and recover from any disaster or emergency, is a critical component of comprehensive community development. Our systematic review delved into the multifaceted landscape of disaster resilience, examining various disaster resilience indicators and measurement frameworks. The review of the existing frameworks has following findings.

Table 2 Thematical classification of the indicator's frameworks

Indicator theme	No of associated framework (n = 32)	% of frameworks	Frameworks
(i)	(ii)	(iii) = (ii)/32	(iv)
Economic capital	20	62.5	BRIC1,CRI1,BRIC2,BRIC3,CDRI2,ResilUS,CDRI3,CDRI6, BRIC4, CDRI4, PEOPLES, CRI5, CRI6, CDRI1, CRI2, CDRST, CRI3, CRI7, CRI8, ResilSim
Social capital	27	84.4	BRIC1, BRIC2, BRIC3, BRIC4, CDRI3, CoBRA, CCRAM, CRI1, CRI2, RRI, CDRI1, PEOPLES, ORF, CDRI2, ResilUS, CDRI6, CRI5, CDRI4, CART, SRI, CDRI5, CDRI, CDRST, CRI3, CRI7, CRI8, ResilSim
Human capital	19	59.4	RRI, CART, CRI1, CRI4, CCRAM, CDRI3, CRI2, CDRI1, CDRI2, ResilUS, CDRI4, CART, CDRI, CDRST, CRI3, CRI7, CRI8, ResilSim, RIMA
Physical capital	17	53.1	BRIC1, BRIC2, CDRI2, ResilUS, CDRI3, PEOPLES, CRI2, CDRI6, CDRI4, CRI6, CRI1, CDRI, CDRST, CRI3, CRI8, ResilSim, RIMA
Natural capital	16	50.0	BRIC1, CRI1, BRIC2, BRIC3, CDRI1, CDRI2, ResilUS, CDRI3, CDRI6, BRIC4, CDRI4, SRI, ORF, CDRI5, CART, CDRI
Cultural capital	3	9.4	CCRAM, CART, CRI6
Political capital	14	43.8	BRIC1, BRIC2, BRIC3, ORF, CDRI6, BRIC4, CRI1, CRI4, CDRI3, PEOPLES, CDRI, CDRST, CRI8, ResilSim
Other theme	8	25.0	CCRAM, CDRI1, ORF, CDRI5, PEOPLES, SRI, CDRST, RIMA

4.1 Comprehensive Frameworks and Varied Disaster Types

The first table presents a comprehensive overview of disaster resilience measurement frameworks, revealing many dimensions in which communities can be assessed. Notably, a significant proportion of these frameworks (50%) adopt an all-hazards approach, recognizing that communities must be prepared to face a broad spectrum of disaster types. This underscores the importance of versatility and adaptability in resilience-building efforts.

Furthermore, the prevalence of indices (53%) suggests a trend toward summarizing multiple indicators into a single value to facilitate a holistic understanding of resilience. Toolkits (44%) and scorecards (3%) also play a substantial role, enabling tailored assessments and checklists to evaluate progress.

4.2 *Thematic Classification of Indicators: A Rich Tapestry*

The second table provides insights into the thematic classification of disaster resilience indicators within these frameworks. Social capital and assets, essential to community cohesion and support networks, take center stage in 84.4% of the frameworks. This emphasizes the pivotal role of community solidarity in building resilience and underscores the interconnectedness of individuals within a community.

While economic dimensions (62.5%) and human capital (59.4%) are also strongly represented, physical capital (53.1%) and environment-related indicators (50%) offer opportunities for further exploration. Recognizing the economic and human elements of resilience aligns with the understanding that financial stability and human resources are integral to a community's ability to withstand and recover from disasters.

Interestingly, political capital (43.8%) and cultural capital (9.4%) are less prominent within the frameworks. This suggests room for deeper exploration of governance structures, decision-making processes, and cultural aspects in the context of disaster resilience.

4.3 *The Way Forward: A Holistic Approach*

Our findings underscore the need for a holistic approach to disaster resilience encompassing various indicators and recognizing the unique cultural and political contexts within which resilience is built. While emphasizing the importance of social and economic aspects, we advocate for a more balanced consideration of physical infrastructure, environmental sustainability, and cultural factors in resilience-building initiatives. Moreover, using mixed methods (25%), combining quantitative and qualitative approaches demonstrates the importance of a multidisciplinary perspective. Such approaches can capture resilience's complexity, integrating quantitative data and qualitative narratives from affected communities.

5 Conclusion

The assessment of disaster resilience through various frameworks plays a crucial role in understanding and enhancing the preparedness and capacity of communities to cope with and recover from disasters. The analysis revealed that social capital emerged as the most emphasized aspect in disaster resilience frameworks, underscoring the significance of community cohesion and support systems in reducing vulnerability. Economic, human, and physical capital were also essential to enhance community preparedness and resilience. However, the relatively lower representation of environmental, political, and cultural capital-related measures suggests potential areas for improvement in resilience assessment. The frameworks were mainly developed with an all-hazards approach, but specific frameworks targeted various natural hazards, indicating the need for tailored strategies to address context-specific risks.

The findings from this study emphasize the importance of comprehensive and context-specific disaster resilience frameworks. Integrating environmental, political, and cultural capital-related indicators can lead to a more holistic understanding of resilience and inform culturally appropriate strategies. Furthermore, considering disaster resilience metrics can enhance preparedness for extreme events. A well-rounded and inclusive approach, incorporating various capitals and considering regional disparities, can enhance community resilience and contribute to more effective disaster management. By acknowledging the diverse dimensions of resilience and implementing targeted interventions, communities can build their capacity to withstand and recover from disasters, fostering a more resilient and sustainable future.

References

1. Ainuddin S, Routray JK (2012) Earthquake hazards and community resilience in Baluchistan. Nat Hazards 63(2):909–937. https://doi.org/10.1007/s11069-012-0201-x
2. Aitsi-Selmi A, Blanchard K, Murray V (2016) Ensuring science is useful, usable and used in global disaster risk reduction and sustainable development: a view through the Sendai framework lens. Palgrave Commun 2(May):1–10. https://doi.org/10.1057/palcomms.2016.16
3. Aksha SK, Emrich CT (2020) Benchmarking community disaster resilience in Nepal. Int J Environ Res Public Health 17(6):17. https://doi.org/10.3390/ijerph17061985
4. Arbon P, Steenkamp M, Cornell V, Cusack L, Gebbie K (2016) Measuring disaster resilience in communities and households: pragmatic tools developed in Australia. Int J Disaster Resil Built Environ 7(2):201–215. https://doi.org/10.1108/IJDRBE-03-2015-0008
5. Ashley C, Carney D (1999) Sustainable livelihoods: lesson from early experience. In the Department for International Development (DFID). https://doi.org/10.1002/9781118663219.wbegss003
6. Béné C, Wood RG, Newsham A, Davies M (2012) Resilience: New utopia or new tyranny? Reflection about the potentials and limits of the concept of resilience in relation to vulnerability reduction programmes. IDS Work Pap 2012(405). https://doi.org/10.1111/j.2040-0209.2012.00405.x
7. Brody SD, Gunn JR (2013) Examining environmental factors contributing to community resilience along the Gulf of Mexico coast. In Kapucu N, Hawkins CV, Rivera FI (eds) Disaster resiliency: interdisciplinary perspectives. Routledge, pp 160–177. https://doi.org/10.4324/9780203102459
8. Burton CG (2012) The development of metrics for community resilience to natural disasters. University of South Carolina
9. Cohen O, Leykin D, Lahad M, Goldberg A, Aharonson-Daniel L (2013) The conjoint community resiliency assessment measure as a baseline for profiling and predicting community resilience for emergencies. Technol Forecast Soc Chang 80(9):1732–1741. https://doi.org/10.1016/j.techfore.2012.12.009
10. Combaz E (2014) Disaster resilience: topic guide. In International development department, University of Birmingham, vol 98, Issue 639. University of Birmingham. https://doi.org/10.4324/9781003123897-13
11. Courtney CA, Ahmed AK, Jackson R, Mckinnie D, Rubinoff P, Stein A, Tighe S, White A (2008) Coastal community resilience in the Indian Ocean Region: a unifying framework, assessment, and lessons learned. In: Solutions to Coastal Disasters Congress 2008, pp 990–1001

12. Cox RS, Hamlen M (2015) Community disaster resilience and the rural resilience index. Am Behav Sci 59(2):220–237. https://doi.org/10.1177/0002764214550297
13. Cutter SL, Ash KD, Emrich CT (2014) The geographies of community disaster resilience. Glob Environ Chang 29:65–77. https://doi.org/10.1016/j.gloenvcha.2014.08.005
14. Cutter SL, Barnes L, Berry M, Burton C, Evans E, Tate E, Webb J (2008) A place-based model for understanding community resilience to natural disasters. Glob Environ Chang 18(4):598–606. https://doi.org/10.1016/j.gloenvcha.2008.07.013
15. Cutter SL, Burton CG, Emrich CT (2010) Disaster resilience indicators for benchmarking baseline conditions. J Homel Secur Emerg Manag 7(1). https://doi.org/10.2202/1547-7355.1732
16. Department for International Development (2011) Defining disaster resilience: a DFID approach paper. In Department for International Development
17. DRLA and UEH (2010) Haiti humanitarian assistance evaluation from a resilience perspective
18. Food and Agriculture Organization of the United Nations (2016) Analyzing resilience for better targeting and action: resilience index measurement and analysis. Food and Agricultural Organization of the United Nations. http://www.fao.org/3/a-i4102e.pdf
19. Gunderson L (2010) Ecological and human community resilience in response to natural disasters. Ecol Soc 15(2):29. https://doi.org/10.5751/ES-03381-150218
20. Gwimbi P (2009) Linking rural community livelihoods to resilience building in flood risk reduction in Zimbabwe. J Disaster Risk Stud 2(1):71–79
21. Holling CS (1973) Resilience and stability of ecological systems. Annu Rev Ecol Syst 4:1–23
22. Hughes K, Bushell H (2013) A Multidimensional approach for measuring resilience. Oxfam GB working paper
23. Irwin S, Schardong A, Simonovic SP, Nirupama N (2016) ResilSIM-a decision support tool for estimating resilience of urban systems. Water 8(377):1–25. https://doi.org/10.3390/w8090377
24. Kafle SK (2012) Measuring disaster-resilient communities: a case study of coastal communities in Indonesia. J Bus Contin Emer Plan 5(4):316–326
25. Kapucu N (2012) Disaster and emergency management systems in urban areas. Cities: Int J Urban Policy Plan 29(S1):41–49. https://doi.org/10.1016/j.cities.2011.11.009
26. Maguire B, Hagan P (2007) Disasters and communities: understanding social resilience. Aust J Emerg Manag 22(2):16–20
27. Mancini AD, Bonanno GA (2009) Predictors and parameters of resilience to loss: toward an individual differences model. J Pers 77(6):1805–1832. https://doi.org/10.1111/j.1467-6494.2009.00601.x
28. Manyena SB (2006) Rural local authorities and disaster resilience in Zimbabwe. Disaster Prev Manag 15(5):810–820. https://doi.org/10.1108/09653560610712757
29. Mavhura E, Manyangadze T, Aryal KR (2021) A composite inherent resilience index for Zimbabwe: an adaptation of the disaster resilience of place model. Int J Disaster Risk Reduct 57:102152. https://doi.org/10.1016/j.ijdrr.2021.102152
30. Mayunga JS (2007) Understanding and applying the concept of community disaster resilience: a capital-based approach. Summer academy for social vulnerability and resilience building (Issue July). http://www.ehs.unu.edu/file/get/3761.pdf
31. Miles SB, Chang SE (2008) ResilUS modeling community capital loss and recovery. In: The 14th world conference on earthquake engineering
32. Nguyen KV, James H (2013) measuring household resilience to floods: a case study in the Vietnamese Mekong River Delta. Ecol Soc 18(3). https://doi.org/10.5751/ES-05427-180313
33. Norris FH, Stevens SP, Pfefferbaum B, Wyche KF, Pfefferbaum RL (2008) Community resilience as a metaphor, theory, set of capacities, and strategy for disaster readiness. Am J Community Psychol 41:127–150. https://doi.org/10.1007/s10464-007-9156-6
34. Olsson P (2003) Building capacity for resilience in social-ecological systems. Stockholm University, Department of systems ecology (Issue December)
35. Orencio PM, Fujii M (2013) A localized disaster-resilience index to assess coastal communities based on an analytic hierarchy process (AHP). Int J Disaster Risk Reduct 3(1):62–75. https://doi.org/10.1016/j.ijdrr.2012.11.006

36. Parsons M, Morley P, Marshall G, Hastings P, Glavac S, Stayner R, Mcneill J, Mcgregor J, Reeve I (2016) The Australian natural disaster resilience index: conceptual framework and indicator approach, Issue February
37. Paton D, Johnston D (2017) Disaster resilience: an integrated approach. In: Paton D, Johnston D (eds) Charles C Thomas Publisher, Ltd. (2nd ed.). Charles C Thomas Publisher, Ltd. https://medium.com/@arifwicaksanaa/pengertian-use-case-a7e576e1b6bf
38. Pfefferbaum RL, Pfefferbaum B, Van Horn RL, Klomp RW, Norris FH, Reissman DB (2013) The communities advancing resilience toolkit (CART): an intervention to build community resilience to disasters. J Public Health Manag Pract 19(3):250–258. https://doi.org/10.1097/PHH.0b013e318268aed8
39. Prashar S, Shaw R, Takeuchi Y (2012) Assessing the resilience of Delhi to climate-related disasters: a comprehensive approach. Nat Hazards 64(2):1609–1624. https://doi.org/10.1007/s11069-012-0320-4
40. Ranjan ES, Abenayake CC (2014) A study on community's perception on disaster resilience concept. Procedia Econ Financ 18(September):88–94. https://doi.org/10.1016/s2212-5671(14)00917-4
41. Renschler CS, Frazier AE, Arendt LA, Cimellaro G, Reinhorn AM, Bruneau M (2010) A Framework for defining and measuring resilience at the community scale: The PEOPLES resilience framework
42. Rivera FI, Kapucu N (2015) Disaster vulnerability, hazards and resilience: perspective from Florida. In: Birkland TA (ed) Springer International Publishing Switzerland
43. Rivera FI, Settembrino MR (2013) Sociological insights on the role of social capital in disaster resilience. In: Kapucu N, Hawkins CV, Rivera FI (eds) Disaster resiliency: interdisciplinary perspectives. Routledge, pp 48–60. https://doi.org/10.4324/9780203102459
44. Sempier T, Swann L, Emmer R, Sempier SH, Schneider M (2010) Coastal community resilience index: a community self-assessment (Issue November)
45. Shaw R, Razafindrabe B, Gulsan P, Surjan A, Takeuchi Y (2009) Climate disaster resilience: focus on coastal urban cities in Asia. Asian J Environ Disaster Manag 01(June):1–15. https://doi.org/10.3850/s179392402009000088
46. Sherrieb K, Norris FH, Galea S (2010) Measuring capacities for community resilience. Soc Indic Res 99(2):227–247. https://doi.org/10.1007/s11205-010-9576-9
47. Siebeneck L, Arlikatti S, Andrew SA (2015) Using provincial baseline indicators to model geographic variations of disaster resilience in Thailand. Nat Hazards 79(2):955–975. https://doi.org/10.1007/s11069-015-1886-4
48. Tilio L, Murgante B, Di Trani F, Vona M, Masi A (2011) Resilient city and seismic risk: a spatial multicriteria approach. ICCSA, LNCS (Part 1) 6782:410–422. https://doi.org/10.1007/978-3-642-21928-3_29
49. Twigg J (2007) Characteristics of a disaster-resilient community: a guidance note 1(August)
50. United Nations Development Programme (2014) Understanding community resilience: findings from community-based resilience analysis (CoBRA) assessments
51. United Nations International Strategy for Disaster Reduction (2017) How to make cities more resilient: a handbook for local government leaders. A contribution to the global campaign 2010–2020: making cities resilient–my city is getting ready! United Nations Office for Disaster Risk Reduction
52. United Nations International Strategy for Disaster Reduction (2005) A guide for implementing the Hyogo framework for action by local stakeholders. http://www.unisdr.org/files/13101_ImplementingtheHFA.pdf
53. United Nations International Strategy for Disaster Reduction (2009) Terminology on disaster risk reduction. In: United Nations international strategy for disaster reduction. United Nation international strategy for disaster reduction. https://doi.org/10.7591/9781501701498-008
54. United Nations Office for Disaster Risk Reduction (2019) Disaster risk reduction in Nepal: status report 2019. In: United Nations office for disaster risk reduction, regional office for Asia and the Pacific. https://www.undrr.org/publication/disaster-risk-reduction-nepal

55. Usamah M, Handmer J, Mitchell D, Ahmed I (2014) Can the vulnerable be resilient? Co-existence of vulnerability and disaster resilience: informal settlements in the Philippines. Int J Disaster Risk Reduct 10:178–189. https://doi.org/10.1016/j.ijdrr.2014.08.007
56. Verrucci E, Rossetto T, Twigg J, Adams BJ (2012) Multi-disciplinary indicators for evaluating the seismic resilience of urban areas. In: 15th world conference on earthquake engineering (15 WCEE), pp 1–10. http://www.iitk.ac.in/nicee/wcee/article/WCEE2012_3490.pdf
57. Yoon DK, Kang JE (2013) A measurement of community resilience in Korea. In: Proceedings of International Symposium on City Planning 2013, pp 1–19

Drought Adaptation in Thailand: A National Governmental Perspective

Puvadol Doydee and Jessada Techamahasaranont

Abstract This paper analyses the national government's capacity for drought risk adaptation in Thailand, emphasizing national laws, regulations, and insights from a national expert workshop on drought risk management conducted in February 2022. Over the past two decades, Thailand has witnessed a significant rise in the occurrence, severity, and magnitude of drought phenomena, largely due to natural climate variability. The effects and vulnerabilities of drought were particularly pronounced during the periods of 2015–2016 and 2019–2020. Additionally, the Vegetation Health Index (VHI) values recorded in March 2020 ranged between 0.15 and 0.85. This paper elucidates the components of drought risk adaptation and management, detailing the roles of various ministries and departments overseeing water resources, and drought risk management in Thailand. As drought impacts span multiple sectors, the success of national government initiatives and action plans largely depend on the coordination of national implementing agencies. It is imperative for these agencies to maintain close communication and cooperation during the execution of national strategies for drought adaptation.

Keywords Drought risk governance · Drought adaptation · Drought vulnerability · Institutional responses to drought · Thailand

1 Introduction

Drought disasters in Thailand are becoming more frequent and severe [19]. Over the past two decades, the country has experienced reduced precipitation and increasing temperatures, which have had an impact on long-term water availability [15]. The Royal Thai Government aims to improve the quality of life for vulnerable groups affected by drought disasters. The Ministry of Interior, through the Department of

P. Doydee (✉) · J. Techamahasaranont
Department of Agriculture and Resources, Faculty of Natural Resources and Agro-Industry, Kasetsart University Chalermphrakiat Sakon Nakhon Province Campus, Sakon Nakhon, Thailand
e-mail: puvadol.d@ku.th

I. Pal et al. (eds.), *Proceedings of the 3rd International Symposium on Disaster Resilience and Sustainable Development—Volume 1*, Lecture Notes in Civil Engineering 623, https://doi.org/10.1007/978-981-96-2139-2_6

Disaster Prevention and Mitigation (DDPM), has outlined strategies to cope with drought situations. There are three types of drought that may affect Thailand: (1) meteorological drought, usually associated with a precipitation deficit [8, 6], (2) agricultural drought, which refers to conditions resulting in unfavorable plant reactions, spanning from diminished crop and forage production to complete crop or forage loss [4, 7]. Local soil and land management practices can also interact with hydrological conditions, leading to agricultural drought [20, 8], and (3) hydrological drought, which occurs due to a shortage in surface and subsurface water flow, primarily in the region's river basins.

All drought originate from a lack of precipitation or meteorological drought, but additional forms of drought and consequences emanate from this insufficiency. For example, agricultural drought occurs primarily due to soil water deficiency, while hydrological drought manifests as a reduction in stream flow, inflow to reservoirs, lakes, and ponds, resulting in livelihood, socioeconomic, and environmental impacts such as reduced wetlands and aquatic habitats (Fig. 1).

Drought is a perturbation, not a stress in a context of agroecosystem properties [1]. This phenomenon is irregular, infrequent, relatively large, and unpredictable. Therefore, people's adaptation to drought is better than their reaction or response to it. Drought can occur in various places, including forests. However, in upland areas,

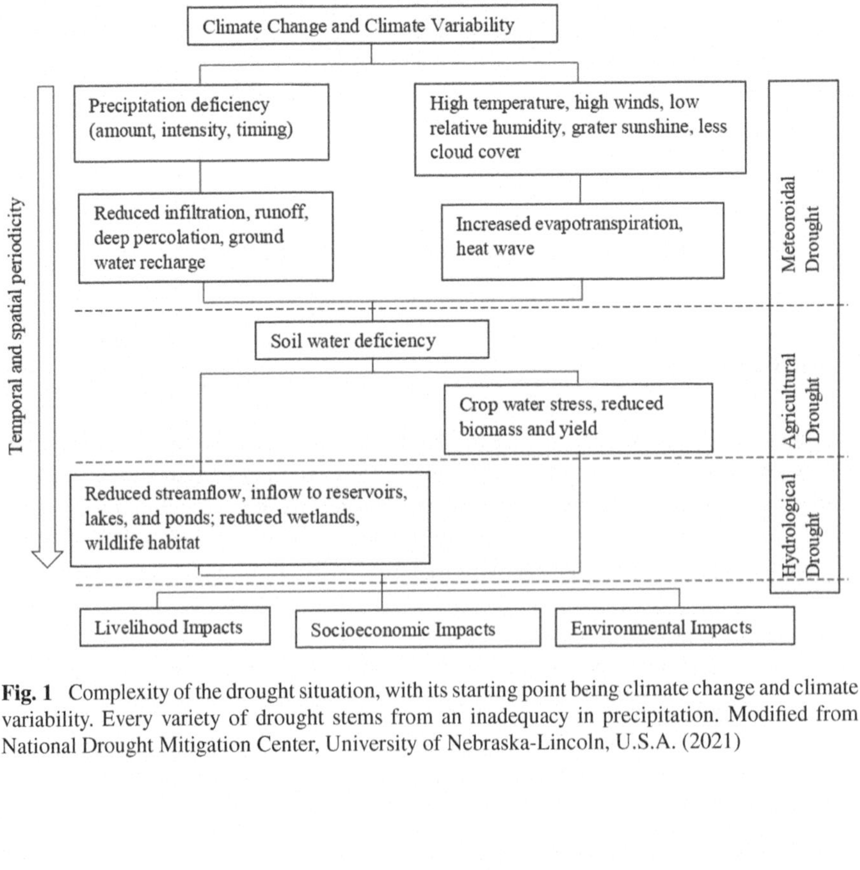

Fig. 1 Complexity of the drought situation, with its starting point being climate change and climate variability. Every variety of drought stems from an inadequacy in precipitation. Modified from National Drought Mitigation Center, University of Nebraska-Lincoln, U.S.A. (2021)

drought sometimes coexists with soil erosion issues. The crops that are able to grow in such conditions are cassava, sugarcane, and watermelon, while rainfed lowland rice is the most common crop ecosystem in Northeast Thailand [9]. Drought is a state of consistent decrease in water quantity, water-flow quantity, or water level, likely to have adverse effects on human beings, animals, and plants in a particular area.

Droughts also increase the risks and impacts of pollution, pests, diseases, heat waves, water contamination, loss and damage to ecosystem services, and ultimately lower the quality of life. According to UNESCAP [16], droughts force people to migrate, often leading to an increase in child malnutrition, stunting, and even adult malnutrition. For example, drought in Thailand has resulted in heat waves that caused a 1.90% increase in excess deaths. Thailand will continue to face vulnerability, with approximately 14.00% of its populace susceptible to drought in the most severe climate change projection. UNISDR [17] proposes that vulnerability evaluation is a procedure that observes and evaluates a region to ready mitigation measures and deter future drought issues arising from land deterioration and desertification.

Drought-related risks pose substantial disruptions to the Thai economy. Remarkably reduced rainfall and sudden declines in groundwater levels within aquifers in catchment zones have resulted in critical water scarcities, affecting agricultural systems, usage, and ecological services. Nevertheless, the Department of Groundwater Resources (DGR) has initiated an extensive groundwater exploitation endeavor employing the Riverbank Filtration (RBF) method to furnish water to regions susceptible to drought across the country. The initial locations encompass Chainat Province, Ratchaburi Province, and Suphanburi Province. The objective of the initiative is to establish a substantial groundwater provision infrastructure and advocate for the combined utilization of surface water and groundwater.

2 Drought Events and Impacts

Thailand is one of the most drought-affected countries in the Asia Pacific region [11]. The drought in Thailand during 2004–2005 resulted in damage of approximately $220 million [18], while the drought of 2015–2017 caused estimated damage of $3.30 billion [2]. This period was recorded as a drought year in Thailand [6]. For example, evidence of persistent drought events during 2015–2016, which was the worst drought in Thailand for 20 years, resulted in considerable losses exceeding $2.50 million (Table 1), severely affecting rice production [13]. Similarly, a study by Pak-Uthai and Faysse [10] described a major drought in 2015 that affected many regions of Thailand, particularly the Chao Phraya River basin, the largest river basin in Thailand. In 2015, more areas in Thailand experienced severe drought, especially the northern parts of the country [13, 15].

The drought of 2019 was estimated to cause losses of $312,000 (Table 1) due to the loss of agricultural crops such as rice, maize, sugarcane, and cassava. The anticipated drought is also likely to impact the production of non-rice field crops, orchards, and freshwater fish farming during the off-season, albeit to a lesser extent. At the macro

Table 1 Drought events and their impacts during 2015–2016, and 2019–2020 in Thailand

2015–2016	2019–2020
Fourteen provinces of Thailand associated with 55 districts, 290 sub-districts, and 2,666 villages were affected by the risk of drought Over 50% of the Mekong watershed area in Northeastern Thailand experienced critical drought conditions Economic losses amounted to \$1.70 billion across the 14 provinces, with insufficient water for agriculture The losses exceeded \$2.50 million in rice production alone The drought during 2015–2016 was considered at the time to be the worst drought in Thailand in the past 20 years	By April 2020, 25 provinces had declared drought disaster areas with coverage of 146 districts and 6,846 villages The Thai Meteorological Department (TMD) declared it as the worst drought in 40 years The Thai government declared a drought emergency Economic losses amounted to \$312,000 due to crop losses in rice, corn, sugarcane, and cassava Seawater intrusion was reported, leading to freshwater shortages and disruptions in agriculture and industry Secondary hazards included exacerbated forest fires in Chiang Mai during March–April 2020, with 5,810 hotspots recorded across the country. Six people were killed while fighting the fires The fires led to increased levels of PM2.5 particulates, reaching 1,000 mg/m^3 (the World Health Organization (WHO) threshold is 25 mg/m^3)

level, the implications of such events lead to a decline in purchasing power among farmer households, increased debt burdens, and contribute to challenges experienced on a larger scale [13].

In March 2020, Thailand experienced drought severity as indicated by the Vegetation Health Index (VHI), which is a measure of the severity of drought based on vegetative health as estimated by satellite images [15]. The VHI combines a vegetative condition index and a temperature condition index. Lower values of the VHI (0.35– < 0.15) in Fig. 2 indicate poor vegetation condition and high temperatures, which are associated with more severe drought.

In January 2021, HII [5] reported that nearly 50% of Thailand was confronted with a drought disaster. The issue of salinity intrusion needs to be considered. Salinity intrusion was experienced in the estuary landscapes as a result of low levels of freshwater caused by the hydrological drought associated with seawater level rise. This phenomenon was particularly noted from the beginning of March. Therefore, monitoring of salinity intrusion at the mouths of major rivers of Thailand under drought conditions is deemed necessary.

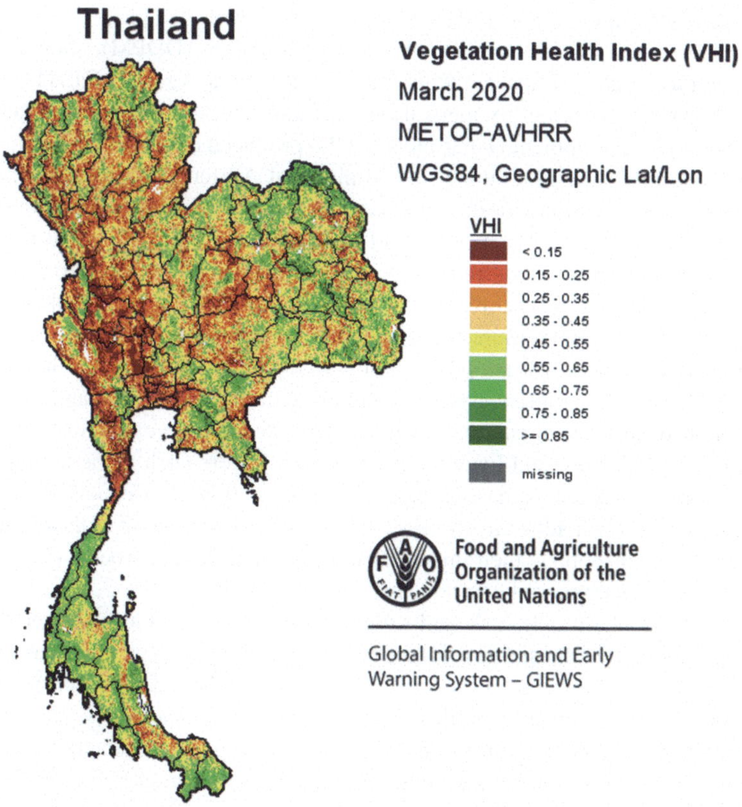

Fig. 2 Map showing the spatial distribution of VHI drought pattern in Thailand, March 2020

3 National Laws and Regulations for Coping with Drought

In Thailand, several national laws, regulations, programs, and disaster risk governance initiatives have been implemented to enhance the understanding of drought disasters and climate risks in the country. These endeavors are spearheaded by diverse governmental bodies, technical organizations, and academia. As an example, the Department of Disaster Prevention and Mitigation (DDPM), in partnership with governmental and technical entities, has formulated databases and initiatives to address drought risks across all tiers, ranging from provincial to village levels. These initiatives align with international agreements such as the Sendai Framework for Disaster Risk Reduction (DRR), the Sustainable Development Goals (SDGs), and the Paris Climate Agreement.

The principal legal mechanism for disaster risk management practices in Thailand is the Disaster Prevention and Mitigation Act of 2007 (DPM Act). This act, along with other related laws, regulations, notifications, and directives, stipulates the DDPM as the core government department responsible for national disaster management. It also empowers local governments to assume responsibility for disaster management in their respective areas, in accordance with the Provincial Plan [12].

The Department of Disaster Prevention and Mitigation (DDPM), supported by the Geo-Informatics and Space Technology Development Agency (GISTDA), has produced drought vulnerability maps using GIS and remote sensing techniques. In 2017, the Land Development Department (LDD) produced drought maps at the sub-district level, while the LDD under the Ministry of Agriculture and Cooperatives developed recurrent nationwide drought maps and regional drought maps for the same year. Drought risk evaluations occur at different administrative levels, overseen by the DDPM, which offers guidelines for provincial and local authorities to formulate drought risk profiles and synchronize them with national strategies and plans.

To mitigate drought, measures have been implemented, including public awareness campaigns to encourage people to employ wise water usage and participate in the implementation of a comprehensive national drought management/adaptation plan, with support from relevant national institutes. Drought risk has cascading effects on human lives, properties, and livelihoods in various factors such as rising temperatures, seawater intrusion, increased hunger, and malnutrition. Consequently, societies and communities must adjust to address present and future drought risks to mitigate these effects and respond to their impacts. Figure 3 illustrates the process of adapting to drought with its important components.

The analysis of drought risk and vulnerability aims to identify target demographics, vulnerable zones, and effects on different domains, encompassing exposure, susceptibility, and adaptability, an evaluation is conducted. This appraisal aids in gauging the extent of vulnerability for target demographics, zones, and sectors. The identification of adaptation options involves analysing current adaptation efforts, evaluating adaptation options, planning, implementing, and monitoring and evaluating (M&E) strategies. This includes conducting an economic cost breakdown analysis to determine appropriate adaptation options, considering local wisdom, research

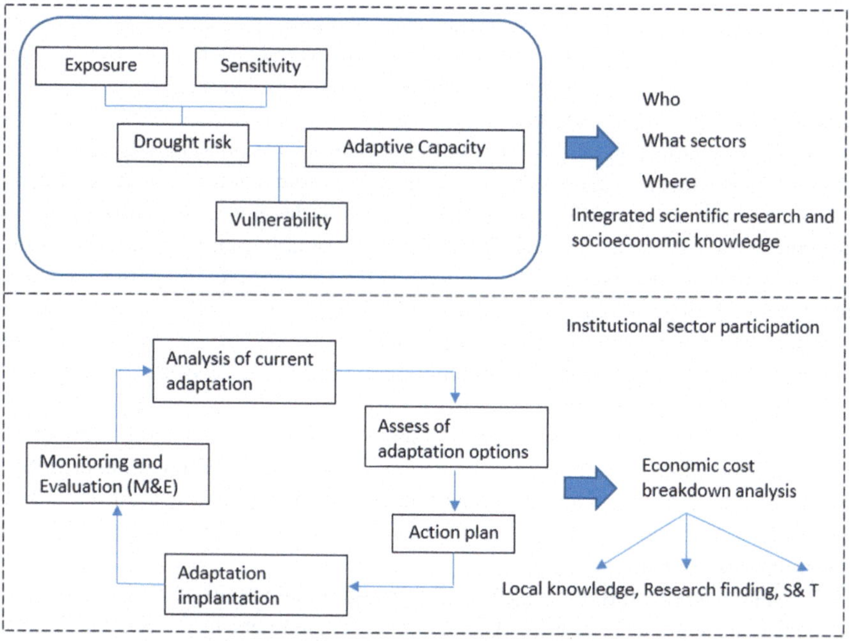

Fig. 3 Element of adaptation and management to drought risk. Modified from T-PLAT (2022)

findings, science, technology, and innovations (STI) that are suitable for the societal and community context. The process of drought adaptation must integrate information and knowledge from scientific and social sciences, with the participation of all stakeholders.

According to German Cooperation [3], Thailand has commenced the development of a nationwide monitoring and evaluation (M&E) framework for the water and agriculture domains. These sectors aim to integrate M&E processes for adaptation strategies aligned with the country's National Adaptation Plan (NAP) and in consonance with the overarching information management framework. The necessity for M&E has been highlighted in the Paris Agreement under the United Nations Framework Convention on Climate Change (UNFCCC), the Sustainable Development Goals (SDGs), and the Sendai Framework for Disaster Risk Reduction 2015–2030.

Setting up a nationwide M&E framework will streamline the documentation of Thailand's progress toward the objectives delineated in these accords. For instance, within the realms of agriculture and water management, initiatives like organic farming, testing resilient crop strains, and soil–water conservation measures have been implemented to bolster resilience against drought. Typically, these sectors have individual M&E systems or frameworks tailored to project requirements.

Thailand has also upheld a structured compilation of data for evaluating drought perils, employed in monitoring drought risks, projecting climate trends, providing seasonal forecasts, and annually forecasting drought hazards nationwide. These initiatives function as aids for sector-specific planning, particularly in provinces susceptible to drought. Data gathering mechanisms incorporate meteorological observation networks featuring automated rain gauges, weather radar systems, and remotely captured data, alongside the automated weather system (AWS) managed by the Thai Meteorological Department (TMD), alongside ongoing surveillance of reservoir water levels. To meet the information requirements for drought, countries need an interagency operational plan. This plan can be developed by a working group that considers how all government agencies can sustainably gather the required information, deliver it effectively, and establish mechanisms for receiving user feedback and evaluation. The working group should be inter-sectoral and have the relevant technical leadership. Solutions need to be scalable and updated, adapting to emerging technologies such as geospatial technology and satellite earth observation, such as MODIS-based vegetation time-series [4].

To execute cross-sectoral strategic measures for drought adaptation, mitigation, and emergency intervention at both the national and subnational levels, approaches such as drought risk and adaptation communication exchange among all stakeholders in Thailand, institutionalization, localization, and impact communication/drought research translation are necessary.

Thailand should further advance the institutionalization of Sex, Age, and Disability Disaggregated Data (SADDD). The detailed data in the Sendai Framework Monitor (SFM) specifically encompass gender (men and women) and age groups (children: 0–14 years, adults: 15–64 years, and seniors: 65 years and above). Disability pertains to individuals with pre-existing disabilities impacted by a disaster, rather than those acquiring disabilities from a hazardous event, such as drought occurrences [14]. SADDD can inform priorities and practices to shape a rights-based and evidence-based approach for drought risk adaptation in all sectors. Combined with other data, such as exposure, socio-economic vulnerability, and intersectional data, critical questions can be answered, such as: Who is most at risk? Which rights are at stake? How can drought risks be reduced for vulnerable groups?

Gathering data for risk, vulnerability, and impact evaluations, and integrating them into drought readiness and planning, mitigation, response, and rehabilitation measures are crucial components pertinent to drought risk governance in Thailand.

Every ministry, in addition to numerous departments, bureaus, offices, and centers, is represented at lower tiers of administration. For example, the Department of Water Resources (DWR) has branches at provincial, district, and sub-district levels, mirroring the structure of the Department of Disaster Prevention and Mitigation (DDMP). Hence, comprehending the intricacies of institutional setups and the multitude of government personnel engaged in the hierarchical Thai administrative system responsible for drought adaptation is paramount (Table 2).

At the national level, the National Disaster Prevention and Mitigation Committee (NDPMC), presided over by the Prime Minister or a designated Deputy Prime Minister, assumes key duties and obligations regarding policymaking in disaster risk governance. At the grassroots level, Provincial Disaster Prevention and Mitigation Committees and Bangkok Metropolitan Committees are established [12]. The Provincial Disaster Relief Committees are responsible for coordinating efforts among

Table 2 Ministries, departments, and their responsibilities involved in water resources and drought risk management in Thailand

Ministries	Departments	Responsibilities
Ministry of Natural Resources and Environment (MONRE)	Department of Water Resources (DWR)	Water resource policies, strategies, and administration outside irrigated zones fall under the purview of the Water Crisis Prevention Center. This entity is tasked with suggesting and harmonizing strategies to tackle water crises and droughts in regions vulnerable to disasters
	Office of the National Water Resources (ONWR)	The obligations involve suggesting policies, crafting strategic outlines, comprehensive plans, and actions for the management of water resources on a national scale, as well as coordinating their implementation. The objectives are to mitigate drought, improve water reservoir efficiency, and ensure the availability of clean water for all people
	Department of Groundwater Resources (DGR)	Policies, strategies, and administration concerning groundwater resources entail executing extensive projects aimed at delivering water to regions susceptible to drought on a nationwide scale
	Office of Natural Resources and Environmental Policy Planning (ONEP)	The formulation, strategizing, and administration of natural resource policies and environmental preservation serve as the nucleus at the national level for overseeing climate change and drought governance
	Department of Environmental Quality Promotion (DEQP)	Encourages public consciousness and engagement in preserving natural resources and the environment, as well as the formulation of environmental policy and planning

(continued)

Table 2 (continued)

Ministries	Departments	Responsibilities
Ministry of Agriculture and Cooperatives (MOAC)	Royal Irrigation Department (RID)	Water resource administration entails the distribution of water for agricultural irrigation purposes, as well as the management of dams and storage reservoirs within its jurisdiction. It provides relief from agricultural drought through the provision of irrigation water
	Department of Royal Rainmaking and Agricultural Aviation (DRRAA)	Research and development in rainmaking technology aim to enhance the efficiency of weather modification and atmospheric water management for reducing drought disasters resulting from global climate change
	Land Development Department (LDD)	Research and management of land utilization function as the primary representative at the national level for the United Nations Convention to Combat Desertification
	Department of Agriculture (DOA)	Crop research and development include initiatives such as direct rice seeding and the cultivation of drought-tolerant species
	Department of Agricultural Extension (DOAE)	Agricultural production and promotion encompass the cultivation of various agricultural crop types, supported by extension services
Ministry of Interior (MOI)	Department of Disaster Prevention and Mitigation (DDPM)	The governing authority and governmental organization for disaster management encompass the duty of mitigating drought risks
	National Disaster Warning Centre (NDWC)	The monitoring and control of warning towers involve disseminating warnings to line agencies
	Provincial Waterworks Authorities (PWA)	Government-owned enterprises oversee the provision and distribution of water supply at the provincial level

(continued)

Table 2 (continued)

Ministries	Departments	Responsibilities
	Department of Local Administration (DLA)	Strengthening local administrative organizations involves supporting communities with the Sufficiency Economy Philosophy application. Involved in collaboration and cooperation with the DDPM at the local level
Ministry of Digital Economy and Society (MDES)	Thai Meteorological Department (TMD)	Weather and climate forecasting include drought forecasting and drought risk mapping
	National Statistical Office (NSO)	Administers data and information in a unified and standardized approach to foster and bolster the nation's competitiveness. Conducts censuses or sample surveys and facilitates the collection of national statistical information databases on socio-economic aspects, including Sex, Age, and Disability Disaggregated Data (SADDD)
Ministry of Higher Education, Science, Research and Innovation (MHESI)	Geo Informatics Space and Technology Development Agency (GISTDA)	The institution furnishes satellite remote sensing and Geographic Information System (GIS) data to both public and private entities for drought hazard monitoring and mapping
	Hydro-Informatics Institute (HII)	The organization conducts research, develops and applies scientific and technological advancements for agricultural and water resources management, including drought management
Ministry of Energy (MOE)	Electricity Generating Authority of Thailand (EGAT)	The publicly-owned utility oversees both electric power generation and transmission, as well as operational activities and maintenance of dams and reservoirs relevant to drought

(continued)

Table 2 (continued)

Ministries	Departments	Responsibilities
Ministry of Social Development and Human Security (MSDHS)	Department of Social Development and Welfare (DSDW)	Organizes and strengthens social development and social welfare initiatives that are implemented based on people's participation and geared towards the goal of developing a quality society, promoting gender equity, and accelerating rural development, including employment and income generation
	Department of Older Persons (DOP)	Develops plans, measures, and innovations, including integrated tasks, to promote and protect the rights and well-being of the elderly. Prepares society for well-being in old age and protects and promotes elder rights in accessing the social welfare system. Promotes networking among organizations to further develop elder-related tasks

sectoral agencies at lower levels, such as districts and sub-districts. The Provincial Emergency Operation Centers (EOC) are in charge of commanding, controlling, supporting, and coordinating drought response and disaster relief measures.

4 Conclusion

Drought risk affects agroecosystem, water, energy, food, land, and the natural environment, leading to destruction of human life and property. Whenever it occurs, the government provides compensation payouts to quickly assist the livelihoods of the victims. The frequency, intensity, and scale of droughts have heightened in Thailand, notably in the last twenty years. Extended periods of severe droughts negatively affect agricultural output, endangering food security and the economic stability of rural households and economically disadvantaged communities.

The assessment of drought effects and vulnerability is a collective obligation shared by stakeholders from all affected sectors. This evaluation aims to identify primary impacts, root causes, scale, severity, risk, and susceptibility levels, along with appraising economic losses within each sector.

This article has been enriched by comments and inputs received from key national stakeholders, including DDPM, DWR, DGR, RID, LDD, DRRAA, DOA, DLA,

TMD, HII, and EGAT. Contributions were provided during the National Consultation Workshop on Drought Risk Management in Thailand held in February 2022.

Acknowledgements We would like to acknowledge the support received from the United Nations Economic and Social Commission for Asia and the Pacific (UNESCAP) for their assistance in coordinating the National Consultation Workshop on Drought Risk Management in Thailand and for providing financial assistance for this study.

References

1. Conway GR (1986) Agroecosystem analysis for research and development. Centre for environmental technology, imperial college of science and technology, London. United Kingdom, p 112
2. EM-DAT (2022) M-DAT: the OFDA/CRED international disaster database. https://www.emdat.be/emdat_db/. Accessed 19 Jan 2022
3. German Cooperation (2021) Thailand, Australia and Germany pool their efforts to strengthen data management and M&E of adaptation in Thailand's water sector. https://www.thaigerman-cooperation.info/en_US/fighting-against-climate-change-thailand-australia-andgermany-pool-their-efforts-to-strengthen-data-management-and-me-of-adaptation-inthailands-water-sector/. Accessed 14 Dec 2021
4. Ha TV, Uereyen S, Kuenzer C (2023) International journal of applied earth observation and geo information agricultural drought conditions over mainland Southeast Asia : spatiotemporal characteristics revealed from MODIS-based vegetation. Int J Appl Earth Obs Geoinformation 121(February), Article no 103378
5. HII (2021) Drought. Hydro informatics modelling system section. Drought bulletin 2(1):2
6. Khadka D, Babel MS, Shrestha S, Virdis SGP, Collins M (2021) Multivariate and multi-temporal analysis of meteorological drought in the northeast of Thailand. Weather Clim Extrem, 34(December), Article no 100399
7. Kim SJ, Park S, Lee SJ, Shaimerdenova A, Kim J, Park E, Lee, WK, Kim GS, Kim N, Kim TH, Lim C-H, Choi Y, Lee W-K (2021) Developing spatial agricultural drought risk index with controllable geo-spatial indicators: a case study for South Korea and Kazakhstan. Int J Disaster Risk Reduct 54(August), Article no 102056
8. Li Y, Lu H, Yang K, Wang W, Tang, Q, Khem S, Yang F, Huang Y (2021) Meteorological and hydrological droughts in Mekong River Basin and surrounding areas under climate change. J Hydrol: RegNal Stud 36(June), Article no 100873
9. Monkham T, Jongdee B, Pantuwan G, Mitchell JH, Sanitchon J, Fukai S (2018) On-farm multi-location evaluation of occurrence of drought types and rice genotypes selected from controlled-water on-station experiments in Northeast Thailand. Field Crop Res 220(May):27–36
10. Pak-Uthai S, Faysse N (2018) The risk of second-best adaptive measures: farmers facing drought in Thailand. Int J Disaster Risk Reduct 28(June):711–719
11. Pandey S, Bhandari H, Hardy B (2007) Economic costs of drought and rice farmers coping mechanisms: a cross-country comparative analysis. Los Baños (Philippines). International rice research institute, p 203
12. Thanyalakmetha S (2020) Thailand country report 2020. Visiting researcher program (FY 2020B). Asian disaster reduction center (ADRC). Department of disaster prevention and mitigation (DDMP), p 9
13. UNDRR (2020) Disaster risk reduction in Thailand: Status report 2020. Bangkok, Thailand. United Nations office for disaster risk reduction (UNDRR), Regional office for asia and the Pacific, p 38

14. UNDRR (2021) Sendai framework monitor (SFM): sex, age and disability data (SADDD). UN office for disaster risk reduction, p 4
15. UNESCAP and ASEAN (2021) Ready for the dry years. Building resilience to drought in South-East Asia (2nd edn) United Nations-economic and social commission for Asia and the Pacific, p 126
16. UNESCAP (2021) ASEAN regional plan of action for adaptation to drought 2021–2025. United Nations-economic and social commission for Asia and the Pacific, p 51
17. UNISDR (2014) Progress and challenges in disaster risk reduction: a contribution towards the development of policy indicators for the post-2015 framework on disaster risk reduction. United Nations Publication, Geneva
18. Wichitarapongsakun P, Sarin C, Klomjek P, Chuenchooklin S (2016) Rainfall prediction and meteorological drought analysis in the Sakae Krang River basin of Thailand. Agric Nat Resour 50:490–498
19. Yang S, Zhao B, Yang D, Wang T, Yang Y, Ma T, Santisirisomboon J (2023) Future changes in water resources, floods and droughts under the joint impact of climate and land-use changes in the Chao Phraya basin, Thailand. J Hydrol 620 Part A (May):129454
20. Yildirim G, Rahman A, Singh VP (2022) Meteorological and hydrological drought hazard, frequency and propagation analysis: a case study in southeast Australia. J Hydrol: RegNal Stud 44(October), Article no 101229

The Impact on PM2.5 in Thailand Society: An Investigation into Causes, Challenges, and Policy Gap in Implementations

Vasuthida Pinrattana and Oulavanh Sinsamphanh

Abstract This research about PM 2.5 pollution is one of the Thai environmental problems that still needs to be fixed, even though it has not recovered much since 2019. The Thai government has an answer to this crisis that has affected people, which is short-term and long-term solutions for people and other organizations to apply. However, it requires time to operate, it is not effective enough, making many people in many areas still face with this problem. This qualitative study explores the number of PM2.5 that happened in Thailand from 2019 to 2023, the cause of PM2.5, and the policies of PM2.5 and analyzes existing policies used in the country with indicators from global organizations. The secondary data was collected from Chulalongkorn University, IQAir from 2020 to 2023, a case study from Beijing, and a case study from Seoul, including UN SDG for analysis. From the information we have found, the causes of PM 2.5 in Thailand come from man-made and environmental such as seasonal, climate change, and burning crops, including the area that gets impacted the most is in northern Thailand such as Chaing Mai. This research shows people's actions and the environment clarify the impact in Thailand and suggests a solution for decreasing PM 2.5 in Thailand.

Keywords Impact · PM2.5 · Public policy · Thailand

1 Introduction

PM2.5 pollution is one of the environmental problems that still has been existing in many developing countries in Southeast Asia. In Thailand, in particular, it is not recovered much and it has been impacted a lot since 2019 [7, 8, 23]. PM2.5 is dust that

V. Pinrattana (✉)
Social Policy and Development, Faculty of Social Administration, Thammasat University, Bangkok, Thailand
e-mail: Vasuthidaza@gmail.com

O. Sinsamphanh
Faculty of Environmental Science, National University of Laos, Vientiane, Laos

I. Pal et al. (eds.), *Proceedings of the 3rd International Symposium on Disaster Resilience and Sustainable Development—Volume 1*, Lecture Notes in Civil Engineering 623, https://doi.org/10.1007/978-981-96-2139-2_7

has a very small size and hazard which is the diameter of PM2.5 is less than 2.5 μm that can enter the human body easily and affect their health, especially the lungs and bloodstream. This small dust comes from many various places, major of this air pollution comes from using diesel engines and most of them are vehicles, traffic jams, emissions from factories that use fuel and coal, and burning from agriculture or outdoor and indoors such as cigarette, candles, and using photocopying machines [15, 29].

In search of effective solution, the World Health Organization or WHO has set guidelines for air pollution to make it easier to create a plan to reduce PM2.5, these levels have seven target levels and have been used since 2021 [37]. According to that guideline, Thailand has met all WHO PM2.5 targets for the framework such as exceeding WHO PM2.5 guidelines by 5 to 7 times, 7 to 10 times, and over 10 times which these three are the most exceeding that the World Health Organization has set for guidelines. Moreover, there is the Paris Agreement which is a commitment from all countries that have to reduce emissions for climate change that is happening, therefore the emission of carbon is also one that needs to be considered as it can be pollution [4, 10]. As a result, it is air pollution that is important that has to be solved as quickly as it can to not affected people's health more than the present time.

However, the direction of the Thai government dealing with this matter is more likely unclear. Such an example, Thai government has an answer to the PM2.5 situation that has impacted their people, by the government has solutions for short-term conditions such as spraying water, and long-term situations such as replacing public transport with electric vehicles instead, which require time to solve this air pollution, and some of them may not compelling enough due to they have spent some time to do it, and people have to adapt themselves to it which make them have many of the PM2.5 in their body. For this reason, they should have more policies for this situation. Moreover, most of the keys to PM2.5 from the government in Thailand help Bangkok and the urban areas. On the contrary, Bangkok and the urban areas are not the only places that impacted by this air pollution, apparently, in 2019 Nakhon Ratchasima was the top polluted province in Thailand [11].

As the matter of fact that, social inequality contributes to variations in the ability to cope with the impacts of PM 2.5 across different populations in society [32]. This becomes particularly evident during the existence of Covid-19, where various groups experience diverse levels of effects [24]. Additionally, several groups in society such as old age [2, 3], people with disability, or other vulnerable groups [5, 31], all face distinct conditions that influence their responses to PM 2.5 damage. Consequently, this chapter aims to examine the causes, challenges, and policy gaps in implementing PM 2.5 management in Thailand. The goal is to provide the Thai government with more effectively informed decisions in addressing the escalating severity of PM 2.5 in the country.

This chapter follows the content review. The contents have emphasized the framework of the UN framework for PM2.5 and the Thai policies to solve the problem of air pollution together with the case study from other countries that had been facing PM2.5 problems. Data were employed from secondary sources including academic papers, website, reports, and NEWS associated with PM2.5.

2 Development of Severe PM2.5 in Thailand

PM2.5 was found severe effect to Thailand much in 2019 since it was higher than the mean of annual PM2.5 [1, 13]. Thailand was twenty-eighth ranks with 24.3 μg per cubic meter in the work country ranking presented by average PM2.5 based on accessible data [18]. Bangkok in Thailand was thirty-third ranking in the world regional capital city ranking with an annual average of 22.8 μg per cubic meter in 2019 which was over from target the WHO for PM2.5 12.8 μg per cubic meter in the mean that the World Health Organization or the WHO set 10 μg per cubic meter for target PM2.5 average [18]. Moreover, Nonthaburi, Nakhon Ratchasima, Chiang Mai, Mae Hong Son, and Khon Kaen have real-time monitoring. These cities had an average of PM2.5 higher in 2019 which was considered moderate and unhealthy for sensitive groups of US AQI level especially, Nakhon Ratchasima, and Khon Kaen. Besides, Nakhon Ratchasima and Saraphi were in the top ten most polluted regional cities in rank eighth and ninth among Southeast Asia countries [18].

In 2020, the COVID-19 pandemic made many countries in the world had healthier air including Thailand world region ranking dropped to thirty-fourth ranks with 21.4 μg per cubic meter of PM2.5 mean in 2020, Bangkok, Thailand dropped to thirty-fourth in the world capital city ranking with 20.6 μg per cubic meters which was over from WHO set for guidelines target about 10.6 μg per cubic meters [19]. Due to the pandemic in 2020, the overall number of cities in Thailand dropped the PM2.5, Nonthaburi, Nakhon Ratchasima, Bangkok, Chiang Mai, Mae Hong Son, and Khon Kaen had a moderate level of US AQI level that cities had PM2.5 had about 20.6–30.5 annual mean which was in 12.1–35.4 μg per cubic meter of PM2.5 [19]. On the other hand, many provinces in Thailand were in the fifth of the most polluted regional cities for annual mean among Southeast Asia countries. Those provinces were Pai with 53 μg per cubic meter which was the second most polluted regional city, Chiang Saen, and Phan with 43.7 and 41.6 μg per cubic meter for the annual mean of PM2.5. In 2020, PM2.5 was discharged a lot in Northern Thailand due to open burning in agriculture, including Southeast Thailand. Furthermore, Northern Thailand reached an unhealthy level in 2020 because of the emission of PM2.5 that came with opening burning for agriculture [19].

In 2021, the PM 2.5 mean of Thailand was forty-fifth country ranking with 20.2 μg per cubic meter, and Bangkok, the capital city, was ranked forty-second with 20 μg per cubic meter which dropped from last year [20]. On the other hand, three cities in Thailand were in the top five of most polluted regional cities in Southeast Asia, which had Lampang with 52.5 μg per cubic meter was the most polluted city in Southeast Asia, exceeding over 10 WHO PM2.5 guidelines, which World Health Organization has updated PM2.5 guidelines by changing the mean of year to 5 μg per cubic meter in 2021, Pai with 47.1 μg per cubic meter, and Nan with 39.5 μg per cubic meter, these two cities were considered to exceed over 7 to 10 times WHO PM2.5 guidelines and it was moderate for US AQI levels [20]. Likewise, many of the provinces in Thailand, including Bangkok for the first four months of 2021 had been the most polluted and had met levels that exceeded over 10 to exceed over 5 to 7 times

WHO PM2.5 guidelines. The issue of PM2.5 in 2021 still came from transportation, industrial, electricity, burning crops, and haze from burning in boundaries around Thailand mostly in the Northern part [20].

In 2022, Thailand was ranked fifty-seventh for the country ranking with 18.1 μg per cubic meter, which was an annual average, and Bangkok was ranked fifty-second for the capital city ranking with 18.0 μg per cubic meter [21]. Bangkok has annual mean dropped around 10 percent from last year and the annual average in Thailand also dropped about 10.4 percent, which implies that Thailand will have cleaner air in 2022 with the US AQI level also considered to have a moderate level. However, some provinces in Thailand that were ranked as the most polluted regional city which there were in the top ten of the most polluted regional city, they were Chaloem Phra Kiat with 36.2 μg per cubic meter which exceeded 7 to 10 times WHO target guidelines, and according to the US AQI level was Unhealthy for Sensitive Groups level for the annual average, Nan, and Pak Nam with 30.4 μg per cubic meter which exceeded 5 to 7 times from WHO PM2.5 guidelines and is considered to be moderate level for the US AQI levels [21]. The root of PM2.5 still came from transportation, industrial, electricity, burning crops, and haze from burning in transboundary around Thailand mostly in the Northern part. Furthermore, burning crops from the area transboundary around the Northern part of Thailand was likely to be one of the majors of worrying that influenced the province in Northern Thailand to face more haze from other places including in their province from open burning crops which happened the most during the dry season in Thailand [21].

In the present time for the first three months of 2023 that now one city in Thailand also has been reported as the most polluted in the world in 2023 which is Chaing Mai on the 11 and 12 of March with 212 μg per cubic meter including the provinces in the northern part of Thailand also have high PM2.5 [36]. In March of 2023 has been reported that Chiang Mai and many provinces in the north have worse air quality that is considered to be very unhealthy to US AQI level which comes from both Thai people's actions and the countries around the northern part of Thailand which are burning farmland including a forest in countries around Thailand. On the contrary, the burning of agriculture is still the main root that makes much of the particles in the atmosphere.

3 The Impact of PM2.5 in Thailand

The impacts of PM2.5 that take place in Thailand have affected people and the economy [36]. Alternatively, the health of people needs to be the most concern due to people who will be the ones who will make everything move, be it economy, agriculture, or travelling. PM2.5 particles are immensely small in that their diameter is less than 2.5 μm, which is smaller than human hair when compared, hence it will be easy to get it in our body through breathing [37]. The effect of PM2.5 on the well-being of people can happen in the short term and long term [12, 22, 35]. The short-term outcome of PM2.5 will take place in the eyes, nose, throat, and lungs irritating such as coughing, sneezing, runny nose, and short breath which will make some people face difficulties in life, besides, people who already are sensitive groups such as allergy will face these symptoms and affect more among other people [22, 27]. The long-term effect on people's health will relate to increasing chronic bronchitis rate and reducing lung function [27]. Besides, PM2.5 will increase more chances of having lung cancer since the particles are affected the most by the lung, and increase the chance of having heart disease as well. The vulnerable group such as people who are considered vulnerable groups such as elderly people, disabled people, pregnant women, and children will also be sensitive to these particles [14, 26]. The State of Global Air reported that around 32,200 Thai people died from PM2.5 in 2019, while Greenpeace reported that people around 29,000 died prematurely from PM2.5 which was considered the number of people who died from the particles was more than the number of people who died by the accidents, drugs or other worse scenarios such as criminal which these were combined in 2021 [34].

It is not only the health that affected by PM2.5 but also the economy that gets this influence. The PM2.5 affected the economy of travelling the most which is one of the three major sources of income in Thailand. Existing studies of the number of foreign tourists in Chiang Mai and Bangkok showed the number of travellers would be different in the time, mostly in April and May, PM2.5 also increased about 5 percent in every month of that year. The decreasing number of tourists showed a lot in Bangkok and Chiang Mai which was reduced to around 659,368 people for Bangkok, and around 106,060 people. This makes Thailand lose the opportunity for economic growth of 4,105.13 and 476.27 million baht [28].

4 Framework in Addressing PM 2.5

4.1 The UN Framework for PM2.5

The United Nations of UN framework for PM2.5 has the 'Clean Air and Sustainable Development Goals' which is 'the 2030 Agenda for Sustainable Development' by the UN that has 17 goals to achieve by 2030, likewise, it has 3 goals that relate to making better air quality to each country members [37], those goals are Goal 3 is Good Health and Well-Being which is target 3.9 and 3.9.1 indicators that pay attention to reduce the number of death and ill by air pollution, Goal 7 is Affordable and Clean Energy, which is target 7.1 and 7.1.2 indicator that will address clean energy that should use in household, and Goal 11 Sustainable Cities and Communities, which is target 11.6 and 11.6.2 indicator which the target 11.6 relates to PM2.5 the most that are reducing the environmental impact of cities, in addition, will focus on air quality, municipals, and other waste treatment, thus 11.6.2 indicator is significant due to its will concern about the PM2.5 average in the cities. These three Sustainable Development Goals are also considered by the World Health Organization or WHO to guarantee the health of people and improve air quality to be better. Both the United

Nations and the World Health Organization use the Sustainable Development Goals as guidance, indicators, or measurements that each country has to achieve and follow by 2023, besides, WHO also applies it as monitoring [37].

4.2 The Policies of PM2.5 by the Thai Government

The Thai government has three policies that will be used as a framework to solve overall PM2.5 in Thailand [6]. The first measurement that the government has come up with makes the management in the area to be more effective, every related agency has to follow and coordinate with the Pollution Control Department and Gistda for making a plan which is before, during, and after air pollution occurs [6].

The second policy is protecting and reducing the cause of pollution. For instance, detecting black smoke from vehicles, maintaining public transportation, cleaning roads, controlling agriculture burn, and controlling and examining emission of pollution from factories. The controlling and reduction will take place in vehicles, agriculture, construction, industrial, and household. Furthermore, reducing the source of PM2.5 will have two terms to operate which are short-term about 2019–2021, and long-term about 2022–2024. In addition, the short-term will be about setting the measurement for each root cause, and the long-term will be concerning applying the measurement and making the measurement to be more worthwhile. While controlling and limiting households will be different from others, it will be the process of supporting households to use clean energy.

The third policy that the government introduced is making pollution management more efficient through a local organization to watch out, follow, and examine air quality. This policy also has both a short period and a long period of the process which will pay attention to developing a system and tools for a mechanism of pollution management that is developing the network for examining air quality, including the local to follow and examine their areas, improving law or measurement to relate situation, supporting research and technology for adapting to handle air quality and publicize, pollution transboundary, and improving the database and develop watch out system [33]. Moreover, some managements have only short-term goals which are setting emission air pollution accounts from their origin which will be periodically, and developing a predicted dust situation system for use as information to improve, and guidance of protection and solution air pollution [33].

5 Analysis of the Thai Policies for the PM2.5 Situation

The policies from the Thai government are still not effective enough although they have both short-term and long-term due to it being hard to apply to different areas, especially with the sources of air pollution which are agriculture, industry, and factory that policy still does not cover that much, the number of PM2.5 also shows that are both decreased and increased since 2019. It is not only not ineffective for some roots, but it is also focusing a lot on the traffic that other parts of Thailand do not have major roots as Bangkok which have all root effects on the city, especially traffic jams in the capital city. Moreover, Thailand still cannot achieve much with the UN and WHO policies for PM2.5 due to the situation of PM2.5 in Thailand from 2019–2022, and the present time show that the condition of air pollution in Thailand is not effective as it should be, although the annual average of PM2.5 that reached since 2019–2022 is likely to decrease as ranking among the country that has monitoring for PM2.5 (Fig. 1), including the capital city as Bangkok city in Thailand that since 2019 to 2022 that the annual mean is decreasing. Therefore Thai policy regarding traffic jams is successful, especially in Bangkok city which is polluted the most by cars, and other vehicles.

However, other cities in Thailand still rank for Southeast Asia, with most of the ranking will tend to rate the ten cities as the most polluted regional cities. Furthermore, the COVID-19 situation that occurred since 2020 would make every country in the world improve the environment, including the root of PM2.5 would decrease emissions [19], but many provinces in Thailand still rank that goes back and forth. For instance, Nakhon Ratchasima with a mean 42.2 μg per cubic meter ranked as the eighth most polluted regional city in 2019 [19], Pai with a mean 53 μg per cubic meter ranked as the second most polluted regional city in 2020 [19], Lampang with a mean 52.2 μg per cubic meter ranking as the most polluted regional city in 2021 [20], and Chaloem Phra Kiat with a mean 36.2 the fifth most polluted regional city in 2022 [21].

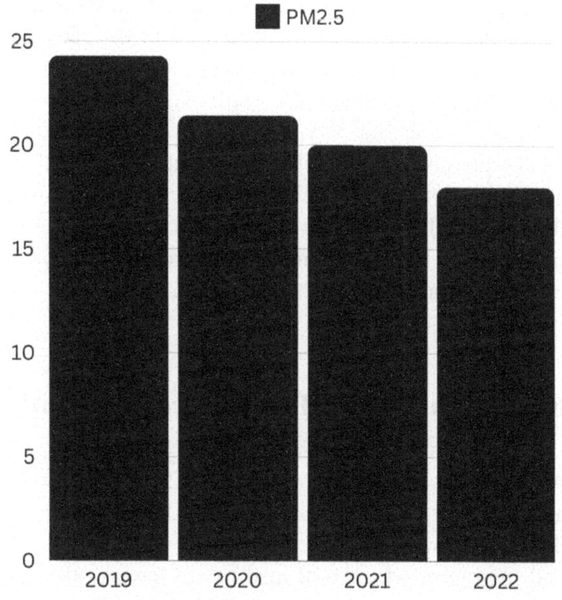

Fig. 1 The statistics of PM2.5 in Thailand in 2019–2022

In addition, in the first three months of 2023, Chaing Mai has ranked as the most polluted city in the world in the current ranking according to Live AQI City Ranking with 243 μg per cubic meter. The WHO framework for PM2.5 also shows that Thailand still exceeds WHO guidance for PM2.5. Moreover, we can see that most of the provinces that have poor air quality will take place in the northern part of Thailand where the people will mainly work in the agriculture sector and travel. The origin of PM2.5 has more in outdoor activity which comes from the action of people, such as traffic jams where smoke will come out from automobile which will happen most in Bangkok that have the most population, construction where dust and chemical will come out during the process of construction, factories, and industries where they are producing will create gas make poor air quality and agriculture burning which is likely to be most source for other provinces after their harvest and prepare to another crop due to they do agriculture as a major work and this action will occur a lot during the dry season, but it also has roots in indoor activities. For example, lighting up incense and candles where the smoke comes from, using photocopying machines from which ozone will come out a lot from this machine, and smoking which people like to do as much as chemicals that respond to oxygen in the air.

However, indoor activities still do not release pollution as much as outdoor activities. In addition, the natural course of PM2.5 is the dry season which will not have much wind from winter to summer that takes place during the dry season. It is not only the dry season that makes the air quality worse, but it is also climate change that plays a big role in making the pressure trap of PM2.5 cannot go away easily which makes PM2.5 in Thailand worse. These PM2.5 affect a lot on the health of Thai people in the short-term and long-term, especially sensitive groups and vulnerable people that will get the impact easier than normal people, and economy in Thailand also will get this impact because most of the economy will be travelling, and travellers may not want to visit Thailand will lose some of their income to it. Likewise, in the present time that PM2.5 still exists a lot in many places in Thailand including Bangkok, the capital city, and northern provinces in Thailand such as Chiang Mai can be possible that people will be getting worse through the harm of air pollution and economy that will make them lose many of consumers to come to the cities. For this reason, Thai policies are still not influential enough that make the number of patients who have problems with the respiratory system and death increase, and the number of travellers not stable and decrease.

6 Conclusion and Policy Recommendations

Air pollution in Thailand which is PM2.5 is still a problem in Thailand that both state and global organizations such as the United Nation want to decrease in the country due to the impacts of the particulars will affect a lot on human health short term and long term which will lead to not have well-being, including the air pollution also impacts on economic, especially travelling that will decrease the growth of economics since travelling is one of major income. Due to PM2.5 in Thailand is still not recovered as much as it should since 2019 to 2023, policy will be the important part to improve people life, by the people themselves also have to participate along with the policy that will make it more effective to solve the environmental problem in Thailand.

It has been observed that the policies in Thailand are still not effective for some of the situations, therefore Thailand still needs more policies that will decrease PM2.5 in the atmosphere with the strictness and awareness that will need to increase for the people who are involved in releasing air pollution. In this emergent situation, the Thai government should emphasize increasing the strictness of PM2.5 in industrial, factory, and agriculture through a tax regime. In terms of incentive, the government may provide tax reduction for those who can lower the PM2.5. This is in line with several countries that aim at reducing PM 2.5 emission such as China [16–17], Norway [9], or Chile [25]. This initiative is to minimize the source of PM2.5 due to the fact that the sources are a cause of the problem that need to get rid of it as much as possible which is the same as the case study of Seoul Metropolitan City, South Korea has seen that solving the problem through the source of PM2.5 will be more worthwhile [30]. Simultaneously, in order to decrease PM 2.5 emissions, the government might suggest imposing taxes as penalties for individuals or entities significantly contributing to PM 2.5 sources. This approach could enhance their motivation to refrain from generating PM 2.5.

In this chapter, its strengthens include several updated important information for PM2.5 in Thailand that can easily acknowledge the cause, impact, and related policies that have been implemented for the PM2.5 problem in Thailand. Moreover, this paper also has analyzed PM2.5 in Thailand from the policy that is used in Thailand and the framework from UN which is the UN Sustainable Development Goals that will be used as criteria to analyze policies for PM2.5 in Thailand. However, the limitation of this research is about writing due to the topic of this research having a lot of information which can lead it to be board for the information to write of this paper which can make it more complicated as have a lot of information, including it will be hard to specify one area as the pollution is happening all over Thailand, as a result, it makes it has much of information to apply in the research.

References

1. Amnuaylojaroen T, Parasin N, Limsakul A (2022) Health risk assessment of exposure near-future PM2. 5 in Northern Thailand. Air Qual Atmos Health 15(11):1963–1979
2. Asavanirandorn C, Pechdin W, Ngamsomsak R, Bhula-Or R (2023) What influences older urban poor's attitude towards online job search? Implications for smart cities development. Smart Cities 6(1):614–625
3. Asavanirandorn C, Pechdin W, Trang NTQ (2022) Identifying factors influencing productivity of older workers in service sector: a case study in pilot companies in Thailand. Behav Sci 12(8):268
4. Blau J (2017) The Paris agreement: climate change, solidarity, and human rights. Springer
5. Chalayonnavin A (2023) Increasing self-esteem for vulnerable young women: a case study of entrepreneurship training for vulnerable young women in Thailand. Br J Spec Educ

6. Chanabawornsakul W, Tuprakay S, Premanoch P, Ratcha M (2022) A study of government measures and guidelines for managing the particulate matter PM2.5 pollution crisis in Thailand. J Assoc Res 27(3). https://so04.tci-thaijo.org/index.php/jar/article/view/259514
7. Chirasophon S, Pochanart P (2020) The long-term characteristics of PM 10 and PM 2.5 in Bangkok, Thailand. Asian J Atmos Environ (AJAE) 14(1)
8. ChooChuay C, Pongpiachan S, Tipmanee D, Suttinun O, Deelaman W, Wang Q, Xing L, Li G, Han Y, Palakun J (2020) Impacts of PM2. 5 sources on variations in particulate chemical compounds in ambient air of Bangkok, Thailand. Atmos Pollut Res 11(9):1657–1667
9. Ciccone A (2018) Environmental effects of a vehicle tax reform: empirical evidence from Norway. Transp Policy 69:141–157
10. Dimitrov RS (2016) The Paris agreement on climate change: behind closed doors. Glob Environ Polit 16(3):1–11
11. Dutta A, Chavalparit O (2023) Assessment of health burden due to the emissions of fine particulate matter from motor vehicles: a case of Nakhon Ratchasima province, Thailand. Sci Total Environ 872:162128
12. Feng S, Gao D, Liao F, Zhou F, Wang X (2016) The health effects of ambient PM2. 5 and potential mechanisms. Ecotoxicol Environ Saf 128:67–74
13. Fold NR, Allison MR, Wood BC, Thao PT, Bonnet S, Garivait S, Kamens R, Pengjan S (2020) An assessment of annual mortality attributable to ambient PM2. 5 in Bangkok, Thailand. Int J Environ Res Public Health 17(19):7298
14. Fu X, Li L, Lei Y, Wu S, Yan D, Luo X, Luo H (2020) The economic loss of health effect damages from PM 2.5 pollution in the central plains urban agglomeration. Environ Sci Pollut Res 27:25434–25449
15. Gao J, Tian H, Cheng K, Lu L, Zheng M, Wang S, Hao J, Wang K, Hua S, Zhu C (2015) The variation of chemical characteristics of PM2. 5 and PM10 and formation causes during two haze pollution events in urban Beijing. China Atmos Environ 107:1–8
16. Han F, Li J (2020) Assessing impacts and determinants of China's environmental protection tax on improving air quality at provincial level based on Bayesian statistics. J Environ Manag 271:111017
17. Han F, Li J (2022) Spatial pattern and spill over of abatement effect of Chinese environmental protection tax law on PM 2.5 pollution. Int J Environ Res Public Health 19(3):1440
18. IQair (2019) 2019 World air quality report. https://www.greenpeace.org/static/planet4-thailand-stateless/2020/02/91ab34b8-2019-world-air-report.pdf
19. IQair (2020) 2020 World air quality report: region and city PM2.5 ranking
20. IQair (2021) 2021 World air quality report: region and city PM2.5 ranking. https://www.iqair.com/world-most-polluted-cities/world-air-quality-report-2021-en.pdf
21. IQair (2022) 2022 World air quality report: region and city PM2.5 ranking. https://www.greenpeace.org/static/planet4-thailand-stateless/2023/03/d1d69c24-2022_world_air_quality_report_en.pdf
22. Janssen N, Fischer P, Marra M, Ameling C, Cassee F (2013) Short-term effects of PM2. 5, PM10 and PM2. 5–10 on daily mortality in the Netherlands. Sci Total Environ 463:20–26
23. Johnston HJ, Mueller W, Steinle S, Vardoulakis S, Tantrakarnapa K, Loh M, Cherrie JW (2019) How harmful is particulate matter emitted from biomass burning? A Thailand perspective. Curr Pollut Rep 5:353–377
24. Klinsrisuk R, Pechdin W (2022) Evidence from Thailand on easing COVID-19's international travel restrictions: an impact on economic production, household income, and sustainable tourism development. Sustainability 14(6):3423
25. Mardones C, Mena C (2020) Economic, environmental and distributive analysis of the taxes to global and local air pollutants in Chile. J Clean Prod 259:120893
26. Maung TZ, Bishop JE, Holt E, Turner AM, Pfrang C (2022) Indoor air pollution and the health of vulnerable groups: a systematic review focused on particulate matter (PM), volatile organic compounds (VOCs) and their effects on children and people with pre-existing lung disease. Int J Environ Res Public Health 19(14):8752

27. Moradi M, Mokhtari A, Mohammadi MJ, Hadei M, Vosoughi M (2022) Estimation of long-term and short-term health effects attributed to PM 2.5 standard pollutants in the air of Ardabil (using Air Q+ model). Environ Sci Pollut Res 1–9
28. Namcome T, Tansuchat R (2021) The impact of fine particulate matter (PM 2.5) pollution on the number of foreign tourists in Chiang Mai and Bangkok. Rajabhat Chiang Mai Res J 22:19–35
29. Orellano P, Reynoso J, Quaranta N, Bardach A, Ciapponi A (2020) Short-term exposure to particulate matter (PM10 and PM2. 5), nitrogen dioxide (NO_2), and ozone (O_3) and all-cause and cause-specific mortality: Systematic review and meta-analysis. Environ Int 142:105876
30. Park S-H, Ko D-W (2018) Investigating the effects of the built environment on PM2. 5 and PM10: a case study of Seoul Metropolitan city, South Korea. Sustainability 10(12):4552
31. Pechdin W, Bunditsakulchai P (2024) Characteristics of individuals at risk of malnutrition in Thailand: an investigation focusing on income insecurity. Front Sustain Food Syst 7:1256119
32. Pechdin W, Sarnkhaowkhom C, Kanthanetr S, Willemse MP (2023) Retelling social inequalities in the era of market competition: review and discussion for sustainable welfare development. Front Sociol 8:1085278
33. Pollution control department (2019) National agenda action plan for addressing air pollution particulate matter issues
34. PpsFoam (2023) Why has the resolution of the PM 2.5 dust problem in Thailand not yet succeeded? (Translated from Thai). Thairath Online. https://www.thairath.co.th/news/society/2664491
35. Sharma S, Chandra M, Kota SH (2020) Health effects associated with PM 2.5: a systematic review. Curr Pollut Rep 6:345–367
36. Shoowong M, Apisitniran L, Sachamuneewongse S (2023) Coming up for air. Bangkok Post. https://www.bangkokpost.com/business/2531910/coming-up-for-air
37. World health organization (2021) WHO global air quality guidelines: particulate matter (PM2. 5 and PM10), ozone, nitrogen dioxide, sulfur dioxide and carbon monoxide. World health organization

Labor Law Implications on Disaster Susceptible Small and Medium Enterprises in Thailand

Weerayut Lasongyang and Shubham Pathak

Abstract Labour law has struggled to provide a just system between employers and employees. International norms, domestic laws, economic setup and several other factors directly affect both parties' minimum wage, resilience and welfare. Consequently, this research includes disaster and their impact on labour law and vulnerable Small and Medium Enterprises which tend to provide employment opportunities in developing countries. This research aims to explore, analyse, and synthesise the labour law scenario in Thailand and how employees and SMEs can be disaster resilient. The method adopted for this research is qualitative, with research tools involving Key Informant Interviews (KII) and rigorous secondary data review that provide literary support to the research. Extensive literature has been provided for the gaps in the existing legal framework in Thailand and the inadequacies of the labour laws and guidelines in Thailand, especially in the local governance level SMEs. The direct impacts were inaccurate minimum wages, which are not paid during disaster scenarios. Equal rights, employment opportunities, and assistance for the professional development of employees and SMEs are neglected during disasters. The lack of involvement from stakeholders, including managerial and governmental assistance, resulted in limited opportunities for the SMEs. This research paves the way for the resilience and sustainability of Thailand and its regional development as Thailand is a leader among developing South Asian countries.

Keywords Labor law · Minimum wage · Professional growth · International law · Sustainable development · Small and Medium Enterprises (SMEs)

W. Lasongyang
Pridi Banomyong Faculty of Law, Dhurakij Pundit University, Bangkok, Thailand

S. Pathak (✉)
School of Accounting and Finance, Center of Excellence in Sustainable Disaster Management (CESDM), Walailak University, Thai Buri, Thailand
e-mail: shubhampathak@gmail.com

I. Pal et al. (eds.), *Proceedings of the 3rd International Symposium on Disaster Resilience and Sustainable Development—Volume 1*, Lecture Notes in Civil Engineering 623, https://doi.org/10.1007/978-981-96-2139-2_8

1 Introduction

Globally, SMEs provide the majority of employment opportunities. However, labour law governs the scenarios contrarily. International norms, universal guidelines, local byelaws, rights and voices of all the concerned parties must be regularly reviewed and revised to ensure that they remain current and relevant [38]. Without constant re-evaluation, outdated guidelines can create an uneven playing field, giving certain individuals or groups an unfair advantage.

When these frameworks are regularly assessed and improved, it ensures that everyone has an equal opportunity to succeed. For instance, reviewing economic zones can help to identify areas that require additional investment and development for migrant workers, which can help to boost local economies [2]. Furthermore, revising domestic regulations can help to protect consumers and ensure that businesses are operating fairly.

By taking the time to regularly review and revise these important guidelines, Thailand can create a more just and equitable society [31, 14]. This not only benefits individuals, but also communities and Thailand as a whole. However, in disaster scenarios, the implementation takes a contrary turn.

As an employer, it's important to recognize and uphold the fundamental rights of your employees. This includes providing fair wages that are commensurate with their skills and experience, as well as decent working hours that allow them to maintain a healthy work-life balance. In addition to these basic requirements, migrant workers as well as domestic employees should also be provided with employment security, which includes benefits such as job security, retirement plans, and healthcare coverage [32].

Furthermore, it's important to acknowledge the family responsibilities of your employees and provide them with the necessary support to balance their work and personal life. This may include flexible work schedules, parental leave, or other accommodations that allow them to fulfill their obligations as caregivers (Fig. 1).

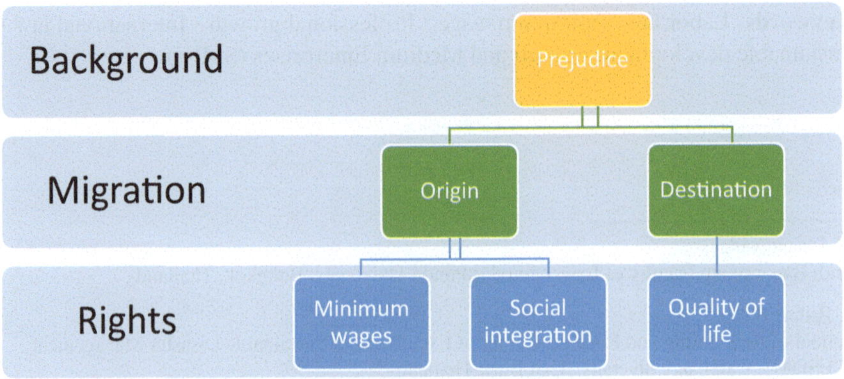

Fig. 1 Research framework for migrant workers and minimum wage

Thai SMEs employers should also take measures to ensure that their employees work in safe environments that are free from hazards and potential risks. This includes providing proper training, equipment, and facilities, as well as adhering to safety regulations and standards [17].

In addition to these basic requirements, Thai employers should also provide social security benefits to their employees [36]. This may include provisions for healthcare, unemployment insurance, disability coverage, and other benefits that offer financial security and peace of mind [25].

Finally, it's crucial to ensure that child and forced labor is not practiced within your organization, and that your employees have the right to form trade unions. These measures promote fair treatment and respect for workers' rights and help to create a more equitable and just workplace. By implementing these principles, employers can create a workplace environment that is fair, equitable, and supportive of their employees' well-being and success.

International Labor Standards (ILO) are conventions or recommendations covering various issues related to labor rights, employment, social security, social policy and human rights. The International Labor Standards (ILO) are comprehensive conventions and recommendations for promoting and protecting workers' rights worldwide. These standards cover various issues related to labor law, employment opportunities, social security schemes, and another rights. They ensure that every worker is handled fairly and has opportunity to decent working conditions, as well as to promote social justice and economic development.

The ILO standards are constantly evolving and adapting to changing social and economic conditions. They are developed through a rigorous process of consultation and negotiation involving governments, employers, workers, and other stakeholders. These standards are widely recognized as the benchmark for labor rights and are increasingly being incorporated into national and international legal frameworks.

These standards cover a broad range of issues, including freedom of association and the right to collective bargaining, elimination of forced labor and child labor, non-discrimination in employment, social security, and protection of the rights of migrant workers. They also cover issues promoting decent working conditions, such as safe and healthy workplaces, fair wages, and reasonable working hours.

Adherence to international labor standards paves the way for a society that is safe and just, where workers are treated with dignity and respect. This commitment can foster greater economic growth, social stability, and human development. International Labor Standards (ILO) are not just a set of rules, but a pathway to a better future, covering a variety of issues related to labor rights, employment, social security, social policy and related human rights (Fig. 2).

Thailand is facing a pressing issue in determining the minimum wages for different sectors. The challenge lies in the fact that the authorities have not yet defined which sectors are included and excluded in this process. This lack of clarity affects a broad range of businesses operating in Thailand, including SME, ME and LE.

A study conducted by Pholphirul and Rukumnuaykit in 2010 highlights the complexity of this problem. The researchers found that the lack of clarity regarding the sectors covered by the minimum wage policy has led to confusion among

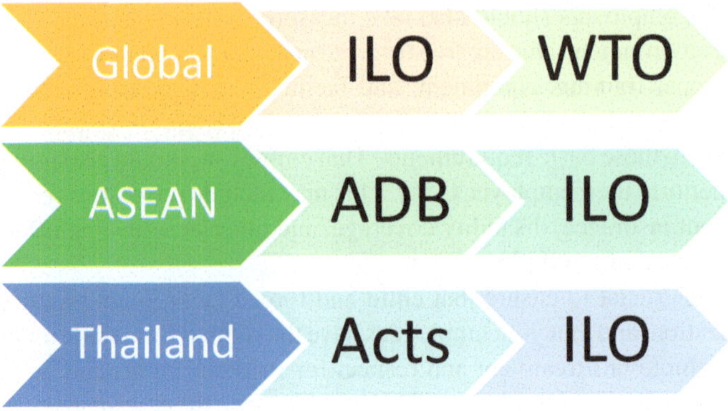

Fig. 2 Framework for labor law

employers and employees alike. Moreover, the study found that the minimum wage policy has not been effectively enforced across all sectors, which has resulted in a significant wage gap between different occupations.

To address this issue, there is a need for greater collaboration between the government, employers, and employees. A comprehensive review of the existing policy is necessary to ensure that it is fair and equitable for all sectors. The review must take into account the varying needs and challenges faced by different sectors and ensure that the minimum wage policy is effectively enforced across the board. Thailand struggles to ascertain the minimum wages provided to these sectors, not defining the covered and uncovered sectors. These are omnipresent in both SMEs and large enterprises (LE) [30].

The regulatory system for just minimum wages is controlled directly by the Ministry of Labour, Department of Labour Protection and Welfare [9, 20]. The regulatory bodies provide definitions, norms, bylaws and guidelines for the public and private sectors of the economy.

The Other bodies, including the Bank of Thailand, the National Economic and Social Development Board, and the National Statistical Office, are responsible for providing transparent information and ensuring collaboration between all involved stakeholders, resulting in the protection of labor law implementations and the rights of the laborers [13].

The disaster in Thailand adversely impacts the standard of employment, minimum wages and rights of all the concerned parties. This leads to inadequacies and is supported by the previous literature to impact directly towards the reduced rights [3, 4, 8, 10, 12, 23, 37]. The theoretical approach involving efficiency wage theory is implied and adopted to ascertain the social, economic and resilience among SMEs [11, 33].

Thai SMEs are facing significant challenges due to the inadequacies in the country's labor laws. The situation is particularly dire during times of disaster, which can exacerbate the existing issues. One of the most pressing problems is the ineffective implementation of laws that are designed to ensure minimum wages, equal employment opportunities, and supportive government assistance. This has led to a situation where SMEs struggle to attract and retain employees, and workers are often paid less than what they deserve.

Another issue is the prevalence of agents and brokers in the job market who charge their own commissions, making it even more difficult for SMEs to operate successfully. This practice often leads to exploitation of workers, who are forced to accept lower wages than they deserve, even though they are performing the same work as their counterparts. This disparity in pay and treatment is a significant problem that must be addressed if we are to create a fair and just labor system.

Furthermore, the complex immigration and labor laws in Thailand pose a significant challenge for migrant workers who face restrictions on their ability to find fair and just employment opportunities. The lack of clarity in these laws can lead to exploitation, and many workers are forced to rely on brokers to find jobs and ensure their safety. Even parents of migrant workers often rely on brokers to ensure the safety of their loved ones, which highlights the severity of the issue.

Therefore, the challenges facing SMEs, workers, and migrant workers in Thailand's labor market are many and varied. It is essential that we address these issues and create a labor system that is fair and just for all. Small and medium-sized enterprises (SMEs) are struggling to cope with the inadequacies in labor laws in Thailand. The situation is particularly challenging during disasters, which can exacerbate the existing issues. One significant problem is that laws designed to ensure minimum wages, equal employment opportunities, and supportive government assistance are not being implemented effectively. This has led to a situation where SMEs and their employees are struggling to make ends meet. Furthermore, the agents and brokers who offer job opportunities often charge their own commissions, making it even more difficult for SMEs to operate successfully. The complex immigration and labor laws in Thailand also pose a significant challenge for migrant workers, who face restrictions on their ability to find fair and just employment opportunities. Even parents of migrant workers rely on brokers to ensure the safety of their loved ones. These challenges must be addressed if we are to create a labor system that is fair and just for all.

2 Methodology

In this study, we utilized qualitative research methods to examine the relationship between employability and several factors that may influence one's ability to secure and retain employment. The data collection tools used in this study included KII and published online resources. We also conducted an extensive literature review to support the consistency and accuracy of our findings.

Our study findings indicate a significant relationship between employability and various factors such as gender, financial status, internship opportunities, and security of the interns. For instance, our analysis shows that individuals from low-income households face more challenges in securing employment due to absence of accessibility to resources for instance education and training. Similarly, the study found that gender can also influence employability outcomes, with women being more likely to experience discrimination and bias in the job market.

The qualitative data obtained through key informant interviews provided us with a deeper understanding of the quantitative findings. For instance, our interviews with industry experts and employers revealed that a lack of relevant work experience and soft skills can also hinder employability outcomes. Furthermore, we found that the security of interns in the workplace can significantly affect their overall employability outcomes.

Our study provides a comprehensive overview of the complex dynamics that influence employability outcomes. It highlights the need for the development of effective policies and interventions that can address the root causes of unemployment and underemployment. Our findings have important implications for policymakers, employers, and individuals seeking to improve their career prospects.

3 Findings and Discussions

The findings from this research have provided an in-depth understanding of the current labour laws and migrant situation, along with the unforeseen implications of the COVID-19 pandemic. The COVID-19 pandemic has had a severe impact on internship opportunities for domestic and migrant professionals who have completed their cooperative education [29]. Despite their academic achievements, many interns are struggling to find full-time employment due to industry-wide hiring freezes and a lack of comprehensive policies to support their transition into the workforce. This situation is further compounded by the fact that many companies are prioritizing experienced professionals over entry-level hires.

Moreover, the absence of clear guidelines and standardized procedures for hiring interns has made the process more challenging for both SMEs and interns [24]. SMEs often miss out on the opportunity to bring fresh perspectives and diverse skill sets into their organizations, while interns are left uncertain about their future professional prospects. As a result, many are forced to take on low-paying jobs or face prolonged periods of unemployment.

It is crucial that businesses take a proactive approach to prioritize hiring interns and provide the necessary support to help them succeed. By establishing clear policies and procedures for intern recruitment, companies can create a more equitable and supportive environment for migrant workers to thrive. Additionally, investing in programs that offer mentorship, training, and career development opportunities can help interns gain valuable skills and experience, setting them up for future success [27]. Therefore, government agencies, laborers and stakeholders need to work together to create a brighter future for all interns and young professionals.

The interns, who had shown immense dedication and potential during their cooperative education, faced significant challenges in securing employment post-pandemic.

This was primarily due to the COVID-19 pandemic and the lack of comprehensive policies at the company level to transition them into full-time roles. One of the respondents expressively spoken,

"Due to COVID-19 pandemic I was not able to be appointed as a full-time employee after my internship program ended."

It was disheartening to discover that while several companies have anti-harassment policies in place, the interns, not being directly under their human resource offices, are left without recourse. This underscores the urgent need for action to ensure the safety and well-being of our interns. This leaves interns at risk of being subjected to harassment without any legal recourse [15].

It's crucial for companies to take immediate action to address this issue. They need to ensure that their anti-harassment policies are comprehensive and cover all employees, including interns. In Thailand, companies should take steps to educate their employees on the importance of preventing harassment in the workplace [18].

By doing so, Thai companies can create a just working environment, where all employees are content. It's time for Thai SMEs to take responsibility and make sure that no employee, including interns, is subjected to harassment. A company executives added,

"There are several work-related pressures and environmental inclinations. However, we tend to manage all employees under the human resource department, but students are beyond the company's authority."

The pressing challenges that the labor sector in Thailand faces is the government's mismanagement and inadequate governance. The lack of effective labor governance can lead to a range of issues, including underpayment, exploitation, and unsafe working conditions for laborers. This can have a devastating impact on their lives and livelihoods, leaving them vulnerable and without adequate protection.

To address this issue, there is a need for a more robust and comprehensive labor governance framework that ensures fair treatment and protection for all laborers. The government must implement policies and practices that promote transparency, accountability, and equity in the workplace. This includes monitoring and regulating labor standards, enforcing labor laws, and providing adequate resources for labor inspection and enforcement.

In summary, the Thai government must take urgent action to improve labor governance to ensure that laborers receive fair treatment, safe working conditions, and protection from exploitation [28]. By doing so, the government can promote a more equitable and just society, where all workers are valued and respected. Inadequate governance of laborers and mismanagement at the company level result in the limited position of laborers in Thailand [19].

The substantial cost of migrating to Thailand, nearly 25,000 THB (equivalent to 720 USD), along with the brokers' fees (10 to 30 percent) and the national verification process, known as 90-day reporting, compels migrant workers to rely on brokers and incur additional expenses.

It is important to acknowledge that certain labourers may not possess identification documentation due to a lack of awareness and understanding surrounding the matter. A prime example of this is the absence of birth registration certificates, which is a prevalent issue among labourers in many countries [26]. This lack of proper identification poses significant challenges for individuals when attempting to access services and opportunities that require identification, such as healthcare, education, and employment.

To address this issue in Thailand, it is imperative to increase awareness and educate these labourers on the importance of obtaining identification documentation and the process involved in doing so [7]. This education must focus on the significance of identification documentation in society, including its importance in safeguarding the rights of individuals and ensuring their access to services. Additionally, it is crucial to provide a clear and concise outline of the process involved in obtaining birth registration certificates and other forms of identification.

By increasing awareness and education on identification documentation, labourers will be able to participate more fully in society and benefit from all available opportunities [21]. This will not only benefit the labourers and their families, but it will also contribute to the development and progress of their respective communities and countries [6]. Additionally, lack of awareness leads to absence of important documents including birth certificates and identification documents.

Migrant workers leave their homes and families to seek better economic opportunities, often landing in unfamiliar lands with a language barrier and limited resources. They are often confined to small rooms, sometimes with no ventilation or natural light, and share their living space with five to six roommates. Despite the cramped living conditions, they find solace in sharing their common issues and concerns with one another [34]. However, their problems often go unheard, and they are left to navigate the complex system on their own.

To address their issues and improve their living conditions, a common platform is required that connects all stakeholders. This platform will provide them with a voice to express their concerns and receive the support they deserve. It will help create a better future for migrant workers by addressing their issues related to housing, healthcare, and workers' rights. By coming together, we can ensure that migrant workers are treated fairly and with dignity, and that their contributions to society are recognized and valued [16]. There are five to eight migrant workers staying inside the same room (Fig. 3).

3.1 SWOT Analysis

When it comes to the situation of migrant workers, there are several key points to consider ensuring their safety and success. Let's take a closer look at each of these points:

- Strengths: One of the most significant advantages of migrant workers is the high demand and supply for their skills in various industries. Their unique expertise and knowledge can contribute significantly to the growth and development of businesses, making them a valuable asset to any company.

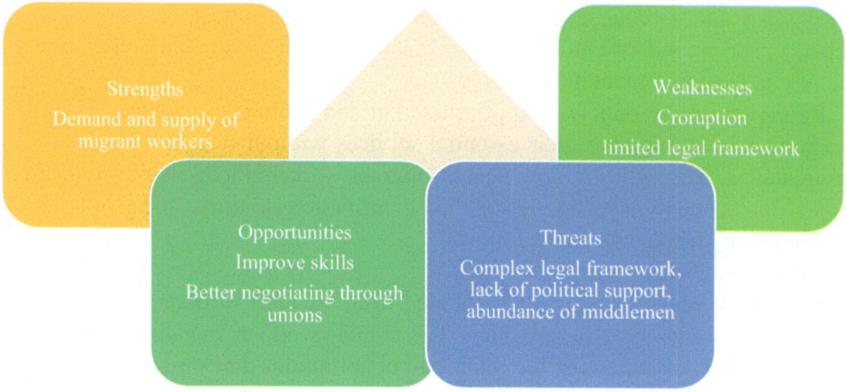

Fig. 3 SWOT analysis for this research

- Weaknesses: However, migrant workers often face significant challenges, such as corruption and an inadequate legal and regulatory framework. These issues can lead to exploitation and mistreatment of migrant workers, resulting in poor working conditions, low wages, and unfair treatment.
- Opportunities: One possible solution to these challenges is the creation of unions. Unions can enhance the skills and negotiating power of migrant workers, ultimately leading to better working conditions and fair treatment. This could help ensure the welfare of the workforce and contribute to their overall success.
- Threats: Despite these opportunities, there are still significant threats to be aware of, such as the legal, political and governance framework. These issues can create significant challenges and obstacles for migrant workers, making it essential to address them head-on to ensure their safety and success.
- Overall, it's clear that there are both advantages and challenges associated with the demand and supply of migrant workers. By understanding these issues and working to address them, we can promote a fair, safe, and prosperous environment for all workers, regardless of their background or origin.

4 Conclusion and Recommendations

For decades, migrant workers contribute to Thailand's economy, providing the country with a vital source of labor and expertise. These individuals come from a variety of backgrounds and bring with them a wealth of knowledge and experience that is essential to the country's success. From agriculture to construction and beyond, migrant workers have helped drive progress and development across multiple sectors.

Despite their crucial role, however, migrant workers often face significant challenges and obstacles. Many are imposed to work overtime in difficult setting, often for low pay and without access to essential benefits and protections. This is unacceptable, and we must do more to ensure that these hardworking individuals receive the just and fair treatment they deserve.

By implementing policies that prioritize fair and equitable treatment of migrant workers, we can improve their essential services without fear of exploitation or mistreatment. This will not only benefit migrant workers themselves, but also the larger Thai economy, which relies on their contributions to thrive and grow.

The recommendations are as follows:

Governance level

The Thai economy has greatly benefited from the hard work and dedication of both registered and unregistered Cambodian migrant workers. Legalizing all migrant laborers would implementing the rights and opportunities they deserve, such as the ability to access healthcare, education, and legal protection. It would also ensure that they are no longer vulnerable to exploitation and abuse.

Collaborating with international organizations is essential in maintaining international norms and standards. This collaboration would involve regular consultations and the exchange of ideas to ensure that our policies are sound and equitable. It would also involve the development of training programs for employers and employees on labor laws and human rights.

To guarantee that migrant workers receive their due wages, the government must take responsibility for depositing minimum wages directly into their bank accounts. This would require the establishment of bank accounts for all migrant workers, which would be managed by government-operated banks. The wages would be securely transferred, and regular audits would be conducted to ensure that everyone is paid fairly.

Creating a legally binding regulatory framework is crucial to ensure the safe and regulated cross-border movement of migrant workers. The framework would involve the establishment of a coordinating agency to oversee all aspects of cross-border migration, including pre-departure orientation, recruitment, placement, and repatriation. The agency would also be responsible for monitoring and enforcing compliance with labor laws and human rights standards. This would ensure that migrant workers are protected from exploitation and abuse, and that they are treated with dignity and respect.

Broker/Agent/Middlemen level

It should be mandatory for brokers to maintain meticulous records of migrant workers and the services they have received. These records should comprise details pertaining to the worker's name, country of origin, work experience, and the services provided to them, such as employment opportunities, accommodation arrangements, and legal advice.

In order to guarantee fair remuneration for their services, brokers must operate within a clear regulatory framework that stipulates their responsibilities towards the concerned parties. This framework should also offer guidance on the fees that brokers may levy and the standards of service that they must uphold.

In addition to their professional duties, brokers must extend assistance to less privileged migrant workers. This assistance may encompass the provision of fundamental necessities such as food and shelter, as well as facilitating access to medical care and other essential services.

Brokers play a pivotal role in facilitating communication between migrant workers and government channels. To this end, brokers must establish a unified platform that enables workers to access information pertaining to their rights and responsibilities, as well as the range of services and support available to them.

Migrant workers level

Migrant laborers ought to undertake the responsibility of forming labor unions to represent their interests and elevate their concerns with brokers and government officials. This proactive approach not only facilitates the negotiation of favorable terms and conditions of employment but also amplifies their bargaining power.

Migrant laborers must be aware of legally mandated minimum wages along with accompanying benefits. Employers may be inclined to exploit migrant laborers by offering wages below the legally stipulated threshold. Laborers must be cognizant of the minimum wage provisions in their respective industries and ensure that they receive the requisite compensation. In the event of a remuneration shortfall, laborers are entitled to lodge complaints with relevant authorities.

Migrant laborers face an array of challenges and impediments when employed in a foreign country. These may include linguistic barriers, cultural disparities, and an unfamiliarity with local laws and customs. It is therefore essential for laborers to communicate their particular needs, risks, and obstacles to relevant stakeholders, including employers, government agencies, and community organizations. This facilitates the provision of requisite support to overcome such obstacles and flourish in their work environment.

Migrant laborers must abide by all applicable laws and regulations to ensure that they receive fair and equitable treatment. This involves obtaining the necessary work permits and visas, complying with local laws and regulations, and adhering to workplace safety protocols. Laborers must be aware of their rights, including the right to lodge complaints if they feel that they are being treated unfairly. By adhering to these guidelines and procedures, migrant laborers can ensure that they are safeguarded and treated equitably.

4.1 Study Limitations

The limitations of this research, including the challenges posed by the language barrier, have been addressed through a comprehensive approach. Official translations in English of the Labor Act and other legal documents were utilized alongside a Thai co-author to cross-check the Thai version of the constitution. This approach was taken to ensure an accurate interpretation of legal and minimalize the risk of inadequate interpretation.

By adopting this approach, we have been able to conduct a rigorous analysis of the legal landscape in Thailand, resulting in a more detailed and nuanced research output. The use of official translations and the involvement of a Thai co-author has enabled us to overcome the limitations posed by the language barrier and ensured the highest level of accuracy and reliability.

We are confident that this research output will provide valuable insights into the legal system and contribute to the wider body of knowledge in this field. Ultimately, this approach highlights our commitment to professionalism and expertise in our research endeavors.

4.2 Way Ahead

The study highlights the pivotal position of Thailand in the ASEAN region, which serves as a central hub for offering vast employment and growth prospects to all member nations. The report underscores the significance of Thailand's contribution towards fostering sustainable development in the region and further reinforces its vital role in propelling the overall growth of ASEAN. According to the study, Thailand is playing a pivotal role in ensuring the growth and sustainable development. As the central hub, Thailand offers significant job opportunities and economic growth prospects to all the member nations. The country's strategic location, coupled with its well-developed infrastructure and skilled workforce, has contributed to its emergence as a key player in the region's economic progress.

The study further explains that Thailand's robust manufacturing sector, coupled with its strong service industry, has enabled it to become a hub for trade and investment. The country's openness to foreign investment, coupled with its favorable business climate, has made it an attractive destination for multinational firms seeking to expand their operations in the region.

Overall, the study underscores the critical role played by Thailand in driving the ASEAN region's economic growth and development. Its ability to provide growth opportunities making it a vital contributor to the region's progress.

Acknowledgements The authors appreciate support from family and friends.

References

1. Agadjanian V, Gorina E, Menjívar C (2014) Economic incorporation, civil inclusion, and social ties: plans to return home among Central Asian migrant women in Moscow, Russia. Int Migr Rev 48(3):577–603
2. Aggarwal A (2007) Impact of special economic zones on employment, poverty and human development (no 194). Working paper

3. Anggadini SD, Luckyardi SS, Urtikanti S (2023) Development of micro small-medium business in Asian countries (Indonesia, Malaysia, Philippines, and Thailand): a comparison in accounting behavior. J East Eur CentL Asian Res (JEECAR) 10(3):487–497
4. Bagchi P (2023) Courting change: how the judiciary shapes labor jurisprudence and transforms the workplace. Int J Law Criminol 3(10):1–5
5. Besnier N (2020) Sport and society in global France
6. Buranajaroenkij D, Doneys P, Kusakabe K, Doane DL (2018) Expansion of women's political participation through social movements: the case of the red and yellow shirts in Thailand. J Asian Afr Stud 53(1):34–48
7. Cahaya FR, Nursalim N, Dhirathiti N (2023) Expectations on labour-related CSR reporting: voices from labour unions in Indonesia and Thailand. Soc Environ Account J 1–24
8. Caraway TL (2023) Labor standards and labor market flexibility in East Asia. In: Global labour in distress, volume II: earnings. In: Decent work and institutions. Cham: Springer International Publishing, pp 437–469
9. Department of labour protection and welfare (2022) Department of labour protection and welfare. https://www.labour.go.th/index.php/en/
10. Diep DM (2023) Current status of internalizing international labor standards on the prevention, removal and the case in Vietnam. J Law Sustain Dev 11(5):32
11. Georgiadis A (2013) Efficiency wages and the economic effects of the minimum wage: evidence from a low-wage labour market. Oxford Bull Econ Stat 75(6):962–979
12. Gumasing MJJ, Ilo CKK (2023) The impact of job satisfaction on creating a sustainable workplace: an empirical analysis of organizational commitment and lifestyle behavior. Sustainability 15(13):10283
13. Imudom W (2000) The role of minimum wage in Thailand. http://www.bot.or.th/BOTHomepage/databank/ArticlesAndPublications/Articles/BulletinArticles/A-Apr-43-E.pdf
14. Kanchoochat V (2023) Siamese twin troubles: structural and regulatory transformations in unequal Thailand. Asian Econ Policy Rev 18(1):47–68
15. Kompipote U (2002) Sexual harassment in the workplace: a report from field research in Thailand., Bangkok: s.n.
16. Kunpeuk W et al (2022) Access to healthcare and social protection among migrant workers in Thailand before and during COVID-19 era: a qualitative study. Int J Environ Res Public Health 19(5):3083
17. Lasongyang W, Pathak S (2024) Labor law implications on migrant workers in Thailand and Cambodia. J Law Sustain Dev 12(2):e3046–e3046
18. Liao X, Lyu B, Abbas J (2023) Sexual harassment in the workplace: rituals as prevention and management strategies in COVID-19 crisis. Heliyon 9(9)
19. Mahalakshmi J, Balamurugan P (2023) Emerging challenges of internal migrant workers in India: a need for law and governance. Int J Intellect Prop Manag 13(3–4):381–396
20. Ministry of labour (2022) Welfare of Labour. https://www.mol.go.th/en/%E0%B8%AA%E0%B8%A7%E0%B8%B1%E0%B8%AA%E0%B8%94%E0%B8%B4%E0%B8%81%E0%B8%B2%E0%B8%A3%E0%B9%81%E0%B8%A3%E0%B8%87%E0%B8%87%E0%B8%B2%E0%B8%99
21. Mossberger K, Tolbert CJ, McNeal RS (2007) Digital citizenship: the Internet, society, and participation. MIt Press, s.l.
22. Mu R, Van de Walle D (2011) Left behind to farm? Women's labor re-allocation in rural China. Labour Econ 18:S83–S97
23. Musikawong S (2022) Understanding the gaps between the bilateral regularization of migration and workers' rights: the case of agricultural migrant workers in Thailand. Theor Inq Law 23(2):289–325
24. Napathorn C (2022) The development of green skills across firms in the institutional context of Thailand. Asia-Pac J Bus Adm 14(4):539–572
25. Napathorn C (2023) HR practices for managing aging employees in organizations: the case of Thailand. Int J Emerg Mark 18(5):1187–1215

26. Olakivi A et al. (2023) Sickness absence among migrant and non-migrant care workers in Finland: a register-based follow-up study. Scand J Public Health 14034948231168434
27. Onyango JO (2024) Supply chain solutions for essential medicine availability during COVID-19 pandemic. J Humltarian Logist Supply Chain Manag 14(1):118–133
28. Pangsapa P (2015) When battlefields become marketplaces: migrant workers and the role of civil society and NGO activism in Thailand. Int Migr 53(3):124–149
29. Pathak S, Laikram S (2002) Cooperative education during Covid-19 pandemic: enhancing legal rights and professional development of interns in Thailand. Int J Disaster Resil Built Environ 13(2):133–149
30. Pholphirul P, Rukumnuaykit P (2010) Economic contribution of migrant workers to Thailand. Int Migr 48(5):174–202
31. Pomeroy RS (1995) Community-based and co-management institutions for sustainable coastal fisheries management in Southeast Asia. Ocean Coast Manag 27(3):143–162
32. Prajanti SDW, Rengga WDP, Kurniawan C (2023) The probability of migrant workers becoming participants of BPJS for employment. J Econ Educ 12(1):84–91
33. Rebitzer JB, Taylor LJ (1995) The consequences of minimum wage laws some new theoretical ideas. J Public Econ 56(2):245–255
34. Sasaki S (2017) Daily experiences of human (in-) securities: a case of migrant women in neighbourhood workshop in Bangkok, Thailand. Gend Technol Dev 21(1–2):63–76
35. Siddiqui T (2004) Efficiency of migrant workers' remittance: the Bangladesh case. Asian Development Bank, Manila
36. Tangcharoensathien V, Supachutikul A, Lertiendumrong J (1999) The social security scheme in Thailand: what lessons can be drawn? Soc Sci Med 48(7):913–923
37. Thomas H, Anner M (2023) Dissensus and deadlock in the evolution of labour governance: global supply chains and the international labour organization (ILO). J Bus Ethics 184(1):33–49
38. UNESCAP (1995) Legislation on equal opportunities and full participation in development for disabled persons: a regional review., s.l.: s.n.

Male Labor Migration and Women Empowerment: A Review Paper

Hari Devkota, Deepak Marahatta, Shankar Tiwari, and Bhim Raj Suwal

Abstract In South Asia, males dominate the migration, leaving females as the primary breadwinners. This results in women taking on the responsibility of family care, agriculture, and household management, which empowers them in various ways. This study, conducted using a comprehensive desk review method, gathered relevant literature from electronic and physical libraries, focusing on scientific journals' peer-reviewed articles. The study examined women's personnel, economic, social, and political empowerment due to male labor migration. The findings reveal that male migration has increased women's participation in financial activities, access to savings and credit, improved gender-friendly tools and technologies, better bargaining power and autonomy, social engagement, and reputation. On the other hand, some studies showed that male migration has increased the financial burden for women, abandonment of farming activities, and social stigmas such as psychological problems, family destruction, and detachment. This study strongly suggests that researchers must conduct further investigations into social and political empowerment, which society usually overlooks but holds more value in achieving sustainable development.

Keywords Migration · Women · Empowerment · Economic · Social · Political

1 Introduction

Male migration and its profound impact on women's empowerment, a topic of increasing interest and research in Nepal, is the focus of this study. Male labor migration is a key driver in the country, has significant implications for the social, economic, and political dynamics within households and communities [27]. The

H. Devkota (✉) · D. Marahatta · S. Tiwari · B. R. Suwal
Center Department of Rural Development, Faculty of Humanities and Social Science, Tribhuvan University, Kiritpur, Kathmandu, Nepal
e-mail: hari.207730@cdrd.tu.edu.np

I. Pal et al. (eds.), *Proceedings of the 3rd International Symposium on Disaster Resilience and Sustainable Development—Volume 1*, Lecture Notes in Civil Engineering 623, https://doi.org/10.1007/978-981-96-2139-2_9

transformative effect of this migration on gender roles and relationships, a particularly intriguing aspect, cannot be overstated. Understanding this complex relationship between male migration and women's empowerment is crucial for addressing gender inequalities and promoting inclusive development. Nepal, a country with a rich history and diverse cultural traditions, has experienced a significant increase in male outmigration over the past few decades. Pursuing better economic opportunities, driven by limited employment prospects, poverty, and social pressures, has led many men to work overseas, leaving their families and communities [24]. This migration phenomenon has transformed the socio-economic landscape of Nepal, reshaping gender roles and relationships and creating both challenges and opportunities for women's empowerment.

Empowering women, encompassing their social, economic, and political agency, is essential for achieving sustainable development and gender equality. Women's empowerment improves their well-being and contributes to the overall progress of families, communities, and the nation. Therefore, understanding how male migration influences women's empowerment is vital for designing effective policies and interventions that promote gender equity and social justice [13].

Various research has investigated the consequences of male migration on various aspects of women's lives in Nepal, shedding light on both the positive and negative consequences. These studies have explored economic empowerment, social changes, healthcare access, family dynamics, and psychological well-being among women left behind by migrant husbands [13]. By synthesizing these studies' findings, we can understand the complexities surrounding women's empowerment, considering male migration. Regarding economic empowerment, research has shown that remittances from migrant husbands can provide women with increased control over household resources and decision-making. Financial autonomy enables women to meet their needs, support their children's education, and invest in income-generating activities. However, the extent of economic empowerment varies depending on factors such as the size and regularity of remittances, the distribution of resources within the household, and women's access to financial benefits and opportunities for income generation [16].

In addition to economic and social dimensions, the discussion on women's empowerment in the context of male migration also encompasses political empowerment. Limited research has explored the extent to which women left behind can engage in political processes, exercise their agency, and participate in decision-making at the community and national levels [38]. Examining women's political participation and representation is crucial for understanding the overall empowerment landscape and identifying avenues for addressing gender disparities in governance and policy-making. While prior research has offered vital insights into the relationship between male migration and women's empowerment, considerable gaps must be filled. These gaps include a limited exploration of women's political engagement and representation, inadequate analysis of gender-based violence and protection mechanisms, insufficient attention to social support networks and community resilience, and a lack of focus on women's access to assets and facilities beyond the household.

Therefore, this paper aims to bridge these gaps and provide a comprehensive overview of the male labor migration and its contribution to women's empowerment. By analyzing and synthesizing the existing literature, the paper seeks to identify key findings related to the issue and highlight the gaps in knowledge to identify the possible research area.

2 Methodology

This study utilized a desk review methods to analyze existing literature on women's social and political empowerment in Nepal's context of male migration as well as some global context. The methods adhered to the guidelines proposed by Snyder [48] for conducting comprehensive literature reviews, ensuring a robust and thorough analysis of the relevant research.

2.1 Search Strategy

To identify pertinent studies, an extensive search was conducted using multiple academic databases, including Science Direct, Web of Science, Scopus, and Google Scholar. These databases were selected for their comprehensive coverage and reliability in academic research. The search process was optimized using a predefined set of keywords, including: 'migration,' 'labor migration,' 'women empowerment,' 'labor migration and women empowerment,' 'impact of migration,' 'components of women empowerment,' 'women empowerment approach' and 'women empowerment related findings in South Asia and Nepal.'

2.2 Inclusion Criteria

To ensure the relevance and quality of the included studies, specific inclusion and exclusion criteria were established:

- Publication Date: Only articles published after 2000 were considered, ensuring the analysis of recent and relevant data.
- Geographical Focus: Studies focusing on South Asia, particularly Nepal, were included to maintain contextual relevance.
- Themes: Research addressing women's empowerment, exceptionally personal, economic, social, and political dimensions, was included.
- Only articles published in English were included to maintain consistency in the review process.

- Peer-reviewed: Only peer-reviewed journal articles, conference papers, and book chapters were included to ensure the reliability and academic rigor of the sources.

2.3 *Data Extraction and Analysis*

Once the relevant articles were identified, data extraction was performed. Key information was systematically extracted and organized, including the study's aims, methodology, findings, and conclusions. The extracted data were then thematically categorized and analyzed to identify common patterns, trends, and gaps in the literature.

The analysis focused on the various dimensions of women's empowerment in the context of male migration, including personal, economic, social, and political aspects. It examined the impact of male migration on these dimensions and synthesized the findings to understand the topic comprehensively.

3 Results

The review shows several types of research on male migration and its impact on women's empowerment. These researchers have identified many issues of women's empowerment, such as access to resources, improved gender-friendly tools and technologies, role change in women, increased access to resources, improved health and education, better social engagement and reputation, and increased non-farm business and engagement. The following session deals with significant findings from scholarly written articles accessed from various sources and some negative consequences of male migration. Based on the review, the study has identified women's empowerment in four major dimensions: personal, economic, social, and political.

3.1 *Personnel Empowerment*

3.1.1 Increased Decision-Making Capacity

Migration has enhanced women's decision-making capacity and power within households and the community. Studies by [3, 7, 10, 13, 20, 23, 28, 30, 31, 36, 42, 43, 44, 49, 50] have all demonstrated this phenomenon. Datta and Mishra [7] found that women were more involved in household decision-making, with 52% of women from the surveyed groups reporting increased involvement. Research by [8] indicated that women in nuclear migrant homes had more control over their husbands' wages than in non-migrant households, indicating greater control over resources and decisions.

Gartaula [13] also found that male labor out-migration increased female-headed households and decision-making power for women.

Maharjan et al. [28] identified that remittances from male migration have reduced the physical work burden on women left behind and increased their decision-making roles, leading to empowerment. This finding is supported by Sadiqi and Ennaji [41], who suggested that male migration expands women's freedom and dignity by assigning them new roles and responsibilities within the household. In nuclear families, male migration has enhanced women's autonomy, decision-making responsibility, and labor demand [28]. Women in migrant households have a more significant role in household decision-making than those in non-migrant households in districts like Syangja and Baitadi, Nepal.

Similarly, the study conducted by [30] indicates that the relocation of male family members has a favourable impact on shifting gender roles and women's autonomy, especially in economic decisions. Other scholars note that women in coffee cultivation and sales gain decision-making power and prestige. Additionally, women from migrant households have greater decision-making authority, control over the household budget, and responsibility for shopping [28]. Few studies have also agreed that women from migrant households experience greater personal freedom and decision-making power than other women. Furthermore, [3] portrays that male outmigration has resulted in several personal changes in rural women's lives.

3.1.2 Increased Women's Workload

Due to migration, the absence of male family members has significantly increased women's workload in managing the household, caring for children, engaging in farm activities, and maintaining social engagements. Numerous studies have highlighted that migration increases the workload for women left behind, especially those who become the family's breadwinners [3, 5, 12, 14]. Women in rural areas are particularly affected by migration, as they often bear the burden of unpaid work within the family and subsistence work [5]. The transformation and expansion of married women's roles in agricultural production have also been observed [12].

Husbands' overseas migration has been linked to increasing participation in farming activities among the remaining women [14]. The increased unpaid work is accompanied by a decreased female paid labor supply [32].

3.1.3 Increased Bargaining Power and Autonomy

Migration also contributes to increased bargaining power and autonomy for women. The presence of female household heads in the community, resulting from male migration, has led to a shift in decision-making dynamics. Kaspar [21] discovered that female household heads made operational decisions jointly with their husbands. Rocha and Gordon [15] revealed that wives felt empowered to decide on family affairs without their husbands. Jabbar [20] demonstrated that migration improved the bargaining power of left-behind women, with women in nuclear families having greater responsibilities and autonomy than those in extended families, a previously observed finding [9]. Durand [10] identified that the departure of a male household head for work in the United States boosted a wife's authority and autonomy, putting her in charge of everyday decisions about family control, productivity, and expenditure. Similarly, [23] found that migration empowered 29.44% of women to a satisfactory extent, with a more significant proportion falling into the moderately empowered category (43.89%) and still a significant percentage (29.67%) in the low empowerment group. The research further identified, women in the studied area demonstrated adequate empowerment in healthcare, education, and agricultural production decisions. Other scholars have also observed that the increased authority of emigrants' wives was influenced mainly by migration and the financial resources it brought, granting them greater power.

3.1.4 Role Change in the Family: Increased Role and Responsibility

Migration has significantly changed the role and responsibilities of women, extending beyond household decision-making to social, political, and community participation. Studies by [40, 42] indicate that migration has an impact on women beyond domestic responsibilities, affecting productivity and community-level decision-making. When husbands migrate, women left behind must take on the responsibilities previously handled by their spouses.

Research by [40] suggests that male outmigration provides opportunities for greater responsibility within the household for wives left behind. Similarly, [41] highlight how women left behind strive for inclusion by entering male-dominated public spaces and becoming breadwinners to support their households. Additionally, [42] find that male outmigration enhances women's engagement in farming activities, as they take on managerial roles in hiring and managing workers, not just limited to household and agricultural work.

Male outmigration increases the role and responsibilities of women left behind, particularly in household management [2, 3, 8, 14, 15, 21, 28, 31, 46]. Women in migrant households experienced increased responsibilities due to migration, including caring for farms and livestock. The migration process also leads to changes in the division of labor in the household, and women assume new roles and responsibilities [3].

Overall, male migration affects the structure of family life and transforms women's social and economic positions, often necessitating adjustments and increased responsibilities for the remaining family members.

3.2 Economic Empowerment

3.2.1 Economic Well-Being/Access to Economic Activities

Migration has proven to enhance economic well-being and development within societies, with access to finance resulting in a multiplier effect on economic progress [1, 13, 30, 36, 39, 44, 49]. For instance, research by [36] in the Philippines, Thailand, and Vietnam demonstrated that remittances aided in debt repayment and covered costs associated with children's education and healthcare for family members left behind. Similarly, a study in Bangladesh revealed that husbands' migration boosted the economic independence, living standards, and social status of women remaining in their home communities [39]. In Nepal specifically, [13] found that remittances had an essential role in improving the living standards of many people, providing a critical source of foreign currency, and contributing to the balance of payments. Furthermore, [25] noted that a significant portion of poverty reduction in Nepal between 1995 and 2004 was attributed to out-migration.

In their studies, several scholars have established a connection between remittances and women's empowerment [2, 11]. For instance, Adhikari and Hobley's [2] research in Laos identified four primary ways in which remittances were utilized: family security and well-being, fungible activities, economic and business ventures, and community development. Their findings indicated that 80% of women experienced empowerment through the utilization of remittances. Although husbands' migration improved the economic security, living conditions, and social standing of left-behind women, those living in their in-laws' families had limited or no access to remittances. As a result, the home's composition affects women's access to financial resources.

3.2.2 Investment in Productive Activities

Migration and remittances have a multiplier effect across all sectors, fostering economic growth and facilitating productive investments [1, 2, 19, 26, 39, 43, 49]. Adhikari and Hobley's [2] study on migration highlighted the positive impact of out-migration on increased household income and subsequent improvements in overall well-being. Mainly, Dalit families, including Dalit women, experienced greater freedom and reduced dependence on higher-caste households for economic support. Similarly, [26] found that husbands' migration enhanced the economic security, living standards, and social status of women left behind in Africa. Additionally, [49] conducted a study in Pakistan, which revealed that while migration had a materially favourable impact on individual migrants and their families, its potential to generate broader social transformation and community development was influenced by existing power structures, particularly gender dynamics and socio-political issues. Thus, migration has the potential to significantly improve the living conditions and social standing of lower and middle-income families, reducing their reliance on higher-caste households for sustenance.

3.2.3 Increase Women's Participation in Financial Activities

Migration has brought about financial freedom for women, granting them strategic choices in life and the ability to make meaningful decisions aligned with their needs and aspirations. Lopez-Ekra et al. [26] conducted research in Guatemala and Colombia, where male out-migration rates were high, demonstrating that women gained access to money for household management and livelihood. Similarly, [42] found that remittance access relaxed women's financial constraints, enabling them to make meaningful and strategic life choices. Studies by [8 and 47] revealed that households directly involved in international migration experienced more than double the average income of non-migrant and internal migrant households, providing greater independence to migrant families in developing countries, particularly in the Third World. Additionally, [11] identified four ways in which remittances were spent in Laos, with 80% of women experiencing empowerment through the utilization of remittances.

3.2.4 Access to Savings and Credit

Migration has increased women's accessibility to managing money, obtaining loans, and better access to financial institutions. Datta and Mishra [7] found that women become more involved in household finances and money management without their husbands. However, women often faced barriers when accessing credit and obtaining favourable terms. Sources of debt for women included moneylenders, neighbours, relatives/caste members, and self-help groups. Maharjan et al. [28] discovered that women in Nepal had greater access to lower-value assets like livestock than high-value assets like land and houses. Gartaula [13] highlighted that migration helped accumulate wealth through remittances, which could be used to purchase land in urban areas. Access to finance and credit increased women's financial independence and ability to access resources such as land, livestock, and other necessary services.

3.3 Social Empowerment

3.3.1 Social Engagement and Reputation

Migration has played a significant role in promoting women's social engagement and reputation. Research shows that migration has led to an increase in women's status within households, mobility, and their roles in the public sphere [2, 7, 13, 20, 25, 28, 31, 39, 41]. Women's well-being and household decision-making power have become more dependent on them, elevating their status and strengthening their positions within the household. Women have gained greater independence in household management and the economy, with increased involvement in agricultural production [41]. Mobility and participation in social networks have also increased for women, challenging conservative norms [7, 20]. Women's attendance at community meetings has also increased [31]. These changes reflect shifts in women's roles and increased autonomy.

3.3.2 Increased Access to Education

Migration has been found to positively impact access to quality education, especially for girls, in various countries [2, 8, 35, 36, 49]. Studies show that migration and remittances help cover the costs of schooling and healthcare for children left behind, increasing girls' educational opportunities. The research highlights the positive effects of migration on education, particularly for economically disadvantaged families and girls in the community.

3.3.3 Improved Health Status

Migration has decreased workloads and improved health outcomes for women in different regions [3, 16, 35]. In extended households, the distribution of domestic and agricultural tasks among women reduces individual burdens. Remittances help improve left-behind communities' health by positively altering food habits, nutritional status, and behavior when seeking health care. Migration also facilitates better access to healthcare knowledge and facilities, as families with migrant members gain exposure to different cultures and become more confident in interacting with healthcare providers. However, research conducted in India has found higher rates of reproductive morbidity and sexually transmitted infections among women married to migrant husbands compared to women married to non-migrants. Female household members also incur an increased burden of work to compensate for the loss of male labor and face more obstacles to receiving healthcare [45].

3.3.4 Better Healthcare Facilities

Migration has facilitated better access to healthcare facilities for migrants and their families in various countries [30, 35]. Studies indicate that migration leads to improved medical facilities and services after the migration of family members. Remittances contributed to better nutrition, access to modern healthcare services, and improved health outcomes for those left behind. Research also explores how communities can enhance their eating habits, nutritional status, and health-seeking behavior. Overall, migration has positively impacted healthcare outcomes through increased access to resources and knowledge.

3.4 Political Empowerment

The researcher defines women's political empowerment as representation in political parties, influence in agencies in decision-making, and finally, representation in different layers of the government mechanisms. In addition to economic and social dimensions, the discussion on women's empowerment in the context of male migration also encompasses political empowerment. Limited research has explored the extent to which women left behind can engage in political processes, exercise their agency, and participate in decision-making at the community and national levels [38]. Scholars have identified economic, social, political, psychological, and legal dimensions as crucial to women's empowerment [29, 37].

Political empowerment has been accorded significant recognition in scholarly literature, with many scholars highlighting its pivotal role in driving societal and community progress [4, 6, 18, 29]. A closer examination reveals that political empowerment has a broader domain, encompassing social, cultural, economic, and legal dimensions. However, in the literature, there is very limited research on political empowerment. Hence, further research is needed to explore how women's political participation and influence can contribute to empowerment and transforming power structures [4].

Migration has fueled some negative impacts as well. It has contributed to family disruption, sexual abuse, psychosocial problems, and social stigma in households left behind. The socioeconomic and demographic disruption caused by migration can have emotional and practical consequences. Women often experience a loss of companionship and sexual intimacy due to their husbands' long absence, which can strain their marriages and disrupt the financial support for themselves and their children [8, 15, 28, 31]. Migration has been associated with conflicts between migrant wives and their in-laws due to remittance issues [8].

Studies have revealed that the absence of husbands owing to migration causes higher stress and harms the wives left behind. There is little indication of an empowerment process resulting from male migration [17, 22]. Some women experience marital separation or abandonment, and many fear such outcomes after migration [31]. Wives left behind often face emotional disruption, stress, sadness, loneliness, and feelings of abandonment [35]. Psychological problems such as loneliness, depression, insecurity, and difficulties in disciplining children are common among women whose husbands migrate. Male outmigration restricts women's mobility and increases their cohabitation with in-laws, further disempowering them [42].

4 Discussion

This review explored the multifaceted impact of male migration on women's empowerment, identifying four primary dimensions: personal, economic, social, and political.

4.1 Personal Empowerment

The review highlights that male migration significantly enhances women's decision-making capacity within households and communities. Studies such as those by [7 and 13] reveal that women in migrant households experience increased autonomy and control over resources, aligning with the broader literature suggesting migration can shift traditional gender roles [28, 41]. This enhanced decision-making capacity empowers women personally and transforms household dynamics by granting them more significant involvement in household management and financial control.

4.2 Economic Empowerment

Economically, male migration has been shown to improve households' financial well-being, primarily through remittances. Research by [36 and 39] underscores the role of remittances in debt repayment, education, and healthcare, leading to higher living standards for women left behind. This economic empowerment is further evidenced by increased participation in financial activities and better access to savings and credit [2, 7]. These findings corroborate studies like those of [13 and 26], highlighting migration's transformative economic impact on women's empowerment.

4.3 Social Empowerment

Socially, male migration fosters greater social engagement and enhances women's community reputation. The review found that women left behind gain increased mobility and social participation, as noted in studies by [7 and 20]. This shift challenges conservative norms and promotes gender equality as women take on more prominent roles in public spheres and community decision-making. Improved access to education and healthcare further contributes to their social empowerment, as demonstrated by research from [2 and 35]. These social changes reflect broader societal transformations driven by migration.

4.4 Political Empowerment

The review also touches on the political empowerment of women, which is an emerging but crucial aspect. Although limited, the existing literature suggests that women's participation in political processes and decision-making at community levels have increased due to male migration. Studies like those of [4 and 38] point out that as women take on more responsibilities at home and in the community, their involvement in political activities also grows. This empowerment is vital for driving societal and community progress, emphasizing the need for further research to understand the full extent of political empowerment resulting from male migration.

4.5 Negative Consequences

Despite these positive outcomes, the review highlights several challenges, including increased workload and emotional stress for women. The absence of male family members often leads to a higher household and farming responsibilities burden, as indicated by studies like those of [5 and 12]. Additionally, emotional and psychological challenges such as loneliness, stress, and marital strain are common, as reported by [35]. Migration has also been associated with conflicts between migrant wives and their in-laws over remittances [8]. These findings underscore the need for supportive measures to address the negative impacts of male migration, ensuring that women's empowerment is holistic and sustainable.

The results suggest that male migration is both an empowering and challenging force for women. Empowerment is evident through enhanced decision-making capacity, economic independence, and increased social engagement. These changes align with existing literature, which posits that migration can shift traditional gender roles and allow women to assert greater control over their lives and resources. However, the increased workload and emotional burden underscore a dual reality where empowerment coexists with significant personal challenges.

4.6 Implications

The findings have important implications for both theory and practice. Theoretically, they support the notion that migration can transform gender dynamics within households and communities. The results suggest empowering women through migration can lead to broader social and economic benefits, such as improved health and education outcomes for children and enhanced community development. This underscores the need for policies and programs that support women in migrant households, ensuring that the benefits of migration are maximized while mitigating its adverse effects.

4.7 Limitations

This review has limitations that must be acknowledged. The heterogeneity of the studies reviewed, spanning diverse cultural and socio-economic contexts, limits the generalizability of the findings. The reliance on secondary data means that real-time dynamics and personal experiences might not be fully captured. Additionally, the political dimension of women's empowerment remains underexplored, representing a significant gap in the current understanding. Recognizing these limitations strengthens the credibility of the review and highlights areas for future research.

4.8 Recommendations

Future research should focus on longitudinal studies to better capture the evolving impacts of male migration on women's empowerment. Specifically, there is a need to explore the political dimension of empowerment in greater detail and understand the role of support networks in mitigating the negative impacts of migration. Policymakers should develop targeted interventions that give women access to credit, education, and healthcare. Community-based programs that foster social engagement and provide emotional support can help alleviate the increased workload and stress women face. Addressing these areas will provide a more comprehensive understanding of the impact of male migration on women's empowerment and inform effective policy and program development.

5 Conclusion

This review demonstrates that male migration significantly impacts women's empowerment across personal, economic, social, and political dimensions. The migration of male family members has increased decision-making capacity, financial independence, and social engagement for women left behind. Numerous studies support these changes, suggesting that male migration can catalyze traditional gender roles and enhance women's status within their households and communities.

However, the positive impacts of male migration are accompanied by substantial challenges. Increased workloads, emotional stress, and conflicts related to remittances present significant hurdles for women. The dual impact of empowerment and additional burdens highlight the complexity of the migration phenomenon. Addressing these challenges requires a multifaceted approach that includes supportive policies and community interventions to mitigate the negative consequences while fostering the positive aspects of women's empowerment. Further research is essential to explore the political empowerment of women in greater depth and to develop strategies that support women in navigating the complexities introduced by male migration.

References

1. Acosta PA (2011) School attendance, child labour, and remittances from international migration in EI Salvador. J Dev Stud 47(6):913–936. https://doi.org/10.1080/00220388.2011.563298
2. Adhikari J, Hobley M (2015) Everyone is leaving. Who will sow our fields?" The livelihood effects on women of male migration from Khotang and Udaypur Districts, Nepal, to the Gulf Countries and Malaysia. Assoc Nepal Himal Stud 35(1):1–15
3. Adra N (2013) The impact of male outmigration on women's roles in agriculture in the Yemen Arab Republic. United Nations (January):1–72
4. Alexander AC, Bolzendahl C, Jalalzai F (2016) Defining women's global political empowerment: theories and evidence. Sociol Compass 10(6):432–441. https://doi.org/10.1111/soc4.12375
5. Binzel C, Assaad R (2011) Egyptian men working abroad: labour supply responses by the women left behind. Labour Econ 18:S98–S114. https://doi.org/10.1016/j.labeco.2011.03.002
6. Chen Y, Tanaka H (2014) Women's empowerment. In: Encyclopaedia of quality of life and well-being research. Springer, Dordrecht
7. Datta A, Mishra SK (2011) Glimpses of women's lives in rural Bihar: impact of male migration. Indian J Labour Econ 54(3):457–477
8. De Haas H (2010) The internal dynamics of migration processes: a theoretical inquiry. J Ethn Migr Stud 36(10):1587–1617. https://doi.org/10.1080/1369183X.2010.489361
9. Desai S, Banerji M (2008) Negotiated identities: male migration and left-behind wives in India. J Popul Res 25(3):337–355. https://doi.org/10.1007/BF03033894
10. Durand J, Massey DS (2001). What we learned from the Mexican migration project. Crossing the border
11. Edgardo J, Gomez S, N, John W, Samuel S (2011) Reaching across the Mekong: local socioeconomic and gender effects of Lao-Thai cross-border linkages. J Curr Southeast Asian Aff 30(3):3–25
12. Gardner BD, Pope RD (2015) Women in agriculture. In: The World Bank Group. The World Bank Group, pp 1–46
13. Gartaula HN (2009) International migration and local development in Nepal. Contrib Nepal Stud 36(1):37–65
14. Ghimire D, Zhang Y, Williams N (2019) Husbands' migration: increased burden on or more autonomy for wives left behind? J Ethn Migr Stud. https://doi.org/10.1080/1369183X.2019.1675502
15. Gordon E (2007) An analysis of the impact of labour migration on the lives of women in Lesotho. J Dev Stud 17(3):59–76. https://doi.org/10.1080/00220388108421798
16. Hadi A (2001) International migration and the change of womens position among the left behind in rural Bangladesh. Int J Popul Geogr 7:53–61
17. Hjorth S (2011) Male migration and female empowerment. Lund University (Issue May)
18. Huis MA, Hansen N, Otten S, Lensink R (2017) A three-dimensional model of women's empowerment: implications in the field of microfinance and future directions. Front Psychol 8(Sep):1–14. https://doi.org/10.3389/fpsyg.2017.01678
19. Islam S, Abdul SP, Abdul (2013) Socio-economic impacts of international migration in Bangladesh. J Econ Sustain Dev 4(4):1–10
20. Jabbar JB (2021) Household size and women empowerment. J Bus Divers 21(2). https://doi.org/10.33423/jbd.v21i2.4323
21. Kaspar H (2005) I am the household head now! Impact of out-migration for laour on gender relations in Nepal. Nepal institute of development studies (NIDS)
22. Kulczycka K (2015) Left behind women and empowerment. Consequences of male labour migration on the economic and social position of left behind women in Nepal. Lund University
23. Kumari K, Singh KM, Ahmad N (2022) Impact of migration on women empowerment: a situational analysis of North-Bihar. Indian J Ext Educ 58(1):101–105. https://doi.org/10.48165/ijee.2022.58123

24. Kunwar LS (2021) Foreign labor migration causes and impacts in Nepal. Patan Pragya 9(02):144–154. https://doi.org/10.3126/pragya.v9i02.42056
25. Lokshin M, Glinskaya E (2008) The effect of male migration of work on employment patterns of women in Nepal. World Bank Econ Rev 23(4757):481–507. https://doi.org/10.1093/wber/lhp011
26. Lopez-Ekra S, Aghazarm C, Kötter H, Mollard B (2011) The impact of remittances on gender roles and opportunities for children in recipient families: research from the international organization for migration. Gend Dev 19(1):69–80. https://doi.org/10.1080/13552074.2011.554025
27. Maharjan A (2013) Migration for labour and its impact on farm production in Nepal
28. Maharjan A, Bauer S, Knerr B (2012) Do rural women who stay behind benefit from male out-migration? A case study in the hills of Nepal. Gend Technol Dev 16(1):95–123. https://doi.org/10.1177/097185241101600105
29. Mahbub M (2021) Women empowerment: definition, theory, process, practice and importance. University of Dhaka (Issue June)
30. Malick B, Sargana S, Dobrova V (2014) Male migration: Decision making autonomy and changing roles among females left behind: a feminist approach. World Appl Sci J 29(4):1–7. https://doi.org/10.5829/idosi.wasj.2014.29.04.13950
31. McEvoy JP (2008) Male out-migration and the women left behind: a case study of a small farming community in southeastern Mexico. All graduate theses and dissertations, 141
32. Mendola M, Carletto C (2009) International migration and gender differences in the home labour market: evidence from Albania. In: Policy research working paper, vol 4900. The World Bank, pp 870–880. https://doi.org/10.1016/j.labeco.2012.08.009
33. Moher D, Liberati A, Tetzlaff J, Altman DG, Group (2009) Preferred reporting items for systematic reviews and meta-analyses : the PRISMA statement. BMJ 339(August):332–336. https://doi.org/10.1136/bmj.b2535
34. Mu R, van de Walle D (2011) Left behind to farm? Women's labor re-allocation in rural China. Labour Econ 18:S83–S97. https://doi.org/10.1016/j.labeco.2011.01.009
35. Nguyen TL, Yeoh BSA, Toyota M (2006) Migration and the well-being of the left behind in Asia: key themes and trends. Asian Popul Stud 2(1):37–44. https://doi.org/10.1080/17441730600700507
36. Paris TR, Luis J, Villanueva D (2009) Labour out migration on rice farming households and gender roles: synthesis of findings in Thailand, the Philippines and Vietnam. Gender pathways out of poverty rural employment, pp 1–28
37. Pratley P (2016) Associations between quantitative measures of women's empowerment and access to care and health status for mothers and their children: a systematic review of evidence from the developing world. Soc Sci Med 169:119–131. https://doi.org/10.1016/j.socscimed.2016.08.001
38. Rajkarnikar PJ (2017) The impacts of foreign labor migration of men on women's empowerment in Nepal. University of Massachusetts Amherst
39. Rashid SR (2013) Bangladeshi women's experiences of their men's migration: rethinking power, agency, and subordination. Asian Surv 53(5):883–908
40. Rustagi P, Nathan D, Datta A, George A (2013) Women and work in South Asia: changes and challenges. In: Women and Work in South Asia. Institute for human development, p 41. https://doi.org/10.4324/9781315887616
41. Sadiqi F, Ennaji M (2004) The impact of male migration from Morocco to Europe on women: a gender approach. Finisterra 39(77):59–76. https://doi.org/10.18055/finis1561
42. Sapkota C, Wie D (2019) The effect of male outmigration on women's empowerment in Nepal. National graduate institute for policy studies, December 19–28
43. Sarker MM (2014) Husbands' international labour migration and the change of wives' position among the left-behind in rural Bangladesh. Res HumIties Soc Sci 4(16):2224–5766
44. Shattuck D, Wasti SP, Limbu N, Chipanta NS, Riley C (2019) Men on the move and the wives left behind: the impact of migration on family planning in Nepal. Sex Reprod Health Matters 27(1):1–14. https://doi.org/10.1080/26410397.2019.1647398

45. Siegmann KA (2010) Strengthening whom? The role of international migration for women and men in northwest Pakistan. Prog Dev Stud 10(4):345–361. https://doi.org/10.1177/146499340901000406
46. Sinha B, Jha S, Negi NS (2012) Migration and empowerment: the experience of women in households in India where migration of a husband has occurred. J Gend Stud 21(1):61–76. https://doi.org/10.1080/09589236.2012.639551
47. Spaan E, van Naerssen T (2018) Migration decision-making and migration industry in the Indonesia-Malaysia corridor. J Ethn Migr Stud 44(4):680–695. https://doi.org/10.1080/1369183X.2017.1315523
48. Snyder H (2019) Literature review as a research methodology: an overview and guidelines. J Bus Res 104(March):333–339. https://doi.org/10.1016/j.jbusres.2019.07.039
49. Wassan MR, Hussain Z, Shah MA, Amin SN (2017) International labor migration and social change in rural Sindh Pakistan. Asian Pac Migr J 26(3):381–402. https://doi.org/10.1177/0117196817732210
50. Yabiku ST, Agadjanian V, Sevoyan A (2010) Husbands' labour migration and wives' autonomy, Mozambique 2000–2006. Popul Stud 64(3):293–306. https://doi.org/10.1080/00324728.2010.510200

Compensation for Losses and Damages Caused by Climate Change: Challenges and Progress

Pratik Poudel and Mokbul Moshed Ahmad

Abstract In 1991, the AOSIS (The Association of Small Island Developing States) brought up the idea of getting compensation from developed countries for destruction caused by climate change. But it wasn't until 2007/08 that other developing countries and non-governmental entities really got the ball rolling. At the Sharm Al Sheikh COP meeting, a plan was created to provide money for responding to losses and damages from climate change. There are tons of tough questions that need to be answered, like who should be held accountable and to what extent. We also have to figure out what events are due to climate change and which aren't, plus how to deal with non-economic losses and damage. Financially and budget-wise, there are still a lot of issues to tackle. The article also mentions tools to assess losses, damages, etc., but there are still some kinks to work out, especially when it comes to assigning responsibility and budgets.

1 Evolution of the Concept of Loss and Damage

Climate change's uneven impact: Climate change harms people's ability to make a living, regardless of whether its effects unfold slowly (like droughts or melting glaciers) or suddenly (like hurricanes or floods). Developing countries, despite contributing the least to greenhouse gas emissions, are among the most vulnerable. This has led developing nations to push for compensation from wealthier countries for the damage caused by climate change.

P. Poudel (✉)
Department of Development Planning Management and Innovation (DPMI), School of Environment, Resources and Development (SERD), Asian Institute of Technology (AIT), Khlong Luang, Thailand
e-mail: st123166@ait.asia

M. M. Ahmad
SERD/DPMI, AIT, Khlong Luang, Thailand
e-mail: morshed@ait.asia

I. Pal et al. (eds.), *Proceedings of the 3rd International Symposium on Disaster Resilience and Sustainable Development—Volume 1*, Lecture Notes in Civil Engineering 623, https://doi.org/10.1007/978-981-96-2139-2_10

The rise of "loss and damage": The concept of "loss and damage" emerged at a 1991 UN climate conference. A proposal by the Association of Small Island Developing States for an insurance pool to protect against rising sea levels wasn't widely adopted [1]. The concept gained momentum in 2007–2008, driven by advocacy from island nations and environmental groups. An article by WWF highlighted the lack of funding for adaptation and mitigation, calling for compensation for climate change adaptation efforts. Legal support initiatives by NGOs like Oxfam and WWF further bolstered the argument for recognizing loss and damage as a global issue.

A contentious debate: The issue of loss and damage remained prominent in UN climate talks (COPs). At COP15, a proposal for a loss and damage mechanism was initially rejected due to pressure from developed countries, though it gained some support. Although New Zealand argued against attributing specific events to climate change, a two-year work program on the issue was established at COP16 in Cancun [2].

Formalization and ongoing discussions: Despite objections from leading countries like the US, the Warsaw International Mechanism for Loss and Damage was established in 2013 under the UN Framework Convention on Climate Change (UNFCCC). Developed nations continued to resist the concept, with a US-backed proposal at the 2015 Paris Summit aiming to remove references to it. However, loss and damage secured its place as a separate article (Article 8) in the Paris Agreement [3].

The path forward: The disagreement over loss and damage reflects the broader tension between developed and developing countries in environmental governance. Developing nations seek compensation, while developed nations like the US avoid responsibility. Calls for a stronger role for loss and damage continued at COP meetings in 2019 and 2021, leading to the Glasgow Dialogue on Finance for Loss and Damage. Finally, at COP27 in 2022, funding arrangements were established to address the losses and damages caused by climate change impacts [1].

2 Key Concepts Behind Loss and Damage

What is Loss and Damage?

Loss and damage refers to the negative effects of climate change that cannot be prevented or fully adapted to. This includes things like lost agricultural productivity, harm to ecosystems, and damage from disasters like floods and cyclones. There's no official definition within UN climate documents, but some experts categorize losses as avoided, unavoidable, and inevitable. Avoided losses, minimized through disaster preparedness and adaptation efforts, aren't part of the discussion. Unavoidable and inevitable losses, however, are caused by limitations in money, government, and economic systems. These losses can be physical or intangible, economic or not, with some overlap between categories [10].

Justice and Adaptation Challenges

The debate around loss and damage involves two main approaches: fairness and compensation. Distributive justice aims for a fair sharing of climate change burdens and benefits between countries and people. Compensatory justice focuses on making up for harm caused by one party to another. This approach helps developing nations struggling to cope with climate impacts, like the recent floods in Pakistan [4].

Another key concept is the limit to adaptation. This limit comes in two forms: hard and soft. Hard limits are situations where adaptation is impossible, making losses unavoidable. Soft limits occur when resources or technology aren't currently available to fully address risks. Examples include populations pushed into poverty by climate shocks in cities or coastal areas. Soft limits may become hard limits if people can't relocate [5].

Assessing and Financing Loss and Damage

Understanding current risks from extreme weather and climate change is crucial for assessing losses. While the frequency or severity of hazards may not change significantly, the rising number of people and assets in harm's way is likely driving observed losses. Studies suggest some increases in weather events like thunderstorms, but there's not enough evidence to definitively say these are caused by climate change. Directly attributing losses to greenhouse gases can be difficult.

The Warsaw International Mechanism works on financial solutions for loss and damage, but progress has been slow. The Paris Agreement mentions loss and damage but doesn't specify funding. Experts have proposed options like insurance, emergency funds, and bonds. Insurance works well for sudden events but not for slow-onset disasters. Special Drawing Rights from the IMF have also been suggested as a way to finance loss and damage, particularly for vulnerable countries.

Non-Economic Losses and Damages

Non-economic losses and damages (NELD) are a vital aspect of climate finance. These are impacts that can't be easily measured in dollars, such as the loss of cultural heritage, sacred sites, human health, and even lives. The Warsaw Mechanism includes raising awareness and developing strategies to reduce NELD risks. Because NELD is specific to each situation and difficult to quantify, it presents challenges for policy-making, but it's still crucial to consider [6].

3 Methods and Tools for Analyzing Loss and Damage

This section dives into how scientists assess and address the risks posed by climate change.

Looking Forward, Looking Back: Tools for Assessing Loss and Damage

Researchers like Ana Lopez use different methods to evaluate climate change impacts, considering both past events and future projections. For future predictions, they rely on complex computer models called General Circulation Models (GCMs). These models simulate future climate scenarios based on greenhouse gas emissions. To get a more detailed picture, scientists use Regional Comprehensive Models (RCMs). These are essentially zoomed-in versions of GCMs, providing higher resolution for specific regions [5].

There are two main ways to create RCMs: dynamical and statistical downscaling. Dynamical downscaling involves nesting RCMs within a global GCM, while statistical downscaling uses historical data to establish relationships between global and regional climate patterns. However, this statistical approach can be less precise.

Another important tool is the Inter Sectoral Impact Model Intercomparison Project (ISI-MIP). This project combines climate projections with socio-economic factors to understand the real-world consequences of climate change. It's important to note that these models can't definitively prove that specific events are caused by human activity [7].

Layering Risks: Prioritizing Resources

One way to measure risk is the "risk layering" approach. This method breaks down risks from natural disasters into four categories: [7]

- **Frequent Events**: Here, the focus is on reducing risk through preventative measures.
- **Medium Events**: This layer combines risk reduction with financial tools like insurance to manage the remaining risk.
- **Rare Catastrophic Events**: When large-scale disasters strike, public or international assistance is crucial.
- **Very Rare Events**: These catastrophic events, occurring once in a century or more, require significant international climate funding.

By analyzing the financial needs of each layer, this approach helps identify countries requiring international support and prioritizes investments in risk reduction.

Case Studies: Putting Risk Assessment into Practice

Wouter Botzen and his colleagues explored how cities can assess climate risks. Their research focused on Jakarta and Ho Chi Minh City, looking at how these cities can manage risks through mitigation and adaptation strategies.

In Jakarta, researchers used models to map potential flooding scenarios, considering factors like land subsidence, sea level rise, and rainfall patterns. They also evaluated the effectiveness of different strategies such as land-use changes or upgrading flood barriers.

In Ho Chi Minh City, researchers developed a framework to identify areas vulnerable to flooding. They used a "Damage Scanner Model" to assess potential casualties and economic losses from floods. They also looked at different flood management measures and how these could reduce future impacts compared to a scenario where no action is taken [8].

Making Tough Decisions: Choosing the Right Approach

The article also discusses various methods for making climate-related decisions:

- **Cost–Benefit Analysis (CBA)**: This method compares the costs of a project with the benefits it provides. This approach is used in the Netherlands to determine new flood protection standards.
- **Cost-Effectiveness Analysis (CEA)**: This method focuses solely on project costs, assuming that benefits are similar across different options.
- **Multi-Criteria Analysis (MCA)**: This method compares benefits and costs that are measured in different units, such as lives saved or environmental quality.
- **Robust Decision-Making Approaches (RDMA)**: This method combines various methods, both qualitative and quantitative, to choose options that perform well under different circumstances. This helps minimize potential future regrets [4].

Insurance and Beyond: Addressing Loss and Damage

Insurance plays a role in managing climate risks, as analyzed by Laura Schäfer and her colleagues. Insurance can warn of approaching adaptation thresholds, encourage better upfront planning for climate resilience, and provide financial safety nets for manageable risks. However, insurance has limitations. It's less effective for slow-onset events like sea level rise or for frequently occurring events where building resilience is a better approach. Additionally, insurance doesn't cover non-monetary losses like cultural heritage loss or biodiversity decline [9].

4 Complications Regarding Loss and Damage

Kirstin Dow and Frans Berkhort identify two fundamental issues regarding Loss and Damage. The first issue revolves around determining who should be held responsible for climate-based compensation, rather than identifying vulnerable countries. The second issue concerns attributing specific events to climate change, raising questions about the extent and manner in which rich countries will be held accountable for their carbon emissions [3]. A key question is: What events should be attributed to climate change? As highlighted by New Zealand in the 2010 climate change conference, the challenge is determining which events should be attributed to climate change. Erin Roberts and Mark Pelling emphasize three key issues that national and international actors must address in the loss and damage debate: attribution, the limits to adaptation, and practical action options [10].

Attribution in climate change is complex due to the mix of slow factors, like snowmelt and drought, and fast factors, like hurricanes and cloudbursts, making it difficult to identify which events are related to climate change. Consequently, countries like the Philippines struggle to secure compensation for recurring natural disasters. James et al. argue that the goal of attribution is not to assign blame but to enhance scientific understanding of the links between Earth system elements and society, providing insights into the drivers of hazards, vulnerability, and exposure without political bias.

Wallimann et al. discuss the ethical challenges in calculating loss and damage for countries. They identify three complications in applying compensatory schemes: determining the contribution level of potential emitters who may not exceed their carbon emission threshold, establishing responsibility for emitters unaware of the harm caused, and compensating recipients who may not be directly affected by carbon emissions. Although compensatory justice is currently politically unfeasible (as noted in Paragraph 52 of decision 1/CP.21), the distributive justice framework seems a practical approach for addressing loss and damage and assigning responsibilities [11].

Understanding attribution requires input from diverse disciplines such as physical science, economics, hydrology, and ecology. For instance, assessing losses and damage during a major flood event involves evaluating meteorological and hydrological hazards, habitat exposure, and vulnerability, including factors like flood resilience of houses, early warning systems, and insurance coverage. Additionally, economic terms like monetary losses and infrastructure damage, ecological impacts, and non-monetary losses (such as identity and heritage loss) must be considered. This broad range of variables complicates attribution in loss and damage [12].

While attribution is crucial, it should not be the sole measure of loss and damage. Other factors, such as adaptation difficulties, must also be considered. Loss and Damage is often linked to the limits of adaptation. The Intergovernmental Panel on Climate Change (IPCC) states that adaptation limits occur when a society cannot meet its objectives, secure market values, or avoid intolerable risk through adaptation measures. These limits are often exacerbated by poverty, past economic development decisions, technology use, and social norms. This raises questions about the intolerable limit of adaptation and when it is crossed. Some scholars categorize limits into soft and hard limits for comprehensiveness, as discussed earlier. Erin Roberts and Mark Pelling propose a comprehensive risk management approach to support national and local climate policies, address current development framework shortcomings, and enhance local risk management capacity [10]. Building on information about approaching adaptation limits and occurring loss and damage, policy frameworks may consider removing soft limits and promoting transformational risk management or adaptation to prevent breaching these limits [13].

Another major complication is climate finance. Gewirtzman et al. argue that while there are mechanisms to finance loss and damage, most are based on voluntary associations, which may not be adequate or reliable [4]. Therefore, voluntary associations need to be reformed into binding agreements requiring financially resourceful countries to contribute a portion of their resources. Another complication is that some financial instruments claim to cover slow-onset events, but reality often falls short. Additionally, private risk instruments are too expensive for poor countries and only cover a limited range of sudden climate shocks. It is estimated that the cost of loss and damage could exceed $1 trillion a year, prompting private insurers to set high premiums. Consequently, poor countries may prioritize other pressing issues over paying premiums for infrequent disasters. Thus, rich countries must assist poor countries in this regard [14].

Non-Economic Loss and Damage (NELD) is a significant concern in climate change impact discussions. NELD includes impacts such as loss of life, health issues, loss of mobility, biodiversity loss, destruction of cultural sites, and loss of traditional knowledge. These impacts profoundly affect individuals, including loss of identity, social cohesion, education, and relationships. The challenge lies in the fact that NELD cannot be quantified in monetary terms and varies greatly based on context, making it difficult for policymakers to address these diverse and unique losses effectively [6]. Additionally, internal and international migration resulting from climate change further complicates loss and damage assessments. Extreme weather events like cyclones, sea level rise, and desertification displace people from their homes, highlighting the complex relationship between natural resources, the environment, extreme weather, and society. Cases like sea level rise in Pacific Island states, Sahara desertification in West Africa, monoculture farming expansion in South America, and Boko Haram's emergence in West Africa underscore these challenges [15].

5 Policy Response for Addressing Loss and Damage

At the policy level, various methods exist to address loss and damage issues, but there is a lack of clarity regarding the underlying concepts. Bangladesh, a country highly vulnerable to climate change, exemplifies this challenge. Despite having strong institutional frameworks to address climate change, Bangladesh faces difficulties due to insufficient integration between disaster risk reduction (DRR) and climate change adaptation. Although mechanisms are in place that utilize existing institutions and frameworks and promote collaboration across different government ministries, the country still struggles with managing losses and damages effectively. Bangladesh's approach includes embedding climate change perspectives into its disaster policy framework to strengthen responses to climate-induced losses, demonstrating a commitment to the Paris Agreement and national targets in the Sendai Framework for Disaster Risk Reduction. However, separate laws, policies, and procedures for DRR and climate adaptation hinder successful management. To address this, Bangladesh has been developing a national mechanism to respond to climate-induced losses and damages, leveraging its existing adaptation and disaster risk finance frameworks, and launching initiatives to deepen knowledge and expertise on these issues [16].

A notable example of addressing loss and damage legally is the Huaraz case in Germany. This was the first case in Germany where a private energy company was sued for its contribution to climate change-related damage. Peruvian farmer Saúl Luciano Lliuya sued the company, seeking contributions to safety measures against flooding caused by glacial retreat linked to human activities. The court accepted the case, marking a precedent where a private company was held responsible for climate change damages. This case has implications for holding large greenhouse gas emitters accountable and communicating related risks to shareholders [17].

In terms of legal frameworks, developing countries have unanimously agreed that developed nations, being major contributors to global greenhouse gas emissions, should compensate for the resulting damage. However, there is no binding legal agreement for such compensation. Instead, the principle of "Common But Differentiated Responsibility" was adopted, recognizing that developed countries have more resources to address climate change. This principle obliges cooperation but does not mandate compensation payments. While financial assistance to prevent climate change has seen some success, compensating those who have suffered losses and damages has remained largely unaddressed until recent years, when advocates pushed this agenda further [18].

6 Conclusion

The Alliance of Small Island States (AOSIS) has been leading the discussion on Loss and Damage since 2005. Alongside the G77 and Least Developed Countries (LDCs), AOSIS aims to secure compensation for developing countries for damages caused by climate change. The Warsaw International Mechanism, established in 2013, partially addressed this demand. Developing countries insist that loss and damage should be treated separately from adaptation and that developed countries should be held accountable. Conversely, developed countries worry that focusing solely on compensation might divert attention from the immediate needs of developing countries and argue that it is challenging to separate loss and damage from other climate adaptation processes.

To address loss and damage comprehensively, a range of approaches is necessary, including risk transfer, social protection policies, migration, recovery, and rehabilitation. Addressing non-economic losses and damages (NELD) is crucial but challenging. A proper framework is needed to handle NELD effectively, requiring dedicated efforts to integrate NELD into both preventive and reactive strategies. Additionally, the displacement of populations due to climate change needs to be addressed, taking into account not only the migration itself but also the impacts on those left behind and the political economy of both sending and receiving nations. Although loss and damage was acknowledged as a separate stand-alone agreement in the Paris Agreement, it lacks a legally binding provision that would obligate developed countries to compensate developing countries.

References

1. Vanhala L (2016) Framing climate change loss and damage in UNFCCC negotiations. Glob Environ Polit 16(4):111–129
2. Calliari E, Surminski S, Mysiak J (2019) The politics of (and behind) the UNFCCC's loss and damage mechanism. Loss and damage from climate change: concepts, methods and policy options, pp 155–178
3. Dow K (2014) Climate change, limits to adaptation and the 'loss and damage'debate. E-international relations
4. Gewirtzman JN (2018) Financing loss and damage: reviewing options under the Warsaw international mechanism. Clim Policy 18(8):1076–1086
5. Lopez AS (2019) The role of the physical sciences in loss and damage decision-making. In Mechler LM-BR, Loss and damage from climate change, Springer, pp 261–285
6. Serdeczny O (2019) Non-economic loss and damage and the Warsaw international mechanism. In: Mechler LM-BR, Loss and damage from climate change, Springer, pp 205–220
7. Schinko TM-S (2019) The risk and policy space for loss and damage: integrating notions of distributive and compensatory justice with comprehensive climate risk management. In: Mechler LM-BR, Loss and damage from climate change, Springer, Cham, pp 83–110
8. Wouter Botzen WJ (2019) Integrated disaster risk management and adaptation. In: Mechler LM-BR, Loss and damage from climate change, Springer, pp 287–315
9. Schäfer LW (2019) Exploring and managing adaptation frontiers with climate risk insurance. In: Mechler LM-BR, Loss and damage from climate change, Springer, pp 317–341
10. Roberts E (2018) Climate change-related loss and damage: translating the global policy agenda for national policy processes. Climate Dev 10(1):4–17
11. Wallimann-Helmer I, Meyer L, Mintz-Woo K, Schinko T, Serdeczny O (2019) The ethical challenges in the context of climate loss and damage. Loss and damage from climate change: concepts, methods and policy options, pp 39–62
12. James RA (2019) Attribution: how is it relevant for loss and damage policy and practice? In: Mechler LM-BR, Loss and damage from climate change, Springer, Cham, pp 113–154
13. Mechler RS (2020) Loss and damage and limits to adaptation: recent IPCC insights and implications for climate science and policy. Sustain Sci 15(4):1245–1251
14. Scott Cato M (2022) The chequered history of climate finance. Sustain Financ pp 39–59
15. Heslin AD-C (2019) Displacement and resettlement: understanding the role of climate change in contemporary migration. In: Mechler LM-BR, Loss and damage from climate change: concepts, methods and policy options, Springer, pp 237–258
16. Haque MP (2019) Towards establishing a national mechanism to address losses and damages: a case study from Bangladesh. In: loss and damage from climate change, Springer, Cham, pp 451–473
17. Frank WB (2019) The case of Huaraz: first climate lawsuit on loss and damage against an energy company before German courts. In Mechler LM-BR, Loss and damage from climate change, Springer, Cham, pp 475–482
18. Simlinger F (2019) Legal responses to climate change induced loss and damage. In: Mechler LM-BR, Loss and damage from climate change: concepts, methods and policy options, Springer, pp 179–203
19. Bouwer LM (2019) Observed and projected impacts from extreme weather events: implications for loss and damage. In: Mechler LMR, Loss and damage from climate change: concepts, methods and policy options, Springer, pp 63–82
20. Mechler RB-B (2019) Loss and damage from climate change: concepts, methods and policy options. In: Mechler RB-B, Loss and damage from climate change: concepts, methods and policy options, Springer, p 557

21. Mechler R, Calliari E, Bouwer LM, Schinko T, Surminski S, Linnerooth-Bayer J, ... Zommers Z (2019) Science for loss and damage. Findings and propositions. Loss and damage from climate change: Concepts, methods and policy options, pp 3–37
22. van der Geest KD (2019) The impacts of climate change on ecosystem services and resulting losses and damages to people and society. In: Mechler LM-BR, Loss and damage from climate change: concepts, methods and policy options, Springer, pp 221–236

Evaluating the Effectiveness of Sustainable Transportation Policies in Mitigating Climate Change in Metro Manila, Philippines: The Public Utility Vehicle Modernization Program (PUVMP)

Tabassam Raza, Hannah Barbara Gonzalez, Maria Belen Vasquez Cresp, Carmelita R. E. U. Liwag, Zuhura Mtoi Kanyawanah, Dayanara Bermudez, Bassam Mohammed Zainudeen, Mark Anthony M. Morales, Shaker Mahmood Mayo, and Dina C. Magnaye

Abstract Energy efficiency in transportation is a crucial component of urban areas, and its impact on climate change is profound. As cities grapple with mounting challenges related to environmental sustainability, the adoption of sustainable transportation modes takes center stage. In this context, the Public Utility Vehicle Modernization Program (PUVMP) in Metro Manila, Philippines, stands out as a noteworthy initiative dedicated to mitigating climate change by modernizing public transportation systems. This study is designed to assess the PUVMP's effectiveness in achieving its objectives and its contribution to sustainable transportation within Metro Manila. To accomplish this, our research adopts a qualitative approach, with an emphasis on an extensive review of pertinent literature as the primary methodology. The literature review entails a systematic and thorough examination of policy documents, reports, and academic articles to gain insights into the policy landscape, strategies for policy development and implementation. The findings gleaned from this literature review are synthesized and subjected to a critical analysis to evaluate how effectively sustainable transportation policies address climate change impact in Metro Manila. The research scrutinizes the strengths and weaknesses of these policies, identifies implementation

T. Raza (✉) · M. B. Vasquez Cresp · C. R. E. U. Liwag · D. Bermudez · M. A. M. Morales · D. C. Magnaye
School of Urban and Regional Planning, University of the Philippines, Quezon City, Philippines
e-mail: tabassam.raza@planades.org

T. Raza · H. B. Gonzalez
Planning and Development Research Foundation, Inc., UP SURP Building, Quezon City, Philippines

Z. M. Kanyawanah · B. M. Zainudeen · S. M. Mayo
Department of City and Regional Planning, University of Engineering and Technology, Lahore, Pakistan

I. Pal et al. (eds.), *Proceedings of the 3rd International Symposium on Disaster Resilience and Sustainable Development—Volume 1*, Lecture Notes in Civil Engineering 623, https://doi.org/10.1007/978-981-96-2139-2_11

gaps, and delves into the challenges encountered in attaining policy objectives. With these anticipated findings in mind, our research aims to formulate innovative recommendations that can guide policymakers and stakeholders in enhancing the efficacy of sustainable transportation policies within the region. By amalgamating empirical evidence and astute insights derived from this research, the resulting conclusions and recommendations are intended to present a compelling narrative that offers actionable strategies for advancing sustainable transportation policies in Metro Manila. Consequently, this research serves as a foundation for reflecting on existing policies and informing optimal strategies to facilitate the successful implementation of sustainable transportation modes in the future.

Keywords Effectiveness · Sustainable transportation · Public utility vehicle · Climate change · Environmental sustainability · Environmental impact · Economic viability

1 Introduction

Transportation plays a crucial role in urban areas, with significant implications for climate change. As cities face increasing challenges related to environmental sustainability, the adoption of sustainable modes of transportation has become a priority. Further, in urban areas, energy-efficient transportation is critical for reducing greenhouse gases emission in addressing climate change adverse impacts. The Public Utility Vehicle Modernization Program (PUVMP) in Metro Manila, Philippines, aims to modernize public transportation to mitigate climate change. The PUVMP in Metro Manila, Philippines, [1–3] represents a notable initiative aimed at addressing climate change impact through the modernization of public transportation systems.

The overall use of PUVMP in Metro Manila is designed to revolutionize the city's public transportation system by replacing outdated and dilapidated vehicles with modern, eco-friendly units. This initiative is aimed at significantly enhancing the safety, comfort, and efficiency of transportation for commuters. By introducing vehicles that comply with contemporary safety and emission standards, the program also seeks to address the pressing issue of air pollution, contributing to a cleaner urban environment. Furthermore, the PUVMP supports economic growth by boosting local vehicle manufacturing and creating new job opportunities within the industry. A key objective of the PUVMP is to professionalize the public transportation sector through the implementation of stricter regulations and the consolidation of individual operators into cooperatives or corporations. This ensures better management and accountability, fostering a more organized and reliable transportation system. The program also provides financial assistance and training to drivers and operators, helping them adapt to the new system and maintain their livelihoods. Overall, the PUVMP aims to create a sustainable, efficient, and well-regulated public transportation network that benefits commuters, operators, and the environment alike.

Therefore, this study aims to assess the effectiveness of the PUVMP in achieving its objectives and contributing to sustainable transportation in Metro Manila (Fig. 1). Essentially, the PUVMP is the government's response to address both the aggravating transport related problems and the future transportation demand of the country. On 19 June 2017, the Department of Transportation (DOTr) issued Department Order No. 2017–011, otherwise known as the Omnibus Guidelines on the Planning and Identification of Public Road Transportation Services and Franchise Issuance or the Omnibus Franchising Guidelines (OFG), which launched the PUVMP.

The program aims to fundamentally transform the public transport system in the country making both commuting and public transportation operations more dignified, humane, and on par with global standards. The program also aims to give a comfortable life for all Filipinos through providing a safer, more efficient, reliable, convenient, affordable, climate-friendly, and environmentally sustainable transportation system in the country [3].

The research employs a qualitative research design, utilizing a comprehensive review of relevant literature as the primary methodology [4]. The literature review involved a systematic and thorough analysis of policy documents, reports, and academic articles to understand the policy landscape, policy and implementation strategies. The findings from the literature review are synthesized and critically analysed to assess the effectiveness of sustainable transportation policies in addressing climate change impact in Metro Manila. Sustainable transportation refers to low- and zero-emission, energy-efficient, and affordable modes of transport [5]. Thus, here the effectiveness refers to the degree to which sustainable transportation policies achieve their intended goals and objectives, such as reducing greenhouse gas emissions, improving air quality, enhancing public transport services, and promoting sustainable mobility options. In fact, the research critically evaluates the strengths and weaknesses of the policies, identifies gaps in implementation, and examines the challenges faced in achieving policy objectives.

Building on the expected findings, the research seeks to generate innovative recommendations that can inform policymakers and stakeholders in enhancing the effectiveness of sustainable transportation policies in the region. By weaving together, the empirical evidence and critical insights obtained from the research, the conclusions and recommendations are envisaged to offer a compelling narrative that presents actionable strategies for advancing sustainable transportation policies in Metro Manila. Subsequently, this will provide a basis to reflect on the existing policies and inform the best strategies to facilitate successful implementation of the sustainable modes of transportation policies in the future.

The PUVMP is the government's response to address both the aggravating transport related problems and the future transportation demand of the country. On 19 June 2017, the Department of Transportation (DOTr) issued Department Order No. 2017–011, otherwise known as the Omnibus Guidelines on the Planning and Identification of Public Road Transportation Services and Franchise Issuance or the Omnibus Franchising Guidelines (OFG), which launched the PUVMP.

The program aims to fundamentally transform the public transport system in the country making both commuting and public transportation operations more dignified,

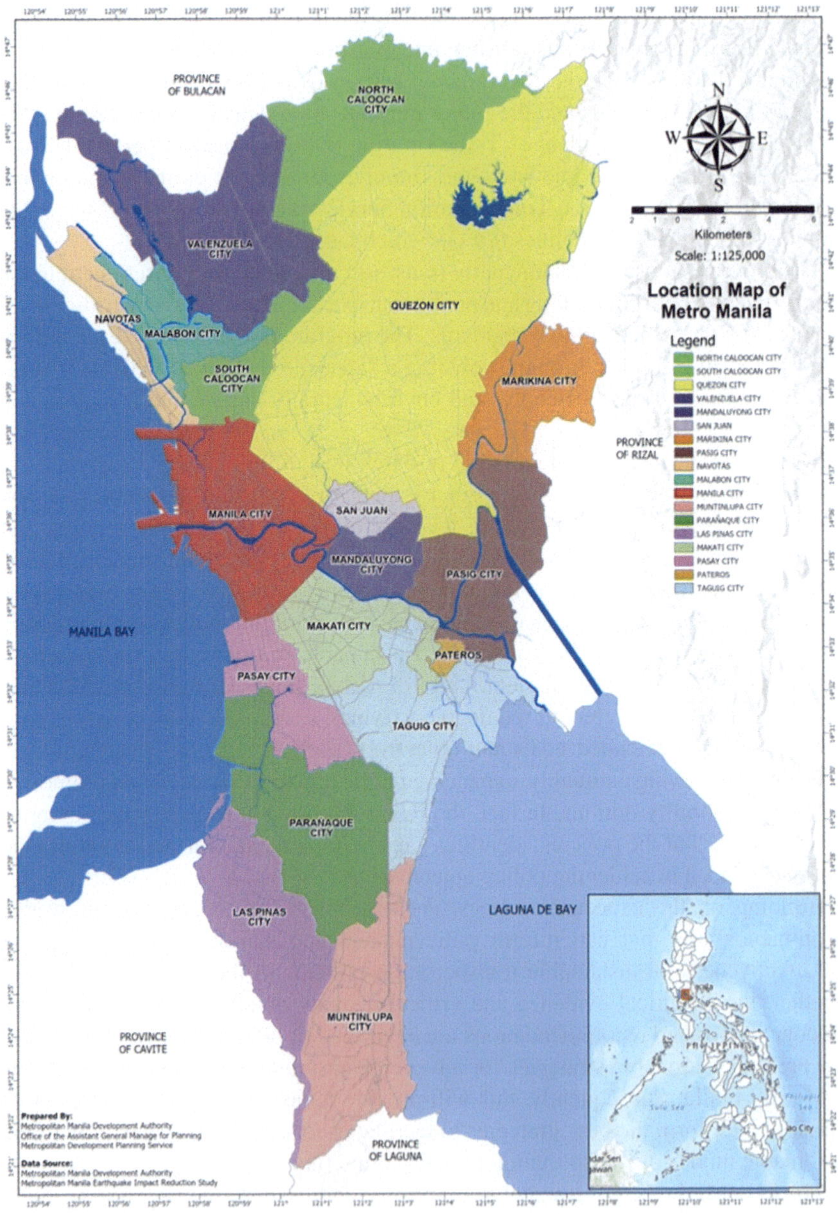

Fig. 1 Location map of greater metropolitan [6]

humane, and on par with global standards. The program also aims to give a comfortable life for all Filipinos through providing a safer, more efficient, reliable, convenient, affordable, climate-friendly, and environmentally sustainable transportation system in the country [3].

The research employs a qualitative research design, utilizing a comprehensive review of relevant literature as the primary methodology [4]. The literature review involved a systematic and thorough analysis of policy documents, reports, and academic articles to understand the policy landscape, policy and implementation strategies. The findings from the literature review are synthesized and critically analysed to assess the effectiveness of sustainable transportation policies in addressing climate change impact in Metro Manila. Sustainable transportation refers to low- and zero-emission, energy-efficient, and affordable modes of transport [5]. Thus, here the effectiveness refers to the degree to which sustainable transportation policies achieve their intended goals and objectives, such as reducing greenhouse gas emissions, improving air quality, enhancing public transport services, and promoting sustainable mobility options. In fact, the research critically evaluates the strengths and weaknesses of the policies, identifies gaps in implementation, and examines the challenges faced in achieving policy objectives.

Building on the expected findings, the research seeks to generate innovative recommendations that can inform policymakers and stakeholders in enhancing the effectiveness of sustainable transportation policies in the region. By weaving together, the empirical evidence and critical insights obtained from the research, the conclusions and recommendations are envisaged to offer a compelling narrative that presents actionable strategies for advancing sustainable transportation policies in Metro Manila. Subsequently, this will provide a basis to reflect on the existing policies and inform the best strategies to facilitate successful implementation of the sustainable modes of transportation policies in the future.

2 Statement of the Problem

While the PUVMP represents a significant step towards sustainable transportation in Metro Manila, there is a need to assess its effectiveness in achieving its intended goals. This study seeks to address the following questions:

- To what extent has the PUVMP been adopted and implemented in Metro Manila?
- What is the environmental impact of the PUVMP in terms of greenhouse gas emissions reduction, air pollution mitigation, and overall carbon footprint?
- What is the economic viability of the sustainable transportation modes introduced under the PUVMP, and what are the socioeconomic benefits for stakeholders involved?
- How has the PUVMP contributed to infrastructure development, including the construction of dedicated lanes, terminals, and charging stations for electric vehicles?

3 Objectives

The primary objective of this study is to assess the effectiveness of the PUVMP in addressing climate change impact in Metro Manila. Specifically, the study aims to:

- Evaluate the adoption and implementation of the PUVMP in Metro Manila.
- Assess the environmental impact of the PUVMP in terms of reducing greenhouse gas emissions, mitigating air pollution, and reducing the overall carbon footprint.
- Analyse the economic viability of sustainable transportation modes introduced under the PUVMP and identify socio-economic benefits.
- Evaluate the infrastructure development resulting from the PUVMP, including the establishment of dedicated lanes, terminals, and charging stations for electric vehicles.

4 Methodology

Through the utilization of a qualitative research design and a comprehensive literature review, this study offers valuable insights into the Public Utility Vehicle Modernization Program (PUVMP) as a sustainable transportation policy implemented in Metro Manila and throughout the country. The Conceptual Framework of the study is presented in the following Fig. 2. The main components of the conceptual framework for this study are visually represented. The PUVMP serve as the starting point for the research. From there, the framework branches out into three main areas of assessment: 1. analysis criteria, 2. challenges to implementation, and 3. recommendations.

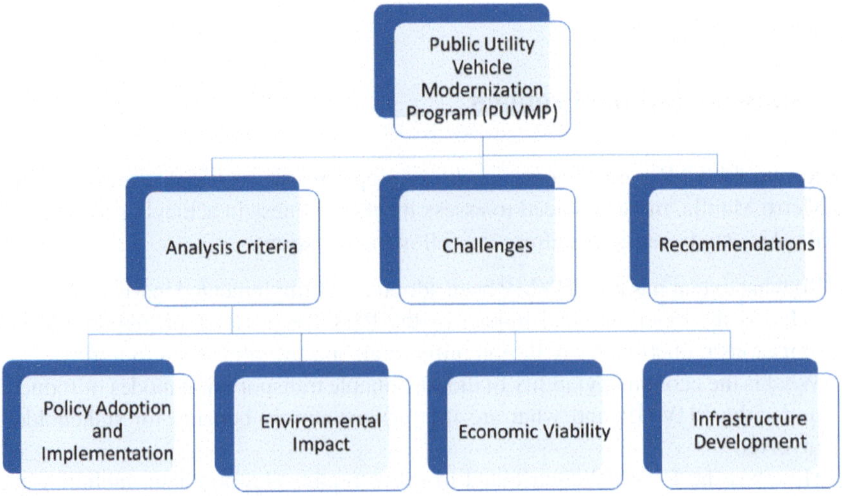

Fig. 2 Conceptual framework of the study

The analysis criteria component subsequently covers four areas: adoption and implementation of policy, the environmental impact of policy, economic viability, and infrastructure development. These factors are essential for ensuring the long-term sustainability and success of the policy. Simultaneously, the framework addresses the challenges that can hinder the effective implementation of sustainable transportation policies. Understanding and overcoming these challenges is crucial for improving the overall effectiveness of the policy. Finally, the framework makes provision for recommendations on how the success of the policy can be improved.

4.1 Brainstorming the Research Topic

Through a collaborative process involving multiple sessions, the research topic "Assessing the Effectiveness of Sustainable Modes of Transportation Policies in Addressing Climate Change Impact in Metro Manila, Philippines" was generated. The sessions employed an open and non-judgmental approach, allowing for unrestricted brainstorming without limitations or constraints. However, key factors such as defining the research purpose, conducting background research through documentary reviews, and considering personal and professional interests were essential in developing a research topic that was acceptable to all authors. The final step involved validation of the research topic using a developed Guidepost for Collecting Secondary and Primary Data for Researches Amidst COVID-19 and in Normal Situation [7].

4.2 Research Issue and Objectives

The first step involved conducting literature reviews to gain a more profound comprehension of the research subject. Subsequently, a concept note for the study was created, encompassing essential elements such as study background, problem statement, research objectives and questions, this methodology, anticipated outcomes, and the study's implications at local, national, and international levels.

4.3 Data Collection

The main approach to gathering data was an extensive examination of literature encompassing policy documents, government reports, and academic articles pertaining to the Public Utility Vehicle Modernization Program (PUVMP) and sustainable transportation policies in Metro Manila.

4.4 Data Analysis

During this stage, the collected data is thoroughly analysed and interpreted to discover significant patterns and insights. The analysis criteria component covers four areas adoption and implementation of policy, the environmental impact of policy, economic viability, and infrastructure development. Throughout the analysis process, four specific criteria were carefully considered, as elaborated below:

- *Adoption and Implementation of the Program*

This specific criterion involves a systematic method for understanding the reception and execution of the program. It entails analysing diverse factors, such as regulatory adjustments, the modernization of public utility vehicles, the consolidation of existing franchises, financing mechanisms for the PUVMP, rationalization of routes, local public transportation route planning, social support programs, vehicle efficiency, and communication channels.

- *Environmental Impact*

With regard to the program's objectives, this criterion aims to gain an understanding of the program's environmental impact. It encompasses evaluating the program's contribution to reducing carbon dioxide emissions and other greenhouse gasses that cause global warming, as well as assessing improvements in air quality, including reductions in particulate matter (PM) and nitrogen oxide (NOx) emissions.

- *Economic Viability and Commercial performance*

In alignment with the program's overarching objectives, this criterion is designed to delve into the program's environmental ramifications with scientific rigor. It entails a multifaceted examination aimed at comprehensively understanding the environmental impact of the program. This includes conducting thorough assessments of the program's role in mitigating carbon dioxide emissions and other greenhouse gases notorious for exacerbating global warming. Additionally, it involves scrutinizing enhancements in air quality brought about by the program, encompassing reductions in particulate matter (PM) and nitrogen oxide (NOx) emissions, both of which are significant contributors to air pollution and its associated health risks. Through a meticulous analysis of these factors, a clearer understanding of the program's ecological footprint and its efficacy in fostering environmental sustainability can be attained.

- *Infrastructure Development*

The objective is to assess the advancement and sufficiency of infrastructure development in supporting the electric transportation system, specifically focusing on aspects like the presence, availability, and convenience of charging stations, as well as their coverage in the Metro Manila area. Additionally, the evaluation will involve analysing the plans, objectives, and investments pertaining to the development of charging infrastructure.

4.5 *Presentation of Findings*

In this section, outcomes and explanations derived from the evaluation conducted on the PUVMP are presented, focusing on factors of adoption and implementation, environmental impact, economic viability, and infrastructure development, as key references used in the analysis. Additionally, this section addresses the obstacles and difficulties encountered in implementing sustainable transportation modes in Metro Manila.

4.6 *Conclusion and Recommendation*

This section presents an overview of the study's purpose, the main findings, and the implications. Additionally, it offers recommendations to improve the adoption of sustainable transportation methods, aligning with climate change objectives established at the local, national, and global levels, with the aim of achieving the desired outcomes.

5 Data Analysis and Findings

The PUV Modernization Program is one of the largest flagship non-infrastructure projects of the government which was launched in 2017 by the Department of Transportation. The PUVMP is composed of ten major components [8] namely: Regulatory Reform, Local Public Transportation Route Plan, Route Rationalization Study, Industry Consolidation, Fleet Modernization,

PUV Financing Modernization, Stakeholder Support Mechanism, Vehicle Useful Life Program, Initial Implementation, and Communications. In this part of the study, the effectiveness of the program in addressing the impacts of climate change is assessed based on four criteria including adoption and implementation of the program, environmental impact. Economic viability and infrastructure development [3]. The effectiveness of the program in addressing the impacts of climate change is assessed based on four criteria including adoption and implementation of the program, environmental impact. Economic viability and infrastructure development as follow:

5.1 *Adoption and Implementation of the Program*

The Land Transportation Franchising and Regulatory Board implemented the PUV Modernization Program last 2017. Based on the LTFRB, PUVMP status updates of

Table 1 Memorandum circulars on key regulatory reforms and other guidelines

Memorandum circular number	Title
2022–047	Guideline on the specifications of Minibus
2022–027	Guideline on the specifications of PUV Class 4 (for PUJ and UVE)
2021–002	Guideline on the specifications of PUJs (Class 1,2,3) and UV Express (Class 3) per type or denomination
2021–084	Extension of time to file application for consolidation pursuant to industry consolidation of PUVMP
LTFRB-LTO JMC	Implementing guidelines for accreditation of scarping facilities

Source [3]

six years, the program was able to accomplish the following as per Accomplishment Report of LTFRB, 2023:

5.1.1 Regulatory Reform

This element introduced the Omnibus Franchising Guidelines, which function as the framework for granting franchises to public utility vehicles in the Philippines. By June 2023, the LTFRB PUV Modernization Program Project Management Office (PUVMP PMO) had issued 214 Memorandum Circulars since its inception in August 2019. These Circulars outline evaluated and approved routes permitted for operation nationwide, along with essential regulatory reforms and other directives pertaining to the PUVMP implementation. Table 1 presents the significant policies and guidelines issued to support the execution and monitoring of the program, relevant to the focus of this paper.

5.1.2 Modernization of Public Utility Vehicles

This component aims to ensure that PUVs follow modern policies as well as focus on environmentally friendly public transportation. As of 26 June 2023, there are a total of 9,275 operational OFG-compliant units (modernized) nationwide with 4,055 located within Metro Manila, the study area for this research.

LTFRB, [3] the study area for this research. The primary factor contributing to the low turnout of OFG-compliant units stems from the outstanding loans held by operators through Government Financial Institutions (GFIs), specifically those acquired for the procurement of modern units such as those offered by institutions like the Development Bank of the Philippines or Landbank of the Philippines. These financial obligations act as significant barriers to the swift adoption of modernized units, impeding the transition towards compliance with the OFG. Furthermore, adherence to dimensional limits outlined for PUV Class 1, 2, 3, 4, and Mini-Bus is mandated

Table 2 Number of modernized operation units

Mode	Modernized operational units
PUJ (class 1, 2, 3)	2,526
UV express (class 3)	240
Mini bus	0
PUB (incl. low floor, low entry city buses)	581
P2P	708
Total	4,055

Source [3]

to align with the established Philippine National Standards, as set forth by the DTI-Bureau of Philippine Standards. This ensures uniformity and standardization across different vehicle classifications, promoting safety, efficiency, and interoperability within the public transportation system.

To provide a comprehensive overview of the progress in modernization efforts, Table 2 presents data detailing the number of modernized PUV units categorized by mode. This quantitative representation offers valuable insights into the distribution and adoption rates of modernized units across various modes of public transportation, facilitating informed decision-making and strategic planning initiatives aimed at accelerating the transition towards a more sustainable and efficient transport infrastructure.

5.1.3 Consolidation of Existing Franchises (Routes and Units)

Under this component, individual operators and drivers opt to form into a consolidated entity (i.e. cooperative, corporation) which give them benefits such as government subsidies and access to credit facilities. Within the confines of Metro Manila, as of May 26, 2023, a cumulative total of 2,029 franchises have undergone consolidation across various modes of public transportation, including PUJ (Public Utility Jeepneys), UVE (Utility Van Express), Mini-Bus, and PUB (Public Utility Bus). This consolidation effort has resulted in a combined total of 75,921 units being amalgamated under these consolidated franchises, thereby streamlining and optimizing the operational framework of public transportation services within the metropolitan area.

The granular breakdown of this consolidation endeavor extends to detailing the number of routes with consolidated franchises and the corresponding count of consolidated units per mode. Such meticulous delineation, as presented in Table 3, facilitates a comprehensive understanding of the distribution and magnitude of consolidation efforts across different modes of public transportation. By elucidating these statistics, stakeholders gain invaluable insights into the efficacy and extent of consolidation initiatives, enabling evidence-based decision-making processes and strategic interventions geared towards enhancing the efficiency and sustainability of public transportation systems within Metro Manila.

Table 3 Breakdown of the number of routes with consolidated franchises and consolidated units per mod

Mode	PUJ		UVE		PUB		MB		Total (all modes)	
Regions	Routes	Units	Routes	Units	Routes	Units	Routes	Units	Routes	Units
Central office	151	5,169	65	1,456	623	13,042	0	0	839	19,667
NCR	291	12,492	55	2,449	0	0	0	0	346	14,941
3	301	19,114	28	358	61	952	4	175	394	20,599
4A	340	19,632	33	225	73	837	4	20	450	20,714
Total	1,083	56,407	181	4,488	757	14,831	8	195	2,029	75,921

Source [3]

Analysis of the baseline data portrayed in Fig. 3 reveals that only 64.3% of the targeted franchise consolidation has been accomplished since the initiation of the program in 2017. Of particular note is the National Capital Region (NCR), a focal area within Metro Manila chosen as the study locale for this research, exhibiting the lowest rate of franchise consolidation at a mere 29.3% (as illustrated in Fig. 4). These statistical insights underscore a pressing need for enhancements in the program's execution to attain its desired objectives effectively.

A significant impediment to the program's success stems from a prevalent misunderstanding among drivers and operators regarding its objectives and implications. Historically, the acquisition of a franchise entailed minimal procedural requirements, allowing operators to play their trade freely without the necessity for rigorous route

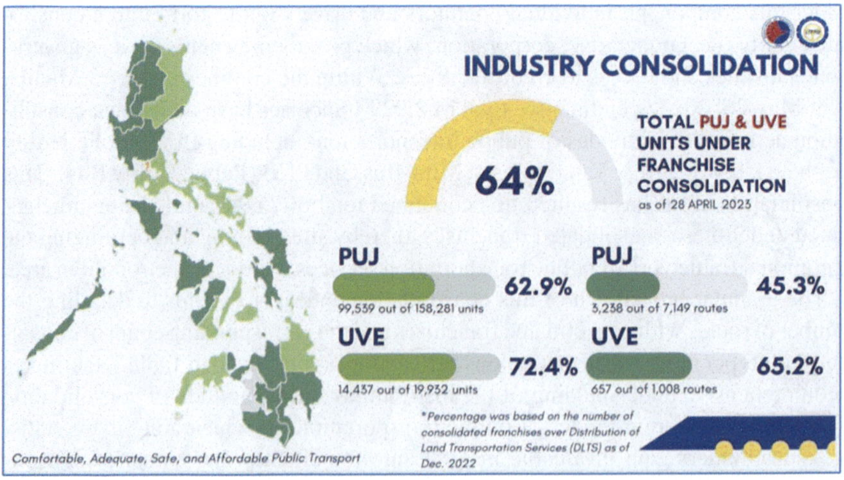

Fig. 3 Number of routes with consolidated franchises and consolidated units per mode (*Source* [3])

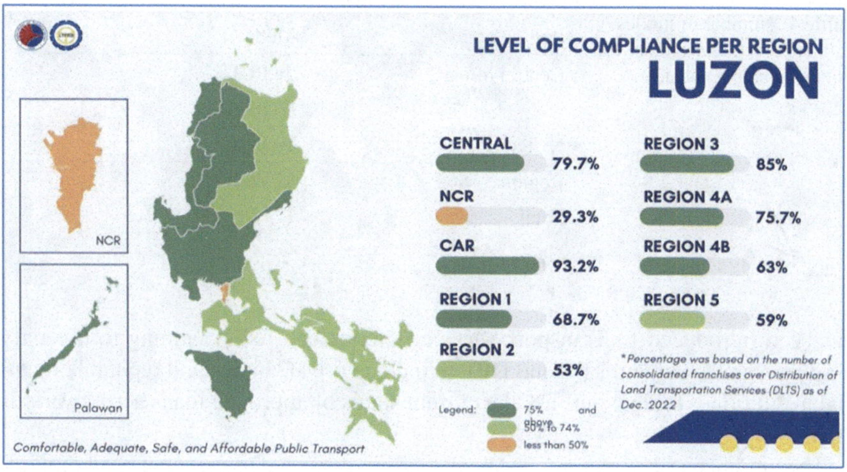

Fig. 4 Number of routes with consolidated franchises and consolidated units per mode in Luzon [3]

assessments. This leniency afforded them unrestricted earning potential, leading to entrenched resistance to the restructuring efforts mandated by the program. Furthermore, apprehensions regarding the financial liabilities associated with the procurement of modernized units exacerbate the reluctance among Transport Service Enterprises (TSEs), as they grapple with the prospect of assuming sole responsibility for these investments.

To address these challenges, a multifaceted approach grounded in scientific rationale is imperative. This entails comprehensive educational campaigns aimed at dispelling misconceptions and fostering a deeper understanding of the program's objectives and long-term benefits among stakeholders. Moreover, proactive measures such as financial incentives and support mechanisms can be implemented to alleviate the financial burden on TSEs, thereby incentivizing compliance and fostering a smoother transition towards modernization. By integrating scientific insights into policy formulation and implementation strategies, the program stands poised to surmount existing barriers and realize its transformative potential within the public transportation landscape.

5.1.4 Financing the PUVMP

This component serves as financial aid to the transport service entities who want to participate in the program by purchasing their modernized units under the government financing institutions (GFIs) such as the Landbank of the Philippines (LBP) and Development Bank of the Philippines (DBP). According to the latest data from LBP and DBP, approved loans amounting to PHP 10.62 billion have been granted for the acquisition of 5,033 modernized units. Additionally, an extended equity subsidy

Table 4 Number of modes with approved loans under the expanded equity subsidy

Regions	Mode
Central office	79 PUJ
	86 PUB
NCR	10 PUJ
Region 3	82 PUJ
Total	171 PUJ; 86 PUB

Source [3]

has been introduced to Transport Service Enterprises (TSEs), aiming to diversify funding sources beyond LBP and DBP to include private banks and reputable financial institutions. Table 4 outlines the current status of approved loans across various modes.

The limited number of approved bank loans is attributed to stringent bank requirements that many TSEs struggle to meet. Moreover, amidst the pandemic, Government Financial Institutions (GFIs) imposed additional prerequisites for route studies such as the LPTRP and Route Rationalization Study, further delaying the loan approval process. However, ongoing difficulties in providing these studies have prompted the exploration of alternative loan options with private banks to ensure the smooth progression of the modernization program without disruptions.

5.1.5 Rationalization of Routes

The primary objective of this study is to devise rationalized and strategically planned transportation routes within the Metro Manila Urban Transport Integration Study (MUCEP) area, encompassing both intra-regional and inter-regional, inter-provincial zones. The MUCEP Area, comprising the entire National Capital Region (NCR) as well as select regions and provinces in LGUs of Region 3 and Region 4A, serves as the focal point for a comprehensive evaluation of existing public transport services. However, the adoption and execution of the Route Rationalization Study have encountered challenges, primarily stemming from internal issues between the third-party consultant and the LTFRB regarding agreed-upon deliverables [3]. Compounding this issue is the expiration of the consultant's contract, necessitating the procurement of a new consultant in accordance with government protocols, thereby further prolonging the completion of the study. This delay in the route rationalization study impedes efforts to optimize public transportation systems, as rationalized routes are essential for enhancing efficiency, reducing congestion, and improving overall service delivery. To address these setbacks, it is imperative to expedite the process of securing a new consultant while ensuring adherence to regulatory frameworks to mitigate further delays. By prioritizing the resolution of internal conflicts and adhering to established protocols, the agency can expedite the completion of the study, thereby facilitating the implementation of optimized transport routes for the benefit of commuters and stakeholders within the MUCEP area.

5.1.6 Local Public Transportation Route Plans (LPTRP)

The local government units, except for LGUs within the MUCEP area, are required to prepare their own public transport route plans that will be implemented within their jurisdiction. From the 1,020 submissions made, 142 were already approved and were issued with a Notice of Compliance (NOC) and Special Notice of Compliance (SNOC) (Table 5).

The diminished count of approved Local Public Transport Route Plans (LPTRPs) stems from a multitude of factors, each contributing to the complexities faced by Local Government Units (LGUs) in finalizing these essential documents. Firstly, the ongoing COVID-19 pandemic has severely strained the resources and capacities of LGUs, diverting their attention and efforts towards urgent public health responses. Consequently, the crafting of LPTRPs has been deprioritized, as LGUs allocate their limited resources towards pandemic-related initiatives. Secondly, the LPTRP may not rank high on the list of priorities for some LGUs, particularly those grappling with pressing socio-economic challenges or infrastructure deficiencies. As a result, the development of LPTRPs may receive inadequate attention and resources, leading to delays in their formulation and approval. Furthermore, changes in administration within LGUs introduce discontinuities in governance and decision-making processes, potentially disrupting ongoing initiatives such as the development of LPTRPs. Additionally, high turnover rates among LGU personnel further compound these challenges, as the loss of institutional knowledge and expertise hampers the continuity and progress of projects. The cumulative effect of these factors hinders the timely completion and approval of LPTRPs, consequently impeding the implementation of the modernization program. Given that approved LPTRPs serve as a prerequisite for securing loans from Government Financial Institutions (GFIs) for the acquisition of modernized units by Transport Service Enterprises (TSEs), the delay in LPTRP approvals poses a significant barrier to the modernization agenda. To address these challenges, concerted efforts are needed to streamline the LPTRP development process, enhance collaboration between LGUs and relevant stakeholders, and provide necessary support and resources to expedite the formulation and approval of LPTRPs. By prioritizing and addressing these underlying issues, the implementation

Table 5 Summary of approved LPTRPs

Issuance	Total
Notice of compliance	60
Notice of compliance (with memorandum circular*)	47
Special notice of compliance	35
Total	142

* LPTRPs with issued Notice of Compliance and Local Sanggunian Ordinance from the LGUs are issued with a Memorandum Circular for the issuance of Certificate of Public Convenience (*Source* [3])

of the modernization program can proceed more smoothly, facilitating the transition towards a more efficient and sustainable public transportation system.

5.1.7 Social Support Program

The objective of this component is to provide support and resources to drivers and operators seeking to improve their skills and those who choose to withdraw from the program. Under the social support framework of the PUVMP, two subprograms have been established: the Tsuper Iskolar Program and the Entrepreneur Program. The Tsuper Iskolar Program aims to facilitate skill enhancement and capacity building among drivers and operators, equipping them with the necessary knowledge and competencies to thrive in the evolving transportation landscape. On the other hand, the Entrepreneur Program is designed to empower individuals transitioning out of the program to explore entrepreneurial opportunities and ventures beyond the realm of public transportation. As of the latest data, a total of 10,867 beneficiaries have been enrolled in the Social Support component of the PUVMP. Among these beneficiaries, 4,906 individuals are enrolled in the Entrepreneur Program, while 5,961 individuals are enrolled in the Tsuper Iskolar Program. These subprograms play a crucial role in supporting individuals within the transportation sector, offering them alternative pathways for personal and professional development. By nurturing skills and fostering entrepreneurship, the PUVMP aims to create a more resilient and empowered workforce capable of adapting to changing industry dynamics and contributing to the socio-economic growth of the nation. See Fig. 5 for sample beneficiaries.

Fig. 5 A total of 261 'tsuper' scholars completed tech-voc courses in Benguet (*Source* [15])

Through systematic tracking and evaluation, the effectiveness and impact of these social support initiatives can be continually assessed and optimized to ensure maximum benefit for program participants.

5.1.8 Vehicle Useful Life Program

This initiative covers a comprehensive framework aimed at enhancing the safety, efficiency, and sustainability of the public transportation sector. It contains various components, including the establishment of a Motor Vehicle Type Approval (VULP) System for new vehicles, implementation of maintenance programs, enhancement of the Motor Vehicle Inspection System, and the introduction of a Scrappage Program for end-of-life vehicles. The Scrappage Program offers a valuable opportunity for Public Utility Vehicle (PUV) operators and drivers to retire their aging units in exchange for monetary compensation. By decommissioning old vehicles, this program not only promotes environmental sustainability but also facilitates the transition towards a modernized and safer fleet. The Department of Trade and Industry (DTI) has compiled a list of 2,681 scrapping facilities across the country. Among these facilities, 63 have been deemed compliant with the provisions outlined in the Joint Memorandum Circular (JMC) issued jointly by the Land Transportation Franchising and Regulatory Board (LTFRB) and the Land Transportation Office (LTO), specifically JMC No. 001, Series of 2021 [9].

This collaboration between government agencies underscores a commitment to enforcing rigorous standards and regulations to ensure the safe and environmentally responsible scrapping of end-of-life vehicles. By adhering to established protocols and working with compliant facilities, the program aims to streamline the scrapping process while upholding the highest standards of safety and environmental protection. Through effective implementation and monitoring, the Scrappage Program contributes to the overarching goals of the PUV modernization effort, fostering a sustainable and resilient transportation ecosystem for the benefit of all stakeholders.

5.1.9 Communication

To ensure effective communication and dissemination of information regarding the PUVMP, this component is dedicated to engaging with key stakeholders through various channels. Its primary objective is to educate stakeholders about the program's objectives, requirements, and benefits. This is achieved through consultation meetings, information dissemination campaigns, and other communication strategies. Since its inception in 2017, numerous stakeholder consultations have been conducted to gather feedback, address concerns, and foster a collaborative approach to program implementation. However, the emergence of the COVID-19 pandemic posed significant challenges to conducting these consultations in recent times, as physical gatherings became impractical and safety precautions were prioritized.

With the COVID-19 situation now under control, the landscape for conducting stakeholder consultations has improved. This presents an opportunity to resume and even enhance engagement efforts, leveraging both virtual and in-person platforms to reach a wider audience and facilitate meaningful dialogue. By leveraging scientific insights into effective communication strategies and utilizing appropriate technologies, such as virtual meeting platforms and multimedia channels, stakeholders can be engaged in a more inclusive and interactive manner. This not only ensures that stakeholders are well-informed about the PUVMP but also fosters a sense of ownership and collaboration, ultimately contributing to the program's success., etc.

5.2 Environmental Impact

Regarding the environmental impact assessment of the PUVMP in Metro Manila, the discernible effects on air quality remain inconclusive due to the limited deployment of modernized units within the study area. Transport is a significant contributor to energy-related greenhouse gas (GHG) emissions, accounting for approximately 30% of the total, and is recognized as the largest source of air pollution in the Philippines. In 2015, GHG emissions from road transport were estimated to be 28.4 million tons of CO2 equivalent (MtCO2e), with the Jeepney sector alone responsible for 15.5% of these emissions. If current trends continue without intervention, emissions from road transport are projected to surge to 87 MtCO2e by 2030. This scenario foresees a substantial increase in the carbon footprint and worsening air pollution in urban areas, exacerbating environmental and public health challenges [10].

However, drawing insights from a comparative analysis of operational characteristics between traditional and modern jeepneys, as evidenced by a GIZ study in 2019 (Fig. 6).

It offers valuable insights into the potential environmental benefits of the PUV modernization initiative. According to the findings, the PUV Modernization Program holds promise in contributing to air quality improvement, primarily attributed to a 12% increase in fuel economy observed in modern jeepneys compared to their traditional counterparts.

Additionally, data from the GIZ study and the DOTr suggest that fleet modernization efforts in Metro Manila could lead to a notable reduction in greenhouse gas (GHG) emissions, estimated at approximately 3.1 to 4.4 million metric tons of CO_2 equivalent (MtCO2e) avoided. Despite these projections, the current operational data indicate that the anticipated fuel efficiency improvements have not materialized in practical operation, as reported by operators. However, it is worth noting that the adoption of modernized vehicles featuring increased passenger capacity, such as the transition from a typical twenty-seat traditional jeepney to a thirty-seat modernized unit, yields significant gains in fuel economy per passenger.

This enhancement in fuel efficiency is expected to translate into a more favorable environmental footprint for operations along the new routes, ultimately leading to a reduction in carbon emissions per passenger compared to older vehicle models. As

Average (range)	Traditional Routes	Modern Routes	Average % ch.
Daily operating hours	14hrs (13-15hrs)	19 hours (11-22 hours)	+36%
Vehicle utilisation	127km (60km-190km)	150km (80-220km Euro IV diesel) (80-120km E-Jeepney)	+18%
Days of operation per week	5.6 days (5.5-6 days)	6 days	+7%
Staff per vehicle/day	1 driver (1-2 drivers)	2 drivers, 1.5 conductors (1-2.5 drivers, 0-2.5 conductors)	+350%
Daily staff earnings	650 PHP (11.57 EUR) (574-755 PHP – 10.21-13.44 EUR) (non-salaried)	600 PHP (10.68 EUR) + benefits (537 M/W-1,000+ PHP – 9.56-17.80+ EUR) (salaried)	~
Vehicle capacity	20 seats (16-24 seats)	30 seats max. (22-24 seated)	+50%
Fuel economy	5.9 km/l (4.2-7.8 km/l)	5.2 km/l (4.2-6.2 km/l)	-12%
Fuel economy per passenger/km	111 km/l	156 km/l	+41%
Daily ridership Pax./day/vehicle	300 (150-350)	460 (Euro IV Jeep 300-750) (E-Jeepney 200-250)	53%

Fig. 6 Comparison of the operational characteristics of traditional and modern Jeepney (*Source* [12])

the deployment of modernized units continues to expand and operational efficiencies are optimized, the PUVMP holds the potential to significantly contribute to the improvement of air quality and environmental sustainability within the MUCEP area.

5.3 *Economic Viability*

Prior to the implementation of the modernization program, drivers relied on the "boundary system" for their earnings, wherein they retained whatever remained after covering rental and fuel expenses. This arrangement provided drivers with a degree of control over their daily earnings, incentivizing them to maximize passenger pickups and extend operating hours. However, this system also fostered on-street competition among drivers, resulting in various adverse effects. Research conducted by Biona [11] in the GIZ study sheds light on the cost structure of jeepney drivers prior to the program's implementation, as depicted in Table 6.

It is evident that longer operating hours translated to greater economic viability for drivers under this system. The transition to a modernized fleet presents a significant challenge for consolidated entities, as the cost of modernized units ranges from PHP 1.6 million to PHP 3.0 million, depending on the class of the jeepney. To facilitate participation in the program, subsidies are provided to Transport Service Enterprises (TSEs) by the government. As of the present date, a total of 1,447 units have been approved under the Land Bank of the Philippines (LBP) and 2,681 units under the Development Bank of the Philippines (DBP) nationwide, aiding existing TSEs in fleet modernization and facilitating loan processes for consolidated entities.

Assessing the commercial performance of operating modern jeepneys necessitates a thorough evaluation of feasibility. Operating modern jeepneys entails higher initial investments due to the cost of new vehicles, expenses for in-vehicle staff, establishment of formal operating entities (e.g., cooperatives or corporations), and fleet management. Comparative analysis of return on investment between modern and traditional jeepney operations indicates that despite the higher upfront costs, modernized operations typically yield higher returns over the 15-year lifespan of the vehicle. This suggests that modernization can enhance commercial performance for operators (Table. 7). Investing in larger vehicles boosts farebox revenues, while formalized operations with structured shifts and fleet management enhance vehicle productivity and generate economies of scale through collective maintenance and parts procurement. This underscores the potential for modernization to not only improve operational efficiency but also bolster financial sustainability for operators in the long run.

Table 6 Jeepney daily cost structure

Average amount	End to end route length		
	< 5 km	5–10 km	> 20 km
Vehicle boundary (rental)	878 PHP	820 PHP	840 PHP
	17.64 USD	16.48 USD	16.88 USD
Fuel cost	634 PHP	900 PHP	915 PHP
	12.73 USD	18.08 USD	18.38 USD
Other expenses	172 PHP	210 PHP	245 PHP
	3.45 USD	4.3 USD	4.92 USD
Driver net task-home pay	900 PHP	660 PHP	933 PHP
	18.08 USD	13.26 USD	18.75 USD

Source [3]

Table 7 Commercial performance of operation over fifteen-year period (PhP/EUR)

Vehicle type	Traditional Jeepney	Modern Jeepney	
	Jeepney and Franchise		
Investment	350,000 PHP (6,200 EUR)	1,925,000 PHP (34,000 EUR)	1,850,000 PHP (32, 750 EUR)
Financing source	Micro financing	DBP finance	DBP finance
Financial interest rate of return (FIRR)	32%	55%	52%
Net present value @12%	629,000 PHP (11,100 EUR)	881,000 PHP (15,600 EUR)	1, 167,000 PHP (20,600 EUR)

Source [12]

Amidst the challenges posed by the heightened community quarantine measures during the peak of the pandemic, the LTFRB extended assistance to Transport Service Enterprises (TSEs) participating in the modernization program. Through enrolment in the Service Contracting Program (SCP) administered by the agency, TSEs were provided financial support by the government to continue operating during the pandemic period. This initiative aimed to prevent transport paralysis and ensure essential services, particularly for healthcare workers who rely on public transportation to fulfil their vital roles in addressing medical needs. The profits generated from the SCP served as a crucial lifeline for TSEs, enabling them to meet their financial obligations, including loan repayments, despite the closure of many routes during the quarantine period. This support mechanism not only sustained the viability of operations but also alleviated financial burdens on operators, fostering resilience amidst unprecedented challenges.

This collaborative effort between the government and TSEs highlights a commitment to ensuring the continuity of essential services and supporting the modernization agenda even in the face of adversity. By providing financial assistance through the SCP, the LTFRB demonstrated proactive measures to mitigate the economic impact of the pandemic on the transportation sector while safeguarding the well-being of frontline workers and the general public.

From a scientific standpoint, this initiative aligns with principles of crisis management and resilience-building, emphasizing the importance of adaptive strategies to maintain essential services during emergencies. Furthermore, it highlights the role of government intervention in supporting sustainable transportation solutions and fostering long-term resilience in the face of unforeseen challenges.

5.4 Infrastructure Development

The Public Utility Vehicle Modernization Program (PUVMP) stands as the largest non-infrastructure initiative spearheaded by the government. However, in accordance

Fig. 7 Example of a common garage and terminal of consolidated transport cooperatives (*Source* [13])

with the Omnibus Franchising Guidelines, the sole infrastructure prerequisite for Transport Service Entities (TSEs) to operate is the establishment of a common garage (as illustrated in 7) for their fleet and a terminal (as depicted in Fig. 7) for operational purposes. Yet, recent legislative developments such as the Electric Vehicle Industry Development Act (EVIDA) or RA 11,697, enacted in 2022, have introduced significant implications for the program. RA 11,697 underscores a commitment to fostering innovation in clean, sustainable, and efficient energy while promoting inclusive and sustainable industrialization, acknowledging the pivotal role of the private sector. Notably, the law mandates that 5% of the total TSE fleet be composed of electric vehicles and necessitates the construction or installation of charging stations for electric vehicles at gasoline stations, alongside dedicated parking slots equipped with charging stations in every establishment.

Presently, the Department of Transportation, in collaboration with partner agencies tasked with implementing RA 11,697, is actively engaged in executing the provisions of the law. This entails conducting consultations with various stakeholders, formulating the necessary implementing rules and regulations (IRR), establishing dedicated offices, and securing budgetary allocations in the annual General Appropriations Act, among other initiatives.

The anticipated impact of RA 11,697 on the transportation landscape of the Philippines, particularly in Metro Manila, is considerable. By facilitating the proliferation of electric vehicles and supporting the development of associated infrastructure, the law holds the potential to revolutionize the transportation industry, ushering in a cleaner, more sustainable, and technologically advanced era of mobility. From a scientific perspective, this legislative framework aligns with global efforts to mitigate climate change and promote environmental sustainability through the adoption of alternative energy sources and innovative transportation solutions.

6 Conclusion

The implementation of the PUVMP in Metro Manila has seen significant advancements across various components since its inception in 2017. The overarching aim of the program is to revolutionize the public transport system, elevating its standards to align with global benchmarks while fostering dignity and humanity in transportation services. However, alongside notable progress, the program also faces several challenges and areas necessitating focused attention to ensure its efficacy and acceptance among stakeholders.

To facilitate the adoption and execution of the program, the LTFRB has proactively issued a excess of policies and guidelines. These initiatives highlight the agency's commitment to supporting program implementation and oversight. Notably, the modernization drive has led to the deployment of 9,144 compliant units, marking a significant stride towards fostering environmentally sustainable transportation practices. Nevertheless, obstacles such as pending loans and financial constraints have shortened the overall spread of compliant units.

The consolidation of existing franchises and the establishment of consolidated entities have yielded encouraging outcomes, with 2,029 franchises amalgamated and 75,921 units integrated within the study area. Financial assistance, amounting to PHP 10.62 billion, has been greenlit by government financial institutions to facilitate the acquisition of modernized units. Nonetheless, challenges pertaining to loan approvals, stringent requirements, and the necessity for supplementary studies have impeded the program's momentum.

The formulation of Local Public Transportation Route Plans (LPTRPs) has encountered obstacles stemming from various factors, including resource reallocation towards COVID-19 response efforts, prioritization dilemmas, administrative turnovers, and high turnover rates. These hurdles have constrained the approval of LPTRPs, consequently hindering loan disbursements for modernized units. Meanwhile, the social support program has enrolled beneficiaries in the Tsuper Iskolar and Entrepreneur programs, offering invaluable assistance and skill enhancement opportunities.

Efforts such as the Vehicle Useful Life Program (VULP) and communication campaigns have been initiated to address different stages of a vehicle's lifecycle and enhance awareness of the program among stakeholders. However, recent stakeholder consultations have been curtailed due to the constraints imposed by the COVID-19 pandemic.

In terms of environmental impact, the current deployment of modernized units within the study area remains relatively low, resulting in limited observable effects on air quality. Nonetheless, a comparative analysis of operational characteristics between traditional and modern jeepneys reveals the potential for improved air quality, evidenced by a 12% change in fuel economy.

Despite the elevated upfront expenditures, modernized operations typically yield greater returns over the 15-year lifespan of the vehicle. This underscores the potential of modernization to bolster commercial performance for operators. Further, the collaborative endeavor between the government and Transport Service Enterprises (TSEs) underscores a shared commitment to ensuring the uninterrupted provision of essential services and advancing the modernization agenda amid challenges.

By promoting the widespread adoption of electric vehicles and facilitating the establishment of necessary infrastructure, the law has the capacity to fundamentally transform the transportation sector. This shift has the potential to attendant in a new era of mobility characterized by enhanced cleanliness, sustainability, and technological advancement.

7 Recommendations

Based on the aforementioned findings, the following recommendations are proposed to augment the effectiveness and acceptance of the PUVMP:

- Streamlining loan processes: simplifying loan procedures and consider alternative financing options, and ensuring timely disbursement of approved loans. This can help increase the number of compliant units on the road.
- Facilitating LPTRP development: Support and resources should be provided to local government units (LGUs) in mainstreaming Climate Disaster Risk Assessment (CDRA) into LPTRP and investment decisions. Further, guidelines and process of integrating the climate risk and vulnerability assessment of the public transportation sector into the current LPTRP process should be developed. In addition, coordination between LGUs and government agencies should be enhanced to expedite the approval process, considering the challenges faced by LGUs, such as limited resources, administrative changes, and high turnover rates.
- Strengthening stakeholder consultations: Transport operators, drivers, and commuters should be engaged to gather valuable feedback and address their concerns, and ensure their active involvement in the program's development and implementation. With the COVID-19 situation stabilized, conducting stakeholder consultations has become more feasible. This offers a chance to resume and strengthen engagement efforts using both virtual and in-person platforms. By employing scientific communication strategies and digital tools like virtual meeting platforms, stakeholders can be engaged more inclusively and interactively. This ensures stakeholders are well-informed about the PUVMP, fostering ownership and collaboration for program success.
- Enhancing communication and awareness: Information dissemination campaigns must be intensified to improve public awareness and understanding of the program's objectives, benefits, and requirements. Various communication channels, such as social media, community engagement, and targeted outreach programs can be utilized to effectively reach and educate stakeholders.
- Continuous monitoring and assessing environmental impacts: Monitoring efforts to assess the environmental impact of the program as more modernized units become operational should be expanded. Air quality indicators, fuel economy

improvements, and other relevant metrics should be regularly evaluated to determine the program's effectiveness in achieving its environmental goals.

- Exploring partnerships with the private sector and NGOs: It is important to collaborate with private banks, financing institutions, and non-governmental organizations (NGOs) to expand financing options and provide additional support to transport service entities. Such partnerships can help address the challenges faced in securing loans and enhance the overall implementation of the modernization program.
- Amplified Return over Initial Outlay: In light of the prospect of amplified returns over extended durations notwithstanding augmented initial outlays, it is recommended that operators focus on the transition to modernized jeepneys. Allocations toward modern vehicle acquisition, in-vehicle staff deployment, and formal operating entity establishment are poised to bolster commercial efficacy and sustainability. Furthermore, synergistic engagement with governmental bodies, as exemplified by initiatives such as the Service Contracting Program (SCP), is imperative. This collaborative endeavor not only alleviates economic adversities but also fortifies the perpetuation of indispensable services and bolsters the modernization agenda, thereby fostering resilience and efficacy within the transportation sector.
- Proliferation of electrical Vehicles and Associated Infrastructure: Given the potential outlined by scientific research, it is recommended to actively facilitate the proliferation of electric vehicles and provide robust support for the development of associated infrastructure. This approach has the capacity to catalyse a transformative shift in the transportation industry, leading to a cleaner, more sustainable, and technologically advanced era of mobility.

By considering these recommendations, the adoption and implementation of the Public Utility Vehicle Modernization Program can be improved, challenges addressed, and this will in the long term contribute to a more sustainable and efficient public transport system in Metro Manila, benefiting both commuters and the environment.

References

1. Asian Development Bank (2019) Urban transport for livable cities forum 2019. The impact of the PUV modernization program on road safety in the Philippines. Manila, Philippines. https://www.adb.org/news/events/urban-transport-livable-cities-forum-2019. Accessed 6 Oct 2023
2. Gatarin RG (2024) Modernising the 'king of the road': pathways for just transitions for the Filipino jeepney, Urban governance 4(1):37–46. https://doi.org/10.1016/j.ugj.2023.11.002. https://www.sciencedirect.com/science/article/pii/S2664328623001092. ISSN 2664–3286. Accessed 12 Dec 2023
3. Land Transportation Franchising and Regulatory Board (LTFRB) (2023) PUVMP status updates as of June, 2023, p 53. https://ltfrb.gov.ph/wp-content/uploads/2023/08/ACCOMPLISHMENT-REPORT-as-of-June-2023-1.pdf. Accessed 8 Oct 2023
4. Raza T, Liwag CREU, Andres AVL et al (2020) Extreme weather disasters challenges for sustainable development: Innovating a science and policy framework for disaster-resilient and

sustainable Quezon City, Philippines. https://doi.org/10.1016/j.pdisas.2020.100066. Published Print: 2020–01, Elsevier Update policy. https://doi.org/10.1016/Elsevier_cm_policy
5. Office of the Energy Efficiency and Renewable Energy (OEFRE) (2024) Sustainable transportation and fuels, ENERGY.GOV, Forrestal Building, 1000 Independence Avenue, SW, Washington, DC 20585. https://www.energy.gov/eere/sustainable-transportation-and-fuels. Accessed 6 Feb 2024
6. Metro Manila Development Authority (MMDA) (2022) Location map of greater metropolitan, data source: metro manila development authority and metropolitan manila earthquake impact reduction study, office of the assistant general manager for planning, metro Manila, Philippines
7. Valdez AS, Raza T, Farolan MI et al (2023) Primary and secondary data collection to conduct researches, write thesis and dissertation amidst COVID-19 pandemic: a guidepost. In: Proceedings of the 2nd international symposium on disaster resilience and sustainable development, Lecture Notes in Civil Engineering, Springer Nature, Singapore 294:269–288. https://doi.org/10.1007/978-981-19-6297-4_20
8. Tacderas MY, NG ACL, Tolentino NJY, Tiglao NC, Herrera CE (2021) Examining the Implementation of the public utility vehicle modernization program (PUVMP) in general Santos city, Philippines: an industry perspective, Philippine. Transp J 4(1). https://ncts.upd.edu.ph/tssp/wp-content/uploads/2022/02/PTJ-V4-N1-TSSP-2021-Vol4-No1-04-Tacderas-et-al.pdf. Accessed 22 Dec 2024
9. Land Transportation Office and Land Transportation Franchising and Regulatory Board (2021) LTO-LTFRB Joint memorandum circular No. 001, s. 202, Guidelines for scrappage facility accreditation and vehicle scrapping. https://jur.ph/laws/summary/implementing-guidelines-to-govern-the-accreditation-of-scrappage-facilities-and-manner-of-scrapping-of-public-utility-vehicles. Accessed 6 Jan 2024
10. Mariano P (2015) Modernising public transport in the Philippines. https://changing-transport.org/modernizing-public-transport-in-the-philippines/. Accessed 8 Jan 2024
11. Biona JB (2015) Jeepney market transformation program. Clean air Asia on behalf of GIZ department of transportation. https://www.facebook.com/photo/?fbid=545086511089739&set=a.184633020468425. Accessed 22 Jun 2023
12. GIZ (2019) Reforming the (semi-)informal minibus system in the Philippines—the 'public utility vehicle modernization program' early route evaluation. https://www.changing-transport.org/wp-content/uploads/2019-11_GIZ_JeepneyModernisation_Early-Evaluation_final.pdf.
13. Binangonan Transport Cooperative (2022). https://www.facebook.com/photo/?fbid=445771960906204&set=a.445771927572874. Accessed 22 Jun 2023
14. Department of Transportation (DOTr) (2017) Department Order 2017–011 Omnibus guidelines on the planning and identification of public road transportation services and franchise issuance, Published on February 17, 2021. https://www.oneplanetnetwork.org/knowledge-centre/policies/department-order-2017-011-omnibus-guidelines-planning-and-identification. Accessed 2 Jan 2024
15. Philippine News Agency (2023) by Agoot L. 261 'tsuper' scholars complete tech-voc courses in Benguet, The Philippine News Agency is a web-based newswire service of the Philippine government under the supervision of the News and Information Bureau (NIB) of the Presidential Communications Office (PCO). https://www.pna.gov.ph/articles/1192834. Accessed 3 Jan 2024

Indigenous Knowledge of Afro-Asian Tribal Communities: An Alternative Science of Natural Disaster Management

Tapas Pal

Abstract Indigenous knowledge of indigenous communities is critical to protecting the Earth. The climate is changing massively due to modernity rather than human intervention, needs, wants, and desires. The flow of modern culture has penetrated almost every planner's mind in recent times. So, sustainable management has become important in contemporary natural disaster management. Tribes have their own indigenous knowledge to predict climate change and natural disasters. They can feel the breath of nature and the sensation of biodiversity. The practical and local oral skills of tribal people can be used to cope with natural disasters, and tribes are close to nature. The time has come when we need to rethink our survival plans, policies, and management to protect the earth in a sustainable way with the application of tribal knowledge. 'Sabka Saath Sabka Vikas' demands the 'rhetoric-to-action' and contemporary environmentalists are planning to combat the climate crisis, promote sustainable development, and implement disaster risk reduction measures. Both science and technology should jointly make smart disaster management possible, but modern'sci-fi 'and 'hi-fi' technological policies are not seen as successful at all. Science and technology can both be "critical friends" for natural disaster management and policymaking, but sustainable resilience is always demanded by the common people. There have been many landslides, tsunamis, earthquakes, and volcanic eruptions, but records show that tribes are affected or damaged very little due to their connection to nature. In fact, we have entered the realm of nature without understanding the sense of 'Pancha Bhoot' and have forgotten that a natural disaster is a natural phenomenon of nature. Tribes in Afro-Asian countries nourish indigenous knowledge to save them from the devastation of natural disasters. This research highlights the indigenous knowledge of tribal communities in Afro-Asian countries to inform the people of the changes in climate, rivers, oceans, soil, ethology, ornithology, and nephology to cope with the level of destruction due to natural disasters.

Keywords Indigenous knowledge system · Qualitative oral skills · Sensation of biodiversity

T. Pal (✉)
Department of Geography, Raiganj University, Raiganj, India
e-mail: geo.tapaspal@gmail.com

I. Pal et al. (eds.), *Proceedings of the 3rd International Symposium on Disaster Resilience and Sustainable Development—Volume 1*, Lecture Notes in Civil Engineering 623, https://doi.org/10.1007/978-981-96-2139-2_12

1 Introduction

The problem is raised when we are trying to penetrate our culture into the Sanskritization [1] of nature. If any misbalances are seen here, they can appear quickly. So, the basic background of the research is raised by the question, like why, how, and when tribal indigenous knowledge is applicable to predicting the probability of the types of natural disasters, the path of hazards, minimizing the level of destruction, and forecasting it. Satellite forecasting and television or radio talk are the scientific platforms, but if the natural method of disaster minimization is applied by common people, then the final result will be more fruitful. Recently, based on their ages, different climate leaders from 160 countries have joined together under US Secretary-General Antonio Guterres (2020), and Archana Soreng, who is from the Santhal community in India, has been elected as a climate leader [2] to highlight indigenous knowledge to reduce the climate crisis and related disasters. Catastrophic destruction for natural disasters has increased, for example, a cloud burst in Uttarakhand (2013) [3], a destructive volcanic eruption in Curah Kobokan village of Indonesia (2021), multiple tornadoes in Kentucky (2021), Hurricane Ida on Louisiana's Port Fourchon of the US (2021), Hurricane Katrina, Cyclone Tauktae (2021) in Gujarat, flash floods in Kerala and Uttarakhand (2021), Cyclone Seroja in Timor (2021), and floods in Western Germany (2021) [4]. Torrential rain fell in Henan Province of China (2021), Super Typhoon Rai in the Philippines (2021), Earthquake in S.W. Haiti (2021), Bushfires of Australia (2019–2020) [5], Swarms of Locusts in Asia-East (2020), Cyclone Amphan in Bangladesh-India (2020), Forest Fires in Uttarakhand (2020), Green Snow of Antarctica (2020) [6], etc. Therefore, it is high time to rethink our indigenous knowledge to deal with, mitigate [7], and predict the impact of natural disasters. Based on the study, it was seen that in Tsunami (2004), not a single Jarawa tribe died due to their indigenous knowledge, but the so-called modern, educated tourists and local people have died [8]. This study highlights indigenous information about regional natural disaster management, such as the 'Megh' (cloud) Puja of the Akunda (dom) tribe along the River Ajay in Birbhum District, West Bengal [9]. Modern inventions and technologies are not able to control and manage the entire natural disaster risk reduction process completely. The UN-2030 Agenda [10] contains 17 different Sustainable Development Goals [11]. Now, there may be SDG-3 ('Good health and well-being'), SDG-13 ('Climate action'), SDG-14 ('Life under water'), and SDG-15 ('Life on land'). Emphasizing sustainable indigenous knowledge for disaster risk reduction is enlightening. Climate leaders are demanding indigenous knowledge to mitigate the climate crisis and related natural disasters. It is proven that indigenous practicing knowledge is fruitful to cope with the level of the effects of natural disasters; for example, indigenous people of the Niger Delta Region can understand the nature of the flood by observing the position of the birds' nests on the bamboo trees adjacent to the river, thus planting bamboo trees along the river helps to control the incidence of floods; tribes of South Cape Province in South Africa receive rain forecasts based on ant behavior and take action accordingly [12]. Tribes in the O.R. Tombo district of Zimbabwe indicate the probability of drought in the horse's erratic behavior [13], or tribes in Kenya watch the number of nests of the weaver (Babui) birds in the trees to forecast the probability of the drought; for example, more nests indicate a humid environment, and fewer nests indicate dry weather [14]. Again the animal behavior forecast the upcoming climatic hazards for example the call of frogs and toads in the rainy season indicates a long rainy season and the call stops indicate a change in weather; an excessive number of insects trapped in spider webs indicates humid weather; changes in the location of fishes in large lakes predict drought; falling leaves of Baobab indicate dry weather or drought; the changing of the nests of the Swallows Birds indicates a drought and their flying in the sky in a circular way indicates a flood [15]; the constant cries of dogs and the overturning of cotton leaves are the indicators of an upcoming cyclone which is seemed by the Santal Tribe [16]; some tribes in the Himalayas get indications of earthquakes from the restless behavior of rats [17] and the Nagoni tribe consider the bee's departure as a sign of a flood [18].

2 Objectives

To find out the relationship, importance, relevance, and cross-country well-being of tribal indigenous practices in natural disaster risk reduction in a sustainable way.

3 Methodology

The research is confined to the specific seven natural disasters (cyclones, earthquakes, tsunamis, droughts, landslides, volcanoes, and floods) of selected seven Asian countries (Philippines, Indonesia, Sumatra, Timor, China, India, and Bangladesh) and six African countries (Nigeria, Zimbabwe, Kenya, Ghana, Malawi, and South Africa). It's qualitative, behavioral, desk based, secondary, and documentary research [19], where the online questionnaire [20] was sent to different academicians in Afro-Asian countries to get information on natural disaster management through indigenous knowledge. Theoretical technique (Cross Country Dataset) [21] and photography (global perspective and secondary dataset) are used. The open ended research questions [22] have been used, like how far and to what extent the selected types of natural disasters are managed with the help of indigenous knowledge [23] and practices of Afro-Asian tribal communities. And another question is whether the indigenous knowledge and practices of tribal knowledge are sustainable for natural disaster management. Different books and articles were reviewed to do this research. Afro-Asian countries are selected because they fall under developing countries [23], and tribal communities still exist there with indigenous identity to some extent [24]. Whereas, in developed countries, it is not possible to maintain tribal culture and indigenous practices properly due to the influx of heavy globalization and modernization [25]. The continents of Asia and Africa together hold 77% (2021) [26] of the world's population, while Europe and the Americas together account for 22.5% (2021) [27]. Therefore, in the continents where the high number of populations and diversity of tribal communities have become the cause of the topic selection.

4 Discussion

Indigenous knowledge and practices of Afro-Asian tribal communities [23] are important in reducing the risk of natural disasters. A lot of indigenous practices of different tribal communities are still being practiced in different countries on the Afro-Asian continent.

4.1 Drought

In China (Xinjiang region), the Uyghur tribe practices carriage irrigation to manage drought [24]. In Kenya, Luo collision, Kip, Nond, Pokoto, and Taveta Nganyi communities forecast drought when they observe wildlife roaming around their houses to search for food and water, abundant orbits in the sky around Lake Victoria, hearing cries of birds like Ivutavutitya Iilum, Maalwi, or Konga, birds become greedy and fly off, the baobab tree loses its leaves and remains leafless even during the long dry season, red moon, and solar eclipse [25]. In Arunachal Pradesh, India, the Konyak tribe foretells drought by drying up mountain springs that indicate a lack of rainfall. This is an indicator to the locals that there is a need to anticipate and prepare for droughts and food shortages, serving as a sign of the need to start storing food grains in earthenware containers [26]. In Ghana, the Yogu tribe predicts drought by following chameleons and snails that always lay eggs and give birth during dry periods [27]. Young chameleons do not see in the rainy season because they must be at least one month old to withstand rain. Again, this tribe knows that the flowering of the 'Dawa Dawa' plant predicted the upcoming drought [28].

4.2 Flood

In Tamil Nadu and Kerala, India, the tribes of Erulas, Kattunayakkans, Kotas, Kurumbas, Panyans, Kattunaikars, and Cholanaikas predict floods by looking at river water levels, the sudden swelling of mosquitoes, ants moving away from river banks, and climbing trees. And buildings, constant cries of wood cats, bats flying in large flocks, unexpected arrival of wild pigs and roosters in the village from upstream areas, ants climbing up the house in a line, rapid movement of ants means heavy rain or flood, mouths of young red ants to move with eggs [29]. And yellow-bellied frogs' calls are a common indicator of rain, and during floods, they spend the night in high forest stands [30]. In Bihar and West Bengal, Santhal, Oraon, Munda, Ho, Kharwar, Kharia, and Lepcha tribes protect themselves from floods and Nor'wester with the technique of 'halla' to warn the community and tell Sariali ho, bad avatani. As a symbol of respect for the hill (Pahar), the Santhal tribe worships the'marangburu' (meaning the great mountain) [31]. The Aku community at the side of the River Ajoy in Birbhum and Burdwan districts of West Bengal used the sheora tree to prepare the roof because it is very flexible and can resist the nor'westers (Kal Baisakhi) [8]. In Southern Africa, Swazi, Emaswati, and Liswati (a Bantu ethnic group) estimate floods by looking at the height of such Emahlokohlok bird nests. The nests of the Emahlokohlok bird are very high in trees when there is a possibility of flooding, and nests are made at a level of medium or low when there is no possibility of flooding [32]. In Malawi, Chewa, Nyanja, Lomwe, Ya, Tumbuka, and Sena tribes assumed flood through observing animal behavior (abundance of ants, hippo migration from the river banks to villages, increased numbers of local bird species, e.g. 'anazeze', 'songwe') and plant behavior (increased flowering of the tree 'Mkhunku' (Acacia Nigrescens), riverine behavior (expanded sounds from the waters moving in the river, changing of watercolor), meteorological furcating (very hot temperatures, increasing rainfall intensity, heavy wind flow) and behavior of heavenly bodies (e.g., sun, moon, stars) (Trogrli) [33]. In Zimbabwe, Ndau, Ndebele, and Shona tribes predict floods based on the characteristics of certain types of trees, The presence of dark black clouds (local name: amaze amnyama) indicates the anticipated arrival of heavy rain, and if the Inkanku bird (rainmaking bird) cries, then heavy rain may occur [34]. Again in Tsholotsho District, Matabeleland, and the Northern Province of Zimbabwe, the Tsholotsho tribe follows the restless behavior of the inkanku (rain-making bird) to predict heavy rains and the ukuhluma kwajihlahla (new leaves growing from trees) to foretell the new rainy season [35]. In Swaziland, South Africa, Swazi, San, and Nguni tribes predict floods based on the arrival of rain based on the presence of a particular species of bird, and the arrival of floods also determines the height at which those birds make their homes [32]. In the Western Niger Delta of the Niger River in Nigeria, Urhobo, Isoko, Itsekiri, Izon, and Ukwuani tribes live on the banks of the river, from where they can observe nature closely and forecast floods. In the trees adjacent to the riverbanks of the Niger Delta Region, the birds normally nest in the branches of the entire tree at the very low river level. So that the birds can easily collect their food from the river. But when the heavy rain and flooding would happen, birds would build up their nests at higher altitudes. Birds can estimate that there is a risk of heavy rains any year and can guess where the river's water level is likely to rise. They planted many crops together on the same land instead of one crop in the fields to stop soil erosion, planting many crops, they were able to maintain the supply of nitrogen in the soil and retain other nutrients [36]. In Ghana, the Chietang tribe follows the Hippopotamuses breed and feed along the shores of the river in Chattanooga and rarely ventures inland, but always moves inland several hours before a flood [27]. In Assam, India, Chutia, Ahoms, Sonowal Kacharis, Koches, Kalitas, Kaibartas Mishings, and Deoris tribes know that melong cries before a devastating flood and massive bamboo flowering before the summer season indicate an upcoming devastating flood [37]. In Sumatra, Indonesia, and the

Sunni Islands, the Asian Simelucans and Moken tribes provide their communities with early warning of any type of natural disaster based on observing animals, birds, insects, plants, clouds, air, and water temperatures [38]. The Khasi tribe of India always followed pre-disaster management, low-carbon generation, betel cultivation and planting new plants in barren lands, and social forestry [39].

4.3 Volcano

In the Philippines, Igorot, Lumad, Badjaos, Ati, Tumandok, and Palawan tribes followed different incidents in nature to predict an upcoming volcanic eruption, like when the riverbed dries up for several months in a row, causing cracks in the riverbed, occasional earthquakes, and species of animals (Baby-demo or wild boar, chicken) started descending from the mountains in groups. When the eruption continues, it will gradually take on a terrible shape, but if the weather is good, there is nothing to fear [40]. During an eruption, they always avoid low lying areas. Since different debris and ash come with the wind, people take shelter in distant places [41]. When the eruption stops, they do not return to the area until they feel the place is safe for habitation. Because the possibility of another eruption in the same location exists even after the eruption [41], Again, the Albay tribe of this country believes that if wild pigs and chickens are running away from the Mayon volcano area, that means a volcanic eruption is coming [42].

4.4 Landslide

In India, Bhutias, Lepchas, Sherpas, and Yolmos predict landslides based on some observations like unusual noises of animals and birds, movement of wild animals closer to the habitat, changing places of ants with their eggs, rolling stones down from the hill alone, going out of the crow's in the sky, starting chirping of haleso (greenish dove), suddenly a bright sun appeared and a faint odor of soil wafted after a lot of rain [26]. The Aka tribe in Arunachal Pradesh, India, considers the Bhojo Phu Mountain a sacred place, believing that anyone who takes any forest material from the sacred mountain will be destroyed. Such belief has helped in the conservation of forest resources in and around hilly areas, which has indirectly helped in mitigating various natural disasters like floods, droughts, and landslides [43]. The Kovalima, Liquika, Timor-Leste, Bunak, Fataluku, and Makase tribes know that landslides will occur when sacred trees (al lulik), such as teak, bamboo, and beachwood) are cut down and sacred stones (fat u lulik), are destroyed or removed [44]. Monu Ain, or Lulik, is a festival held by a traditional leader to take stones, sand, and trees from nature, apologize for killing snakes, and promise not to do it again. First, an animal (e.g., a chicken, pig, or dog) is killed, and its blood is spread on the affected area with the help of betel leaves. Then, the trees are transplanted in sacred places such as around water sources, riverbanks, beaches, and highlands or hilly areas [45].

4.5 Tsunami

In India, the Great Andamanese, Onge, Jarawa, Sentinelese, and Shompens tribes measure the depth of the ocean by the smell of the air, or the sound of the oars of their boats; tribal people can understand the tsunami in advance [46]. They take shelter in high places during tsunamis. At that time, they ate coconut and honey and caught fish with the help of a bow and arrow [47]. Again, the Moken, Jarwas, and Onge tribes of the Andaman Islands believe that if a high wall of water appears after a great earthquake in the land, they go deep into their forests to higher ground and escape the wrath of the wave [48]. In Indonesia, the Simelu Island community predicts tsunamis by going to the coast to monitor tidal movements when earthquakes occur. If the tide ebbs or recedes immediately after an earthquake, run for higher ground, as the tide will follow with huge waves. Immediately after the earthquake, they rushed to higher ground and stayed there. Traditional storytelling methods are used to reduce the impact of tsunamis and earthquakes. The 'ancestral experience' was in 1907, when an earthquake created a tsunami and killed a large number of islanders. Stories of the cataclysmic events of 1907 were translated into stories, testimonial monuments, and memoirs, which were then passed on to an unformed and young generation in an ad-hoc pattern. It saved so many lives in the 2004 tsunami [49].

4.6 Storm

Tribal fishermen in Barangay Tulang Dayot, Philippines, look to the leaves of the star apple tree to predict an impending typhoon; If seen upside down from the tree, a natural hazard (e.g., a typhoon) will occur, and the Bayanihan procedure will follow [50]. The Acehnese tribe of Indonesia follows the dark cumulus clouds (cumulus nimbus) on the horizon and their upward movement from the wind, the position of the hives in the trees, the calm sea conditions during the passage, and the smell of the ocean. To predict storms, high waves, and coastal erosion. They believe the rancid smell is an indication of excess evaporation at the sea surface and the formation of massive clouds. They planted atom pasi (which absorb salt vapor) to prevent and mitigate the effects of high waves, strong winds, and coastal erosion [51]. Tikopia Island and Solomon Island tribes take shelter under overhanging rocks on high ground when typhoons strike [52]. The Taboys tribne Gmelina, Talisay, Pile, Marukbarok, Tamarind, Santol, and Nauru of Rapu-Rapu Island in the Philippines predict typhoons based on the branches of trees and banana leaves falling to the ground even when there is no strong wind. Two days after such an observation, heavy rains, storms, or strong winds will hit the community. For food security, after a typhoon, they dig a hole in the ground, put root crops (such as cassava) inside the hole) and fill it with soil. Stored root crops are prevented from rotting and can last up to a month, thus providing food security during storms [53]. The Chakma, Marma, Tripura, Tanchangya, indigenous Assamese, Keot, Chak, Pangkho, Mro, Murang, Bom, Lushei, Khiang, and Khumi tribes of Bangladesh predict cyclones to see weather patterns; for example, the sky darkens and indicates cloudy weather followed by rain. It will be extremely hot and humid. They follow animal behavior such as cattle becoming restless and stopping eating, cattle and dogs crying continuously at night, or birds flying aimlessly, which means a cyclone may be coming soon [54]. Smoky and cloudy types of water in ponds, and rivers become hot, if a cyclone is coming [55]. Santals, Marma (in Chittagong), Tripura tribes, and Tanchengya tribes of Bangladesh forecast Cyclone to follow the dogs if they are constantly crying before the cyclone hits, if the leaves of cotton and mender are all turned upside down, if flies, mosquitoes, and ants become restless, or if the wind blows hard [56]. The Vanuatu tribe of the Republic of Vanuatu manages Cyclone by Nimaleten (cyclone shelters that are made of wood, creepers, palm, and Pandanus leaves) [57].

4.7 Earthquake

Inagta Alabat tribe of Alabat Municipality of the Philippines always conscious of unusual appearances of oarfish) surfacing in great numbers to predict earthquakes [58]. In Sumatra, Indonesia, the Nias community built the house as one of the world's best examples of earthquake-resistant architecture. Traditional houses in Nias (earthquake-resistant wooden house construction) are made of wood, where the structural members are slotted together instead of nailed. This technique adds flexibility to the home. The houses are supported by vertical pillars (enomo), which rest on stone blocks, and several diagonal beams (ndriwa). Ndriwa is placed between massive walls, across and alongside the building, serving as longitudinal and lateral bracing. They lean against each other at their base and top, where they fit into the horizontal beams on the floor of the building. The technique of diagonally bracing the ndriwa in both directions allows the structure to withstand the strongest of tremors [59] (Figs. 1–32).

5 Importance and Relevance

Tribal Indigenous Knowledge is zero financial investment disaster management. In this case, it is enough to know about generation-wise indigenous knowledge. If there is documentation about tribal indigenous knowledge, modern people can easily pre-observe natural disasters with local geographical knowledge. As a result, local knowledge is important to mitigate local natural disasters. Geographically, where modern technological knowledge has not been reached, tribal indigenous knowledge

Fig. 1 When a large number of common swallows (Hirundo angolensis) are flying in the sky over Lake Victoria in Kenya, that means the probability of heavy rain is coming [38]

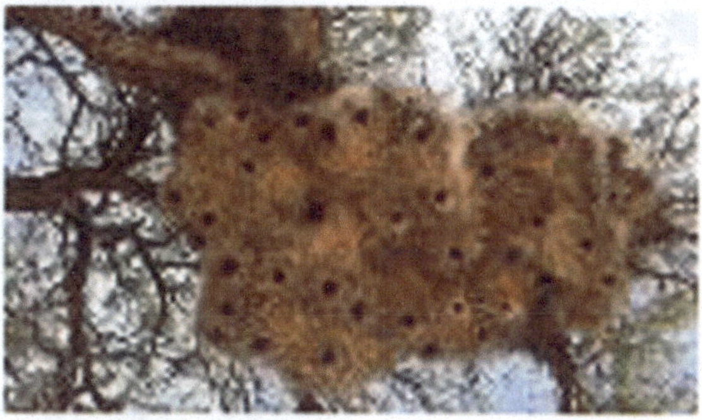

Fig. 2 Here, weaver birds prepared many nests, which means the upcoming weather may be humid [60]

Fig. 3 Pattern of the forming of nests by weaver birds (Nguni or Gema Cocks) in different seasons indicates the coming nature of the season, like if they form a single nest, which means drought weather conditions may come [43]

Fig. 4 Rampuria village in Darjeeling, where a soil conservation project was undertaken by WWF-India with the indigenous knowledge of the local tribal community [61]

Fig. 5 When the bird Ivutavutiya Iilum of Maalwi cries, that means the drought condition may come to that geographical area [62]

Fig. 6 If the leaves of the baobab tree fall suddenly or a huge number of fruits arrive on this tree, that means the drought season may be coming [63]

is useful and can reduce the amount of damage in disaster management. People need a proper formal education to have modern technological knowledge, but the world map has only an 87% global literacy rate [89]. Again, very few people have indigenous or technological knowledge. As a result, all kinds of disaster forecasting and management using technology are not possible (Figs. 33–35 and Table 1).

Fig. 7 The view of the red moon forecasts the flood or drought condition in any geographical area [64]

Fig. 8 Chameleon eggs, indicates the upcoming drought in Ghana [65]

Fig. 9 When snails lay and hatch their eggs during a dry spell, it predicts an impending drought in Ghana [66]

6 Findings

Tribal indigenous knowledge is practiced in tribal geographical territories, which is important for natural disaster management. Sustainable survival is very difficult for modern civilization in respect of natural disasters, hazards, risks, and vulnerability. Whereas, tribes can predict natural disasters in advance through their own indigenous management system, they can forecast the type of natural disasters and reduce

Fig. 10 Red ants climb a tree is the sign of upcoming rainy days [67]

Fig. 11 Squirrels squawk in trees forecasts the irregularity in the local weather [68]

Fig. 12 Bats are flying in large flocks, which means the flood may come in the area [69]

Fig. 13 The unexpected arrival of wild pigs from upstream areas to villages in the Philippines raises concerns about an impending flood disaster [70]

Fig. 14 Red ants are busy transferring their eggs before the upcoming flood [71]

Fig. 15 The absence of yellow toads predicts the upcoming drought season in an area. Again, its presence indicates the changing season [72]

the intensity of damages [92]. The natural disaster cannot be managed completely with modern technology [93], so if tribal indigenous knowledge is connected with technological advancement, then total disaster management will be improved a lot. Tribal lands are isolated, so, tribal knowledge for natural disaster management is scattered. A hybrid disaster management program [94] can be launched by bringing

Fig. 16 Sheora Gachh (Trophis aspera) is used by the Aku community at the side of River Ajoy in Birbhum and Burdwan districts of West Bengal to prepare the house-roof because it is very flexible to resist the nor'westers (Kal Baisakhi) [73]

Fig. 17 Abundance hippo migration from the river banks to villages gives the signal of an upcoming flood in Malawi [74]

Fig. 18 Excessive flowering in 'Mkhundu' (Acacia nigrescens) tree predicts floods in Malawi [75]

Fig. 19 A dark cloud forecasts heavy rain [33]

Fig. 20 Bamboo flowering indicates the upcoming flood disaster [76]

Fig. 21 Before rolling the rocks in landslide-prone areas, doves (Haleso) start chirping [77]

Fig. 22 Himalayan alder is fruitful to protect landslide [78]

Fig. 23 Rhododendron treks is highly connected with the landslide forecasting [79]

Fig. 24 When the leaves of the star apple (caimito) are inverted, they predict the upcoming typhoon [80]

Fig. 25 Cumulonimbus clouds in the sky foretell impending heavy rain [81]

Fig. 26 When the branches of the Gmelina arborea Roxb tree fall down without a strong wing, it indicates that a typhoon is on its way [82]

Fig. 27 Indigenous cyclone shelter, which is prepared by the Vanuatu tribe [83]

Fig. 28 Excessive flowering in tamarind trees (Tamarindus indica) forecasts the upcoming flood in Pakistan [84]

Fig. 29 Santol fruit falling to the ground without any storm or strong wind indicates the upcoming typhoon in the Philippines [85]

Fig. 30 Cassava is a root crop that is used as a post-disaster food security crop in the Philippines after a long period of drought [86]

all this indigenous disaster management knowledge together with modern disaster management. So, reasoning, finding, and comparing with existing studies show that tribal communities are very aware of changes in climate, rivers, oceans, soil, animal behavior (ethology), bird behavior (ornithology), cloud science (nephology), etc. If there are any imbalances or changes here, they can appear quickly. Thus, the main

Fig. 31 If the giant dwarfish (or oarfish) 28 is seen at the site of the coast that forecasts the upcoming earth quake and tsunamis in Japan [87]

Fig. 32 To alleviate drought conditions, the tribal community of the Xinjiang region of seventeenth-century China used a carriage system as part of their irrigation system [88]

background of the research is raised by the question, that is, why, how, and when tribal indigenous knowledge is applicable to disaster type probability, hazard pathway, destruction level reduction, and disaster type prediction. From satellite forecasting or television to radio talk scientific platforms, if natural methods of disaster reduction are applied by common people, the final results will be more fruitful. Recently, based on their age, different climate leaders from 160 countries came together under US Secretary General Antonio Guterres (2020) and elected Archana Soreng of the Santal community. This is evidence that climate leaders are demanding indigenous knowledge to mitigate the climate crisis and related disasters. Concepts of naturalism and traditionalism among tribal communities can be used in disaster management.

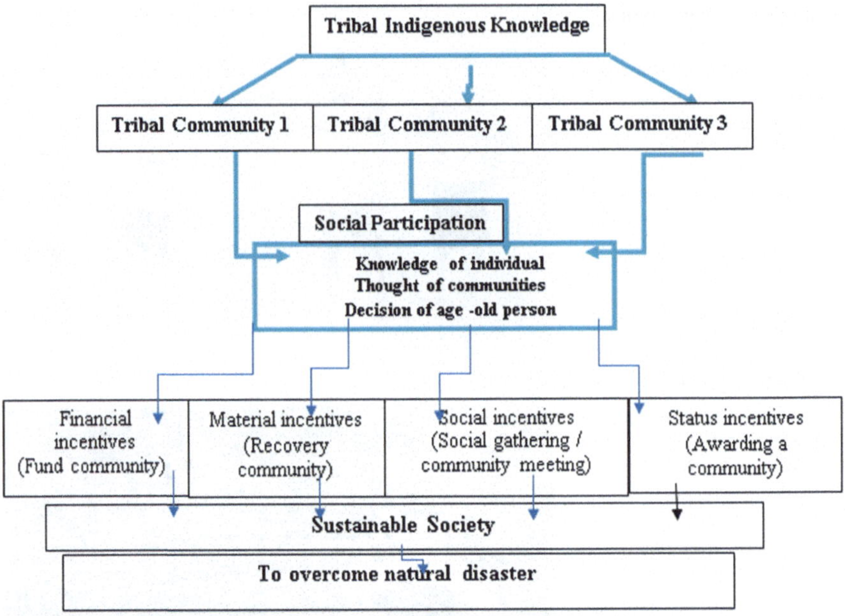

Fig. 33 Overcome the natural disasters by tribal communities to form sustainable society [90]

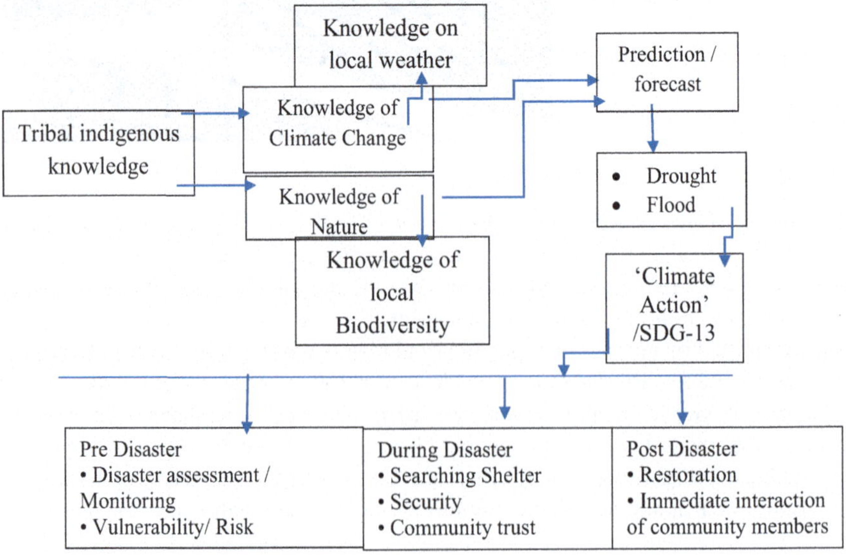

Fig. 34 Relation between pre, during and post natural disasters management by indigenous knowledge of tribal communities [91]

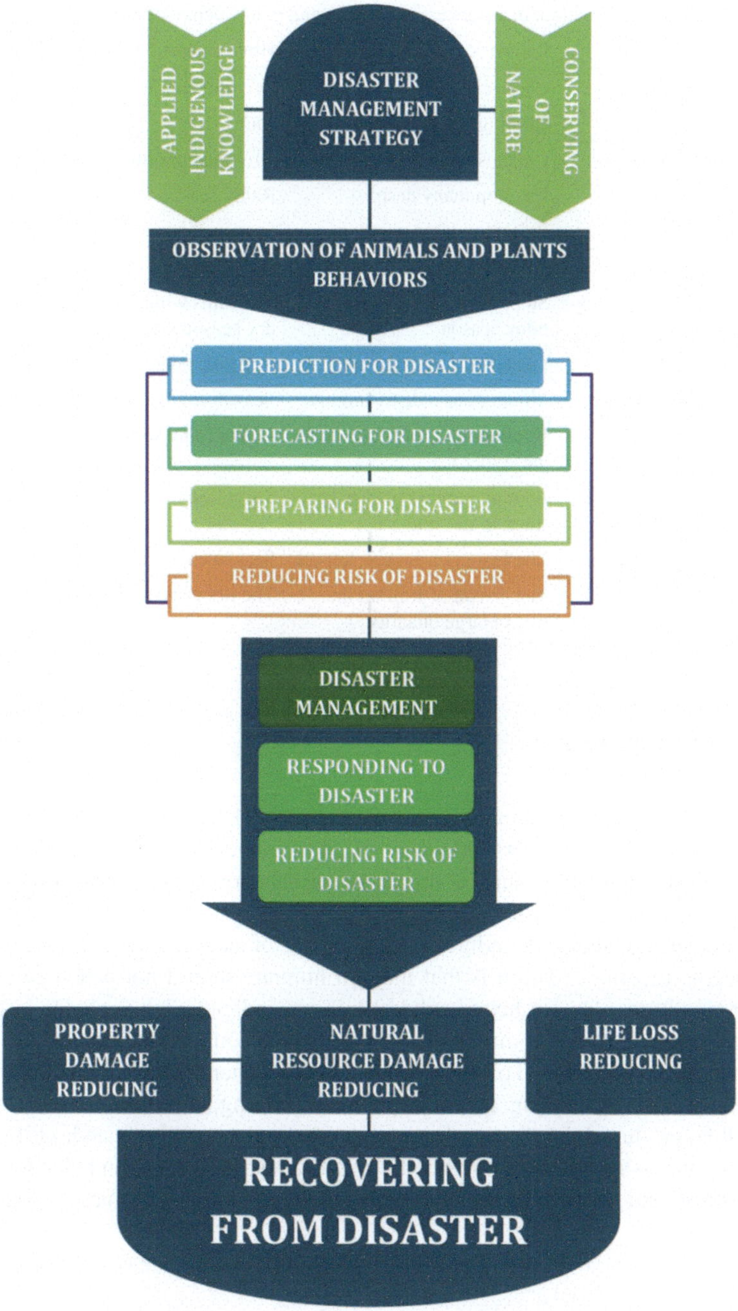

Fig. 35 Applied Indigenous knowledge and disaster management system for local people [91]

Table 1 Forecasting of natural disasters using animals and plants behavior in Kenya

Category	Name	Behavioral indications	Hint of natural disasters
Bird	Swallows	Change the nest	Drought condition
		Circling in the sky	Rainy day
	Weaver	Create single nests	Drought condition
		Develop many nests	Rainy season
Amphibian	Frogs	Present in large numbers	Season change
		Absence	Dry season
		Stopped croaking during the rainy season	The rains would soon end, and the dry season would arrive
	Toads	Present in large numbers	Season change
Insects	Lake flies	Moving one side to another	Season change
	Spiders	If too many flies get stuck in the spider net	Heavy rainy season
		A small number of flies get stuck in the spider net	Short rainy season
Plants	Baobab	The leaves of the tree fall off	Season change and the dry season
		The arrival of fruit on the tree in large quantity	The dry season is coming

Source [91]

7 Recommendations

Indigenous communities should respect indigenous knowledge in dealing with natural disasters. Modern peers should not abandon indigenous and traditional knowledge. Scientists should find out how much science is behind indigenous knowledge in disaster management. Disaster management planners should nurture modern technological knowledge alongside indigenous knowledge of natural disaster management. The specific recommendation is that modern humans should not penetrate primitive tribal cultures. Due to heavy cultural mixing, indigenous knowledge is being lost among various tribal communities of Afro-Asian countries like the Ongi of the Andaman Islands, the Santal of West Bengal, and Jharkhand. Due to climate change, particularly climate warming, the atmosphere is warming, which means more moisture will be produced, increasing the amount of rain during storms and, ultimately, increasing the risk of catastrophic flooding. Thus, destructive human behavior must be examined, and sustainable thinking is important to maintain climate philosophy.

8 Conclusion

Indigenous knowledge is sustainable in managing natural disasters and is sustainable in tribal culture because it is directly linked to nature. The livelihood of tribal culture is useful for saving the earth. Tribes in Afro-Asian countries are providing traditional places to sustain indigenous practices to cope with the risk of natural disasters. It is very clear that the disaster management knowledge of the natives is still alive in this geographical space. And finally, modern disaster policy needs to be integrated with traditional knowledge of disaster management. Integration of indigenous and traditional knowledge philosophies with modern IT-based disaster management policies is now important. Disaster planners are developing disaster management knowledge based on modern technology, but tribal communities are practicing indigenous knowledge of disaster management, where tribal environmental philosophy is finally highlighted. The strengths and limitations of this indigenous knowledge are zero-sum investments in disaster management. In this case, it is enough to know about generational indigenous knowledge. Modern humans can use it to predict natural disasters if they have documentation of indigenous knowledge. Indigenous knowledge involves local geographic knowledge. As a result, local knowledge is important in mitigating local natural disasters. Again, very few people have IT knowledge. As a result, all types of disaster forecasting and management are not possible using technology. Indigenous knowledge is important for natural disaster management. Tribal geographical areas have preserved their indigenous knowledge. But survival is very difficult for modern civilization. Through their own indigenous management systems, tribes can predict and reduce the severity of natural disasters. Indigenous geographies retain their identity through indigenous knowledge. Tribal society cannot control natural disasters completely, while natural disasters cannot be controlled completely with modern technology. It emerged from this study that combining tribal traditional knowledge with technological advancements would greatly improve overall disaster management. Alienation of tribal lands: that is, tribal knowledge for natural disaster management is sparse. A hybrid disaster management program can be launched by combining all this indigenous disaster management knowledge with modern disaster management. The future research areas that could leverage the contributions of this study are to establish the relationship between traditional knowledge of tribal communities and sustainable societies, where they are able to mitigate the most adverse impacts of natural disasters.

Acknowledgements I am very grateful to Dr. Siben Paul for providing the relevant information.

References

1. Yoshita S (2020) Activist Archana Soreng in UN Chief's new youth advisory group on climate change. Environment: the wire. https://thewire.in/environment/un-secretary-general-antonio-guterres-archana-soreng-climate-change
2. BBC (2023) Kedarnath: 'Survivors took refuge in trees—and died of hunger'. https://www.bbc.com/news/world-asia-india-65912680
3. Brianna N (2021) 10 of the deadliest natural disasters in 2021. Extreme weather joined earthquakes and volcano eruptions to claim thousands of lives around the world. U.S. News & World Report L.P. https://www.usnews.com/news/best-countries/slideshows/here-are-10-of-the-deadliest-natural-disasters-in-2021
4. Bushfires in Australia can occur all year-round. For what "season" means, see seasonality of bushfires in Australia
5. Samurović K (2020) Why is Antarctica's snow turning green? Geography realm. https://www.geographyrealm.com/why-is-antarcticas-snow-turning-green/
6. Misra N, Sanya R (2005) Ancient tribe survives tsunami. The Associated Press. https://www.cbsnews.com/news/ancient-tribe-survives-tsunami/
7. Subasinghe C (2021) Global to glocal. In: rebuilding for resilience: a barrier island case. Springer International Publishing, Cham, pp 9–34
8. Paul S (2024) The use of indigenous knowledge and practices for the sustainable natural disaster management among Afro-Asian tribal communities (PhD thesis). Awarded in department of geography, Raiganj University under the supervision of Dr. Tapas Pal
9. Bali Swain R, Yang-Wallentin F (2020) Achieving sustainable development goals: predicaments and strategies. Int J Sust Dev World 27(2):96–106
10. Filho Leal W, Wall T, Azul AM, Brandli L, Özuyar PG (2020) Good health and well-being, Springer Science and Business Media LLC
11. Wongbusarakum S, Loper C (2011) Indicators to assess community level social vulnerability to climate change: an addendum to SocMon and SEM-Pasifika regional socioeconomic monitoring guidelines. National oceanic and atmospheric administration (NOAA) and Apia, Samoa: secretariat of the Pacific regional environmental programme (SPREP). http://socmon.org
12. Dube MA, Musi PJ (2002) Analysis of indigenous knowledge in Swaziland: implications for sustainable agricultural development, working paper series no. 34, African technology policy studies network, Nairobi, Kenya
13. Ngara R, Mangizvo RV (2013) Indigenous knowledge systems and the conservation of natural resources in the Shangwe community in Gokwe district, Zimbabwe. Int J Asian Soc Sci 3(1):20–28
14. Howell P (2001) Knowledge is power? Obstacles to disaster preparedness on the coastal chars of Bangladesh, humanitarian exchange, vol 18. Humanitarian Practice Network, London
15. Strong likelihood of big earthquake in Himalayan region, need to prepare better: Scientists SECTIONS (2022) Strong likelihood of big earthquake in Himalayan region, need to prepare better: scientists PTI
16. Thompson TJ (1995) Christianity in Northern Malaŵi: donald fraser's missionary methods and Ngoni culture. Leiden, Netherlands, Brill, pp 1–29. ISBN 978-90-04-31996-7
17. Guerin B, Janta B, van Gorp A (2018) Desk-based research and literature review. Eval Interv Prev Count Violent Extrem 63:RR-2094
18. Lumsden J (2007) Online-questionnaire design guidelines. In: Reynolds A, Woods R, Baker J (eds) Handbook of research on electronic surveys and measurements. Idea Group Reference, London, pp 44–64. https://doi.org/10.4018/978-1-59140-792-8.ch005
19. de la Fuente Á, Doménech R (2021) Cross-country data on skills and the quality of schooling: a selective survey. BBVA research and Universidad de Valencia, BBVA research
20. Dube S. Open-ended questions vs. closed-ended questions in user research, INVESP. https://www.invespcro.com/blog/writing-open-ended-and-closed-ended-questions-for-user-research/
21. Xaxa V (1999) Tribes as indigenous people of India. Econ Polit Wkly 34(51):3589–3595

22. Daes EIA (2004) The impact of globalization on indigenous intellectual property and cultures. Australian Human Rights Commission
23. Panda GK, Chatterjee U, Bandyopadhyay N (2023) Indigenous knowledge and disaster risk reduction. Springer Science and Business Media LLC
24. Wang Y, Zhou J, Guo X (2008) Present situation and its development of the Karez Well in the Xinjiang. Ground Water 30(6):49–52
25. DMCN (2004) Traditional indicators are used for climate monitoring and prediction by some rural communities in Kenya. A contribution to the integration of traditional and modern scientific climate prediction methods in Kenya. Drought monitoring center, Nairobi, Pilot application project
26. Jha V, Jha A (2011) Traditional knowledge on disaster management: a preliminary study of the Lepcha community of Sikkim, India. Indian Knowl Tradit Knowl 10(1):173
27. Ngwese NM (2018) Traditional and local knowledge practices for disaster risk reduction in northern Ghana. Sustainability 10:825. https://doi.org/10.3390/su10030825
28. Zubeesorganickitchen. Benefits of Dawa Dawa plant. https://zubeesorganickitchen.com/benefits-of-dawadawa/
29. Acharya A, Prakash A (2019) When the river talks to its people: local knowledgebased flood forecasting in Gandak River basin, India. Environmental development
30. Senthilkumar K, Gobalakrishna C (2016) Socio-cultural Prole of Irular Community in Tamil Nadu. Indian J Appl Res 6(11). ISSN–2249–555X, IF: 3.919, IC Value: 74.5
31. Debnath MK (2011) Living on the edge: the predicament of a rural indigenous Santal Community in Bangladesh (Doctor of Education thesis)
32. Dlamini BT (2003) Indigenous knowledge systems: what do young scientists in Swaziland make of their relationship with school science? In: Putsoa B, Dlamini M, Kelly V (eds) Proceedings of the 11th annual SAARMSTE conference held in waterford Kamhlaba, UWC, University of Swaziland, pp 589–592
33. WIKIMEDIA COMMONS. Acacia nigrescens, bloeiwyse, Pretoria NBT.jpg. https://commons.wikimedia.org/wiki/File:Acacia_nigrescens,_bloeiwyse,_Pretoria_NBT.jpg
34. Dube E, Munsaka E (2018) The contribution of indigenous knowledge of disaster risk reduction activities in Zimbabwe: a big call to practitioners. J Disaster Risk Stud. Courtesy of AOSIS: Jamba from https://www.ncbi.nlm.nih.gov
35. Matowanyika J (1994) What are the issues with indigenous knowledge systems in southern Africa? In: indigenous knowledge systems and natural resource management in southern Africa. Report of the Southern Africa Regional Workshop, Harare, Zimbabwe, IUCN-ROSA, Zimbabwe
36. Dlamini D (2009) The management of indigenous knowledge initiatives in Swaziland. University of KwaZulu-Natal (Thesis, Information Science), Pietermaritzburg, p 119
37. Bordoloi R, Muhaddadi AU (2015) Indigenous technical knowledge is associated with disaster management and fisheries-related activities in the highest flood effected district (Dhemaji) of Assam, India. Indian J Tradit Knowl 14(3):407–415
38. Rahman A, Sakurai A, Munadi K (2016) Indigenous knowledge management to enhance community resilience to tsunami risk: lessons learned from Smong traditions in Simeulue Island, Indonesia. IOP Conf Ser Earth Environ Sci 56. https://doi.org/10.1088/1755-1315/56/1/012018
39. Poffenberger M (2015) Restoring and Conserving khasi forests: a community-based REDD strategy from northeast India. Community Forestry International, 1620 Gasquet Flat Rd, Gasquet, CA 95543, USA
40. Reyes J, Ayo K, Baluyan MPM, Balaguer A (2020) Indigenous knowledge in disaster risk reduction: the tales of three islands (San Miguel, Camotes, and Alabat) in the Philippines. https://doi.org/10.22439/cjas.v37i1.5908
41. Reyes JAL, Ayo KAR, Baluyan MPM, Balaguer ARB (2019) Indigenous knowledge in disaster risk reduction: the tales of three islands (San Miguel, Camotes, and Alabat) in the Philippines. Cph J Asian Stud. https://rauli.cbs.dk
42. Disaster relief–Mayon volcano eruption (2001) World Animal Protection, OCHA. https://reliefweb.int/report/philippines/disaster-relief-mayon-volcano-eruption

43. Chirp!. Doves. https://www.chirpmagazine.online/dovesVojoPhu
44. Traube G, Andrew McWilliam E (2011) Land and life in timor-leste: ethnographic essays. https://doi.org/10.26530/OAPEN_459352. ISBN: 9781921862601
45. Hiwasaki L, Luna E, Syamsidik, Shaw R (2014) Local and indigenous knowledge for community resilience: hydro-meteorological disaster risk reduction and climate change adaptation in coastal and small island communities. UNESCO Office in Jakarta. ISBN:978-602-9416-11-4
46. Misra N (2005) Stone age cultures survive tsunami waves. The associated press, NBC NEWS. https://www.nbcnews.com/id/wbna6786476
47. Behera, Kumar J (2018) Tsunami, tribes and media. The Tribal Tribune 1(5). https://www.etribaltribune.com/index.php/volume-1/mv1i5/tsunami-tribes-media. ISSN: 2249 3433
48. Bhaumik S (2005) Tsunami folklore 'saved islanders'. The BBC's Subir Bhaumik. BBC News. http://news.bbc.co.uk/2/hi/south_asia/4181855.stm
49. Rai P, Khawas V (2019) Traditional knowledge system in disaster risk reduction: exploration, acknowledgement and proposition, Jàmbá J Disaster Risk Stud 11(1):a484. https://doi.org/10.4102/jamba.v11i1.484
50. Hiwasaki L, Luna E, Syamsidik, Shaw R (2014) Local and indigenous knowledge for community resilience: hydro-meteorological disaster risk reduction and climate change adaptation in coastal and small island communities. Jakarta, UNESCO, pp 60
51. Reyes JAL, Ayo KAR, Baluyan MPM, Balaguer ASRB (2019) indigenous knowledge in disaster risk reduction: the tales of three islands (San Miguel, Camotes and Alabat) in the Philippines. Cph J Asian Stud 37(1):103–132
52. Mercer J (2010) Framework for integrating indigenous and scientific knowledge for disaster risk reduction. Disasters, Wiley Online Library
53. Sander H (2021) Impact based forecasting for typhoons in the Philippines
54. Source(s) Anticipation Hub. https://www.preventionweb.net/news/impact-based-forecasting-typhoons-philippines
55. Faquire ABMRK (2022) The mystery behind creation of Chakma speech variety. The financial express. Accessed 18 Feb 2022
56. Howell P (2003) Indigenous early warning indicators of cyclones: potential application in coastal Bangladesh, IGAD
57. Gabrie T, Anish S, Ziwei C, Caroline L, Jennie K, Julie K, Lopez-Carmen VA, Brendan E. The experiences and challenges of Tripura indigenous youth in Chittagong Hill Tracts, Bangladesh, during COVID-19. In: The global indigenous youth caucus, United Nations
58. UNDRR (2015) Local knowledge saves lives in Vanuatu. United nations office for disaster risk reduction–regional office for Asia and Pacific. United nations office for disaster risk reduction–sub-regional office for the Pacific
59. Reyes J, Ayo K, Baluyan M, PM, Balaguer A (2019) Indigenous knowledge in disaster risk reduction: the tales of three islands (San Miguel, Camotes, and Alabat) in the Philippines. Cph J Asian Stud 37(1):103–132. https://doi.org/10.22439/cjas.v37i1.5908
60. Gpsastry. Konga Japam (2014). http://gpsastry.blogspot.com/2014/01/konga-japam.html
61. Peter M (eds) (2008) Indigenous knowledge in disaster management in Africa. UNEP (United Nations Environment Programme), Nairobi, Kenya. P.O. Box 30552
62. WWF-India/Pemba TshBhutia
63. Gpsastry. KongaJapam (2014). http://gpsastry.blogspot.com/2014/01/konga-japam.html
64. WorldAtlas. The species of baobab trees. https://www.worldatlas.com/articles/the-species-of-baobab-trees.htm
65. Shutterstock. Red full moon. https://www.shutterstock.com/image-photo/red-full-moon-669863983
66. Shutterstock. Chameleon eggs. https://www.shutterstock.com/search/chameleon-eggs
67. Shutterstock. Snail eggs. https://www.shutterstock.com/search/snail-eggs
68. Shutterstock. Ants climb a tree. https://www.shutterstock.com/search/ants-climb-a-tree
69. Silbajoris A (2018) Why do squirrels squawk in trees? Sciencing. https://sciencing.com/squirrels-squawk-tree-10057973.html
70. National Park Service. Bat flight program. https://www.nps.gov/thingstodo/bat-flight-program.htm

71. THEWEEK (2021) Wild pigs release the same emissions as 1 million cars each year. https://www.theweek.in/news/sci-tech/2021/07/22/wild-pigs-release-the-same-emissions-as-1-million-cars-each-year.html
72. Arevalo HA (2007) Featured creatures entomology and nematology. University of Florida. https://entnemdept.ufl.edu/creatures/urban/ants/myrmica_ruba.htm
73. Amphibian Fact. Yellow-bellied toad. https://www.amphibianfact.com/yellow-bellied-toad.asp
74. I See Flora (2012) Sheora Gachh-trophis aspera. https://icflora.blogspot.com/2012/01/sheora-gachh-trophis-aspera.html
75. Tasoff H (2019) The Hippo's hidden world. UC Santa Barbara. https://www.news.ucsb.edu/2019/019687/hippo-s-hidden-world
76. Shutterstock. Dark cloud. https://www.shutterstock.com/search/black-clouds
77. Bamboo batu the house of bamboo. Bamboo flowering fully. https://bambubatu.com/bamboo-flowering-a-botanical-phenomenon/bamboo-flowering-fully/Wildboar
78. Chirp! Doves. https://www.chirpmagazine.online/dovesVojoPhu
79. Don D. Alnus nepalensis. Himalayan alder. https://observation.org/photos/25269422/
80. Chatrapathy S (2022) Top 4 Rhododendron treks in the Indian himalayas. https://indiahikes.com/top-rhododendron-treks-in-india-himalayas
81. Shutterstock. Leaves of star apple isolated on white background. https://www.shutterstock.com/image-photo/leaves-star-apple-isolated-on-white-550772326
82. What's this cloud. Cumulonimbus clouds: thunderstorms. https://whatsthiscloud.com/cloud-types/cumulonimbus/AtomPasie
83. Khan M (2022) Gmelina arborea Roxb. Forestrypedia. https://forestrypedia.com/gmelina-arborea-roxb-%EF%BF%BC/
84. RNZI Courtesy of Care International (2016) Disaster prep seen to have paid off in Vanuatu. https://www.rnz.co.nz/international/pacific-news/315553/disaster-prep-seen-to-have-paid-off-in-vanuatu
85. Tamarind tree (Tamarindus indica) with pods. https://en.wikipedia.org/wiki/Tamarind#/media/File:Tamarindus_indica_pods.JPG
86. Freepik. Santol fruit on summer in Thailand ripe santol on the santol tree tropical fruit. https://www.freepik.com/premium-photo/santol-fruit-summer-thailand-ripe-santol-santol-tree-tropical-fruit_28440939.htm
87. Grant B, Cassava L. Plant care—information on how to grow cassavas. https://www.gardeningknowhow.com/edible/vegetables/cassava/growing-cassava-yuca.htm
88. Our Breathing Planet. Giant Oarfish. https://www.ourbreathingplanet.com/giant-oarfish/7029711292 anindita naskar
89. Sun Q, Tao W, Kobori I, Liang L (2009) Case study V: Karez in the Turpan region of China, seeing traditional technologies in a new light: using traditional approaches for water management in drylands. In: Bigas H, Adeel Z, Schuster B (ed). Paris, UNESCO, pp 12–13
90. Buchholz K (2022) This is how much the global literacy rate grew over 200 years. Statista. World Economic Forum
91. Paul S (2004) The use of indigenous knowledge and practices for the sustainable natural disaster management among afro-asian tribal communities (PhD thesis). Awarded in department of geography, Raiganj University under the supervision of Dr. Tapas Pal
92. Peter M (eds) (2008) Indigenous knowledge in disaster management in Africa. UNEP (United Nations Environment Programme), Nairobi, Kenya. P.O. Box 30552
93. Stigter CJ, Dawei Z, Onyewotu LOZ, Xurong M (May) Using traditional methods and indigenous technologies for coping with climate variability, Springer Nature 70:255–271
94. Krichen M, Abdalzaher MS, Elwekeil M, Fouda MM (2024) Managing natural disasters: An analysis of technological advancements, opportunities, and challenges. Internet Things Cyber Phys Syst 4:99–109
95. Asghar S, Alahakoon A, Churilov L (2004) A hybrid decision support system model for disaster management. In: IEEE fourth international conference on hybrid intelligent systems (HIS'04). https://doi.org/10.1109/ICHIS.2004.5. ISBN:0-7695-2291-2

Resilience Starts Locally: Enhancing Sustainability Through Localization in Humanitarian Response

Aniruddha Dey, Sheikh Tawhidul Islam, Swarnabha Bandyopadhaya, Nandini Sanyal, Al Jubaer, and Dipanjan Mishra

Abstract The paper addresses the necessity of localization in areas of humanitarian response for communities in disaster-prone areas, emphasizing the need to empower local youths and first responders through need-based, targeted training. It examines frameworks provided by UNISDR [13] and ActionAid [1] and presents a case study, where CARN-AP and its partner organizations conducted training sessions in Bangla language to equip participants with essential skills relating to humanitarian response. In-person training at capacity building training took place in March 2024 in India that included practical exercises using tools like Kobotoolbox and culminated in an exposure visit for field data collection. The effectiveness of this training was demonstrated when local youths successfully conducted a Joint Rapid Needs Assessment (JRNA) after Cyclone Remal in May 2024. Despite these successes, the paper highlights persistent challenges in fully embracing localization, including the lack of recognition and resources for local volunteers. Ultimately, the study underscores the critical role of localized training in building resilient communities and addressing global challenges through local empowerment.

A. Dey (✉) · D. Mishra
PRISM, Kolkata, India
e-mail: deyaniruddha@gmail.com

A. Dey
CARN-AP, Kolkata, India

S. T. Islam
Professor, Regional Centre for Space Science and Technology Education in Asia and the Pacific, International Innovation Institute, Beihang University, Hangzhou, China
e-mail: stislam@buaa.edu.cn

S. Bandyopadhaya
Prayukti International, Kolkata, India

A. Jubaer
Honorary Research Associate, Institute of Remote Sensing and GIS, Jahangirnagar University, Savar, Bangladesh

N. Sanyal
Science Connect Ltd, Dhaka, Bangladesh

I. Pal et al. (eds.), *Proceedings of the 3rd International Symposium on Disaster Resilience and Sustainable Development—Volume 1*, Lecture Notes in Civil Engineering 623, https://doi.org/10.1007/978-981-96-2139-2_13

Keywords Disaster risk reduction · Climate change adaptation · Localization · Disaster risk management · Humanitarian response · Resilience · Sustainability

1 Introduction

The communities living in disaster and climate hotspots suffer from various kinds of challenges occurring at various scales (e.g., every day challenges, accidents as small events, and emergencies disaster impacts including, catastrophes as large scale events). These disturbances cause to destroy their primary asset base and leave them into long term consequential and cumulative impact conditions. Sometimes new episodes of disasters strike on them while they are still struggling to recover from the earlier disaster impacts. In such conditions, state agencies and external partners like NGOs, donor agencies come forward with different support provisions for the communities to improve their conditions. In most of the conditions the administrators, professionals and experts working in the disaster effected fields (e.g., post disaster needs assessment, relief and rescue operations, community mobilization for rebuilding the resources etc.) are not local, rather they come from other different parts of the country or foreign countries with little or no prior knowledge on the local environmental conditions and socio-economic systems [10]. These external support services, in many instances, raise expectations among the local people to receive resources and create a dependency syndrome, which undermines local resilience based on their own strength [11]. Local people, especially the youths remain to be neglected, underutilized in these processes. Thus the local needs assessment results appeared to be noncomprehensive, non-contextual, fragmented and misleading due to the disconnection of the local people, especially the young members of the community. In contrast, the incorporation of local communities, by providing need-based capacity building training, awareness raising and motivations, may help disaster risk reduction actions more effective, efficient, sustainable and impactful [12]. This paper, in this background context, discusses how local communities can play better roles in disaster risk reduction aspects as well as humanitarian actions. At the outset, conceptual discussions are given to locate the niche of localization within broad disaster risk management concept. Thereafter a case study is discussed to show how localization is effectively promoted in both India and Bangladesh contexts in reducing disaster impacts. The learning from the case study may help to replicate and promote localization processes in other parts of the world with necessary modifications to fit the process within their contexts.

2 Localization: Rationale for Bring the Issue Forward

Localization is a dynamic process aimed at empowering local communities by equipping them with the knowledge, technical skills, leadership building skills and a strategic mindset to harness local resources collectively [6, 8]. This approach serves a dual purpose: preventing and mitigating disasters while enabling proactive engagement in disaster response efforts. Through localization, communities shift from passive recipients of relief and support to active agents of societal change, ultimately bolstering their capacity to withstand and rebound from disasters. However, despite substantial investments and institutional initiatives in areas prone to both rapid and slow-onset disaster impacts, communities continue to endure with recurrent challenges [7]. This raises questions about the efficacy of institutional efforts, particularly at the local level. Our firsthand, field-based experiences suggest that the localization process has often been misunderstood and poorly institutionalized and that eventually resulted in the failure to transform local communities into catalysts for change within their own societies. Furthermore, externally-driven institutional endeavors, such as those led by non-governmental organizations, have frequently overlooked the imperative of fostering genuine community participation in disaster risk reduction. Consequently, numerous interventions in sectors like water, sanitation, hygiene, income generation, health, and education are found to be less effective. Once project timelines conclude, resource mobilization ceases tho results in these interventions become abandoned and non-operational. The external initiatives have contributed, in various means, in the emergence of new local elites who amass wealth and influence, while the broader populace remains largely unchanged [14]. The gap hinders the ability of the local populace to respond efficiently to disasters and communicate critical information to affected communities.

The aim to promote localization process which is presented here as a case study is to make rationale for engagement of local communities, mainly the younger generation who come from disaster hot-spots and economically poor regions. Capacity development of the youths is needed in this connection so that they can understand the broader perspectives of disaster risk management issues and can fit their actions within the broader framework. The capacity development activities were performed in three areas–firstly, to develop knowledge of the youths on how to frame questions on issues of local concerns, how to develop research concepts and questions to assess local needs in emergencies and humanitarian contexts, connecting the root cause with the consequences using a system thinking approach and report writing. Secondly, training was provided to develop understanding on spatial and non-spatial data. Knowledge on digital data collection process can be provided to the local youths using no-cost, open source, easily usable online resources like Kobotoolbox for socio economic survey, open-source GIS (e.g., QGIS) and GPS (Global Positioning System) for preparing geo-spatial database relating to local disaster management issues and resilience building. Third, training was provided to develop knowledge and understanding on First Aid including CPR (Cardio Pulmonary Resuscitation) and Rescue operations.

3 Disaster Resilience and Its Link with Localization

The concept of resilience has its roots in a number of disciplines, including engineering, psychology and ecology. The original meaning was largely constructed in the field of ecology and was understood as a measure of the ability of ecological systems to persist in the face of disturbance and maintain relationships between different elements of the system; this idea has been recently adapted by many other disciplines like disaster risk management. According to United Nations International Strategy on Disaster Reduction (UNISDR) terminology on disaster risk reduction (2009)–"Resilience means the ability of a system, community or society exposed to hazards to resist, absorb, accommodate to and recover from the effects of a hazard in a timely and efficient manner, including through the preservation and restoration of its essential basic structures and functions". This definition emphasizes the two periods "before the disaster" and "during times of the disaster" in a way that introduces resilience as the ability to "resile from" or "spring back from" a shock [13].

Resilience building calls for an integrated approach and that encapsulates and goes beyond the DRR (Disaster risk reduction) and CCA (Climate Change Adaptation) concepts. It also takes into consideration of regular development actions including the social safety net programs used for targeted vulnerable communities. 'Resilience' thinking advocates for 'transformation, where necessary' in order to make the systems flexible enough to absorb shocks and stresses. However, the most challenging parts of 'resilience' concept is to define strategy to operationalize it on ground since it cuts across development and humanitarian works and also overlaps with a wide range of existing conceptual frameworks such as SDGs. The wide variations in the power structure of the society (poor and rich; young and old), the contextual settings (e.g. urban, rural), gender relations (women and men; boys and girls) also make the 'resilience' concept difficult to operationalize. ActionAid's [1] definition of 'resilience' is cross-cutting and works across rural and urban contexts. They recommended three elements into it—these are (i) absorptive capacity, (ii) adaptive capacity and (iii) transformative capacity of communities irrespective of their living conditions whether it is in the rural or urban contexts. However, ActionAid's 'resilience' definition read "the ability of people to recognize, challenge and transform the unjust and unequal power relations that dictate their vulnerability, to adapt positively to changing circumstances, and to mitigate, prepare for and rapidly recover from shocks and stresses such that their wellbeing and enjoyment of human rights is safeguarded" [1].

The discussion above suggests that the activities relating to prevention, preparedness, response and recovery from disasters are primarily local phenomenon. These are better understood by local people since they are the people who suffered from the impacts. They know the problem contexts like local terrains, fluvial and climatic processes, local drainage systems much better than the people coming from other regions. They know profoundly the importance of physical assets and other resources for generating incomes in order to secure their livelihoods. The local people are deeply familiar with socio-cultural contexts/elements, local governance and political aspects. These familiarity and knowledge are crucial for successful implementation of interventions. The definition provided by [13] and ActionAid [1] on resilience framework are thus strongly aligned with the need for promoting local people in disaster risk management and resilience building processes.

4 Localization in Humanitarian Response: Elements and Structural Framework

Localization is a process-led function where different actions are performed step-by-step. The first element in the process that may end-up recognizing the fact that local community members and the youth volunteers can perform better in local resilience building process if they are provided with need-based training and resources. This process includes arranging physical or online meetings or/and workshops where local capacity needs are valued and capacity needs of the local community members are assessed. This assessment helps to define the scope of the thematic areas where the capacity building trainings will be provided and resources to be mobilized. These actions will help to attain the first outcome (Outcome 1—Local Capacity Enhanced, Fig. 1). Once the capacity development actions are performed, the local resource persons and the volunteers can perform a number of activities and results will be generated. This will also demonstrate the capacities of the local volunteers. At this stage the second outcome will be developed (Outcome 2) which reads 'local problems are addressed and people can build-back-better' the adversities caused from disaster impacts. The attainment of Outcome 2 is the significant milepost for the localization process where local youths are expected to contribute in the long term recovery processes (post disaster) and also to perform other actions towards attaining a number of national and international (e.g., SDG) targets.

5 Localization: Lessons from India-Bangladesh Case Study

The sections above discussed about the contextual backgrounds of localization process including the DRR and Resilience framework provided by [13] and ActionAid [1] respectively. Sections above 4 provided the conceptual structure on how the localization process should be taken forward towards building community resilience in the contexts of humanitarian response. The process was systematically exercised with an aim to examine the effectiveness of the process by involving a number of organizations and local participants from India and Bangladesh. The process is described in the following sections to describe its robustness and also the limitations (Fig. 2).

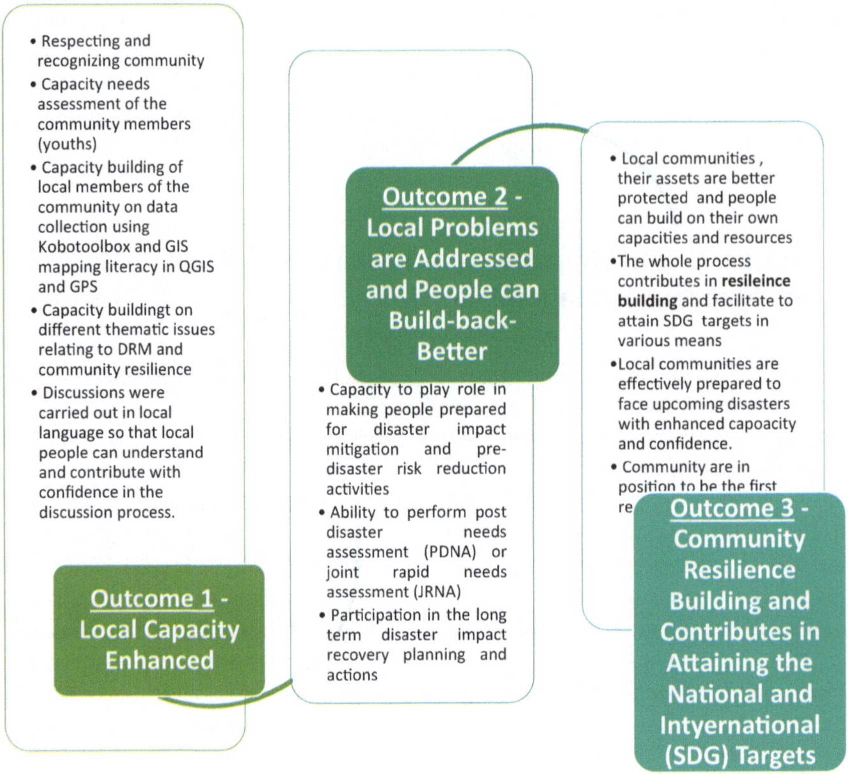

Fig. 1 Structural framework of localization process

The network organization called CARN-AP (Climate Action Resource Network-Asia Pacific, https://carn-ap.net/) in collaboration with Sphere India (https://www.sphereindia.org.in/), Climate Action Network-South Asia (CANSA, https://cansouthasia.net/), State IAG-West Bengal, PRISM and Rural Development Forum of the Institution of Engineers (India), Science Connect Ltd. (from Bangladesh) organized a number of online sessions to discuss the need of promotion of localization in the fields of humanitarian response. The initial meetings decided to arrange twelve thematic training sessions on disaster risk management aspects for local young people. The sessions were organized in Bangla language titled "Building Resilience Together: Empowering First Responders in Regional Languages" held from 20 July, 2023 to 04 October, 2023. The aim was to assess the needs and scope of capacity building training activities in order to abridge the existing capacity gaps. The sessions helped to empower the local youths and the first responders with the necessary knowledge and technical skills (especially in open GIS such as QGIS and GPS) to effectively respond to disasters and communicate vital information to communities and also to disaster management agencies working in different tires of disaster risk management. The twelve sessions helped in sensitizing the local

Fig. 2 Capacity building training in the localization workshop (upper left and upper right), participants in the fieldwork (lower left), and trained local youths in the Sundarban mangrove forest areas collecting loss and damage data after the strike of Cyclone Remal on May 26, 2024

young people and about one thousand youths from different local areas from India and Bangladesh participated in the sessions. Discussions with the participants during the training sessions informed that online sessions are good but not very effective. They requested to arrange in-person training for durations of 5 or 7 days. In response, CARN-AP and its partner organizations organized in-house capacity development training programs held from 12 March to 16 March, 2024, at the Uttaran campus of PRISM in Madarihat Panchayat, Alipurduar, West Bengal, India. A total of thirty participants from various districts of West Bengal and Bangladesh attended the in-house training program. Among them fifty seven percent participants were female. The titles of twelve thematic sessions were (i) Humanitarian charter, protection principles and core humanitarian standards; (ii) Tiers of disaster response mechanisms of governments; (iii) Shelter and settlement management; (iv) Family level disaster preparedness; (v) Child protection in emergency and the need to create child friendly space; (vi) Roles of different task forces in disaster risk management; (vii) Water, sanitation and hygiene promotion; (viii) Livelihood development and income augmentation; (ix) Biodiversity conservation and ecosystem services in sustainable disaster risk management; (x) Psycho-social counselling; (xi) Climate smart solutions in the resilience building process; (xii) Geospatial data collection, processing and map making using QGIS, GPS and Kobotoolbox and related innovations in disaster risk management.

In the in-person training program on the use of Kobotoolbox was performed so that the trainees can gather spatially referenced data with latitude and longitude values from the field. Basic training on GIS (used QGIS) and GPS (Global Positioning System) were provided, as indicated before. The sessions terminated everyday with question-and-answer component with the help of Kobotoolbox. On the 4th day, the entire team went for an exposure visit. During the visit, the participants were divided into theme-based groups and sent to different locations of Totopara tribal village in Alipurduar District of West Bengal, India. The team members had discussions with local villagers about their problems, got answers and inserted the response in the Kobotoolbox. On the 5th day based on the facts and their own field based observations each team wrote reports, containing three thousand words using a report template prepared beforehand by themselves. The report had sections such as introduction, methodology, result and analysis and finally recommendations. The findings were presented by the team members during the certificate distribution ceremony. A person was assigned to write in detail the experiences and processes using a pre-defined paper framework so that enough materials are amassed which can be converted into a paper to be published in the peer reviewed academic journal and this article is prepared based on that initial draft. The total process was also captured in audio-visual systems both in Bangla and English languages so that a short documentary film on the whole process can be made for showing the usefulness and step-by-step process of localization to the people living in other parts of South Asia. The video documentary can be accessed through social media channel Youtube at https://youtu.be/M_zz0UGr2_E?si=AF6g7rdPVmxXuc48. It is imperative to mention that experts on different thematic topics from academic institutions and professional agencies in India, Bangladesh, United States, United Kingdom, Sri Lanka and Canada joined online and gave lectures in related respective the sessions. Two months after the training, a cyclone named Remal struck on 26th of May 2024) in the coasts of Bay of Bengal in India and Bangladesh caused severe damages of assets and properties of the communities. In such contexts, the local youths who received training played useful roles by gathering loss and damage data of the impacts by using the Kobotoolbox installed in their mobile phones and later analzed using QGIS for making map and other results. Based on the maps and data generated the Joint Rapid Needs Assessment (JRNA) was developed by the local youths who are familiar with the local contexts. This report was accessed and used by the National Institute of Disaster management in India (NIDM), SPHERE India, UNICEF (India), State Inter Agency Group -West Bengal (IAG-WB) which testimonies the success of the capacity building process of local youths through localization framework (Table 1).

Limitations of the capacity building training on localization: Governments and other organizations often fail to appropriately embrace the concept of localization. As a result, the efforts of local volunteers frequently go unrecognized and underappreciated. Furthermore, funding agencies are often hesitant to fulfill their commitments, which exacerbates the issue. Consequently, local volunteers are left without the necessary resources to effectively participate in a variety of disaster risk reduction activities.

Table 1 Summary activities of capacity building training workshop

Focal theme of the training	Learning Outcomes	Recommendations
- Empowering local communities by equipping them with the knowledge, technical skills, and strategic mindset to collectively harness local resources - Transforming local communities into catalysts for change within their own societies - Addressing global challenges necessitates tackling local issues first, with active involvement of local communities, resulting in increased resilience of the at-risk societies. While regional, national, and international experts and resources can play supportive roles, the heart of these efforts must always be grounded in the actions and agency of local young people, communities, and organizations - The sessions helped participants acquire knowledge and technical skills to effectively respond to disasters and communicate vital information to communities	Day 1: On the first day of training 30 participants were divided into three teams. In the morning session they were introduced with the functions of climate-smart solutions. During the evening session they learned to use the Kobotoolbox application Day 2: Participants focused on learning the fundamentals of First Aid, practicing CPR (Cardiopulmonary Resuscitation) and engaging in search and rescue related activities Day 3: Activities from Days 1 and 2 were consolidated and practiced. Participants also learned how to create a hazard map to identify vulnerable areas using QGIS, GPS installed in their mobile phones Day 4: The entire team embarked on an exposure visit. Split into theme-based groups across various locations; they interacted with villagers to assess gaps and gather information. Using Kobotoolbox and GPS technologies, they documented critical insights and captured photographic images Day 5: Certificate giving ceremony took place where the participants presented their experiences, technical skills gained, motivation received and initial findings	The conceptual framework of localization including the operational procedure to implement it have been demonstrated in this article. The results of the whole exercise seem to be highly promising and hence recommended to promote and adopt localization where the actions are locally related (e.g., impact assessment or project implementation etc.). - Any survey conducted in disaster-prone areas should involve the widespread deployment of tools like Kobotoolbox Toolbox, such as JRNA or PDNA - Organizations should strategically plan initiatives to enhance public awareness of first aid principles, primary prevention, and rescue procedures - These programs will be more effective and sustainable when local communities are actively engaged because they are often the first responders. Therefore, organizations and governments should promote the concept of localization

6 Concluding Remarks: The Implications of Localization and Global Imperatives

The process of localization emerges as an indispensable strategy for building resilient communities, with a particular focus on empowering youth as active agents of change [5]. The work shows how the concept of localization was developed from the scratch and then divided the process into a number of programmatic components so that the concept of localization can be taken forwards for application. The step-by-step localization process begins by providing local communities capacity building trainings on the tools and methods so that they can take actions with necessary ownership on disaster risk reduction, participatory planning, disaster preparedness and mitigation efforts. The results show that the trainings enabled them to craft locally grounded solutions that resonate with their unique needs and cultural appropriateness, fostering social acceptance and a sense of ownership [4]. By actively involving local communities, the local initiatives gain authenticity and sustainability, reducing the likelihood of interventions becoming unattended and non-functional once external support concludes. Furthermore, in an era marked by global challenges like environmental pollution, climate change, energy crises and pandemics, localization proves its worth by recognizing that these issues often transcend local origins. Local communities in these cases remain central in coping with these challenges, both in the short and long term. Ultimately, the importance of localization lies in its potential to address global challenges through the empowerment of local communities. By embracing localization, communities can take the reins of their onus, making resilience a shared endeavor. While regional, national, and international expertise and resources play crucial roles, the heart of the matter always remains the actions, agency, and engagement of local people, communities, and organizations. In this symbiotic relationship, localization stands as a beacon, guiding the way toward a more resilient and interconnected world [2, 3]. In these background contexts, a structural framework of localization is proposed and the whole process was exercised by involving local youths and experts. The results of the process exercised showed that the training provided developed necessary capacity building and thus the local youths and volunteers with initial limited knowledge has become local resource persons to take actions on different problem contexts.

References

1. ActionsAid (2016) Resilience handbook: a guide to integrated resilience programming, Johannesburg, ActionAid international secretariat. https://actionaid.org/sites/default/files/2016_resl ience_handbook.pdf
2. Al-Abdeh M, Patel C (2019) 'Localising' humanitarian action: reflections on delivering women's rights-based and feminist services in an ongoing crisis. Gend Dev 27(2):237–352
3. Ayobi Y, Black A, Kenni L et al (2017) Going local: achieving a more appropriate and fit-for-purpose humanitarian ecosystem in the Pacific. Melbourne, Australian Red

Cross. www.redcross.org.au/getmedia/fa37f8eb-51e7-4ecd-ba2f-d1587574d6d5/ARC-Localisation-report-Electronic-301017.pdf.aspx
4. Barakat S, Milton S (2020) Localisation across the humanitarian-development-peace nexus. J Peacebuilding Dev 15(2):147–163
5. Cohen MJ, Ferguson K, Gingerich TR, Scribner S (2016) Righting the wrong: strengthening local humanitarian leadership to save lives and strengthen communities. Boston MA, Oxfam America. https://oxfamilibrary.openrepository.com/bitstream/handle/10546/595015/rr-righting-the-wrong-260116-en.pdf;jsessionid=E841A7771FAFE5831384BEF16F081C33?sequence=1
6. Fisher A, Fukuda-Parr S (2019) Introduction—data, knowledge, politics and localizing the SDGs. J Hum Dev Capab 20(4):375–385
7. Healy S, Tiller S (2014) Where is everyone? Responding to emergencies in the most difficult places. Doctors without borders. https://www.msf.org/sites/default/files/msf-whereiseveryone_-def-lr_-_july.pdf
8. Khan AK, Kontinen T (2022) Impediments to localization agenda: humanitarian space in the Rohingya response in Bangladesh. J Hum Dev Capab 7(14): 1–15
9. Roepstorff K (2020) A call for critical reflection on the localisation agenda in humanitarian action. Third World Quarterly 41(2):284- 301
10. Roepstorff K (2019) A call for critical reflection on the localization agenda in humanitarian action. Third World Q 41(2):284–301
11. Saguin K (2018) Why the poor do not benefit from community-driven development: lessons from participatory budgeting. World Dev 112:220–232
12. Slim H (2017) Nexus thinking in humanitarian policy: how does everything fit together on the ground? Int Comm Red Cross. https://www.icrc.org/en/document/nexus-thinking-humanitarian-policy-how-does-everything-fit-together-ground
13. UNISDR (2009) United Nations International Strategy for Disaster Reduction, INISDR terminology on disaster risk reduction, UNISDR, Geneva. file:///C:/Users/IRS%20&%20GIS/Desktop/Localization%20paper/7817_UNISDRTerminologyEnglish.pdf
14. Van Brabant K, Patel S (2018) Localisation in practice: seven dimensions of localisation emerging indicators and practical recommendations. DEPP and the global mentoring initiative. https://reliefweb.int/sites/reliefweb.int/files/resources/Localisation-In-Practice-Full-Report-v4.pdf

Disaster and Climate Resilience in Quezon City: Evaluating Local Development Planning for Sustainable Development Goals

Tabassam Raza, Indrajit Pal, Carmelita R. E. U. Liwag, Jade S. Agapinan, Isabella M. Prodigalidad, Danica Marie V. Arriesgado, Abegail Rose L. Valenzuela, Bassam Mohammed Zainudeen, Julia Angela Mae E. Collado, and Mark Anthony M. Morales

Abstract The Philippines' susceptibility to diverse disasters and their implications for national development necessitate integrating Climate Change Adaptation (CCA) and Disaster Risk Reduction (DRR) into policy, development plans, and governmental tools at all levels. Legislation like the Climate Change Act of 2009 and Disaster Risk Reduction and Management Act of 2010 underlines the critical role of CCA and DRR, particularly within local development planning, to address vulnerabilities and enhance disaster resilience. These policies enable Local Government Units (LGUs) to include CCA and DRR in their territorial development plans, conducting comprehensive risk assessments and implementing mitigation and adaptation measures. Recognizing the interplay between CCA, DRR, and Sustainable Development Goals (SDGs), it's clear that they mutually reinforce one another. The absence of CCA and DRR measures can impede SDG attainment. This study delves into the impact of mainstreaming CCA and DRR in local development planning on SDG achievement, specifically in Quezon City, Philippines. An extensive literature review, including documents like the Comprehensive Land Use Plan (CLUP), Comprehensive Development Plan (CDP), Local Disaster Risk Reduction and Management Plan (LDRRMP), and Local Climate Change Action Plan (LCCAP), examines how these initiatives contribute to the 17 SDGs. While

T. Raza
Planning and Development Research Foundation, Inc, School of Urban and Regional Planning, University of the Philippines, Quezon City, Philippines

I. Pal
Disaster Preparedness, Mitigation and Management (DPMM), Asian Institute of Technology, Khlong Luang, Thailand

C. R. E. U. Liwag · J. S. Agapinan (✉) · I. M. Prodigalidad · D. M. V. Arriesgado · A. R. L. Valenzuela · B. M. Zainudeen · J. A. M. E. Collado · M. A. M. Morales
School of Urban and Regional Planning, University of the Philippines, Quezon City, Philippines
e-mail: jsagapinan@up.edu.ph

I. Pal et al. (eds.), *Proceedings of the 3rd International Symposium on Disaster Resilience and Sustainable Development—Volume 1*, Lecture Notes in Civil Engineering 623, https://doi.org/10.1007/978-981-96-2139-2_14

the findings are specific to Quezon City, they have broader implications, particularly for LGUs facing climate and non-climate hazards and vulnerable populations. The findings and recommendations of this study are expected to have implications for the achievement of Sustainable Development Goals (SDGs) (especially, Goal 11—Sustainable Cities and Communities and Goal 13—Climate Action) related to climate action and resilient cities. Integrating CCA and DRR into local development planning is vital for community safety, especially in vulnerable areas like Quezon City, aligning with the city's SDGs.

Keywords Climate change adaptation · Sustainable development goals · Disaster risk reduction · Sustainable cities and communities · Climate action

1 Introduction

The Philippines has always been among the most disaster-prone countries in the world. The World Food Programme in 2015 claimed that Due to climate change, however, the disasters that hit the country, particularly the typhoons, have become more frequent and intense, resulting in the loss of tens of thousands of lives and the destruction of millions in property, livelihood, and infrastructure in the past couple of years [1]. In addition, the Climate Change (CC) related disasters threaten to reverse the Philippines' gains toward attaining its Sustainable Development Goals (SDGs), including endeavors to reduce poverty and hunger, among other critical objectives. Indeed, this demonstrates that climate change is a critical development issue that endangers the economic progress made over past decades. For example, agriculture, which depends on stable and predictable weather patterns, will suffer significantly if these patterns are disrupted by climate change [2].

In 2022, severe tropical storm "Paeng" devastated and placed the State of Calamity in more than 160 localities in Luzon, Visayas, and Mindanao. According to the Situational Report published by the National Disaster Risk Reduction and Management Council (NDRRMC) on November 19, 2022, an estimated 1,473,826 families or approximately 5,982,377 persons were affected nationwide. In addition, the storm inflicted significant economic losses, with approximately 7.1 billion in damages to the agricultural sector and 5.9 billion in infrastructure [3].

The vulnerability of the Philippines to hazards and the implications of such to the country's development goals make the mainstreaming of CCA and DRR necessary in various phases of policy formulation, land use and development plans, poverty reduction strategies, and other development tools by the government at all levels. The Climate Change Act of 2009 and the Philippine Disaster Risk Reduction and Management Act of 2010 acknowledge the significance of incorporating CCA and DRR into local development planning as a fundamental approach to addressing the unique risks and vulnerabilities faced by communities and building their resilience to the impacts of disasters [4, 5].

Thus, the policies provided the framework and mandate for local government units (LGUs) to integrate CCA and DRR in the local development planning of their respective territories, which entails assessing the risks and vulnerabilities of LGUs and the application of appropriate mitigation and adaptation strategies. Over the years, the Department of Human Settlements and Urban Development (DHSUD) and the Department of the Interior and Local Government (DILG) developed several guidelines to help in the preparation or updating of risk-sensitive Comprehensive Land Use Plan (CLUP) and Comprehensive Development Plan (CDP), such as the "CLUP Guidebook: Supplemental Guidelines on Mainstreaming Climate Change and Disaster Risks in the CLUP," and "CDP + : Guidelines on Mainstreaming DRR and CCA in the Comprehensive Development Plan," as well as the "Climate and Disaster Risk Assessment (CDRA)." as an evaluation tool to determine the risks posed by natural hazards and climate change. Consequently, these guidelines from the Housing and Land Use Regulatory Board (2015) enable informed decision-makers to address the identified risks effectively [6].

2 The Case Study: Quezon City, Philippines

The Quezon City is a highly urbanized metropolis located in the Philippines' National Capital Region (NCR), well-known for its pioneering sustainable development initiatives and ordinances. The city has implemented a comprehensive development plan that integrates disaster risk reduction and climate change adaptation, ensuring alignment with the Sustainable Development Goals (SDGs). To enhance its resilience, Quezon City has focused on sustainable urban planning, the development of green infrastructure, and the promotion of eco-friendly building practices. These initiatives aim to mitigate the impact of climate-related disasters and reduce the city's carbon footprint. Quezon City developed its plans by the implementation of abovementioned guidelines. Through these efforts, Quezon City is committed to creating a safer, more sustainable urban environment for its residents.

The current Quezon City Disaster Risk Reduction and Management Council is organized by Office of the Mayor in 2019 through the Executive Order No. 7, S-2019, entitled "Reconstituting the Quezon City Disaster Risk Reduction and Management Council (QCDRRMC), and Reaffirming with Modifications the Composition, Organization, Functions, and Funding Thereof." It is organized according to the four (4) DRRM thematic areas (Fig. 1).

The CDPM identifies and mitigates the hazards and their potential impacts on communities. At the same time, the CDP ensures preparedness and improves the institutional capacities of the LGU to respond to the hazards and their impacts. The CDR provides the basic subsistence needs of families and communities affected by disasters, while the CDRR restores and improves facilities, livelihood, and living conditions of affected communities and reduce disaster risks following the "building back better" principle. Additionally, the city prioritizes social services, such as enhancing healthcare and education, to strengthen community resilience. These efforts are

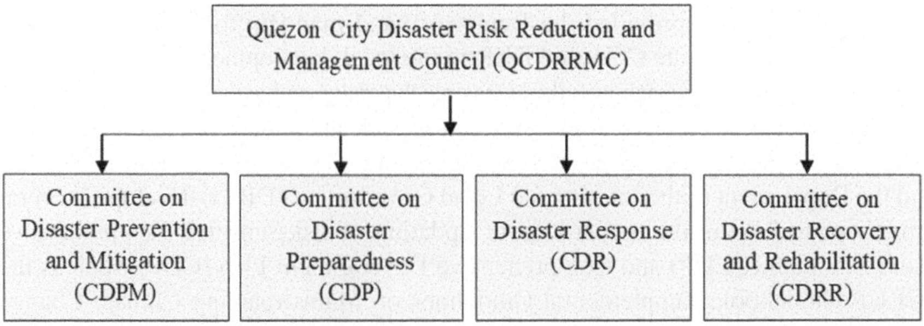

Fig. 1 Quezon city DRRMC four thematic areas

directed towards creating a safer, more sustainable urban environment capable of withstanding the challenges posed by climate change and natural disasters.

3 Review of Related Literature

A handful of studies have been done in the past on mainstreaming Climate Change Adaptation (CCA) and Disaster Risk Reduction (DRR) in local development planning in the Philippines and elsewhere. In the Philippines, case studies on integrating CCA, DRR, or both in the Comprehensive Land Use Plan (CLUP), the Comprehensive Development Plan (CDP), or both have been done in the past. Such was [7] study on the incorporation of CCA and DRR in the CLUP and CDP of Sorsogon City and [8] study on the inclusion of CCA and DRR on the CDP of Nueva Ecija Province.

Other studies discussed the issues and challenges experienced by LGUs in mainstreaming CCA and DRR. Raza, in 2018, [9] identified weak adaptive capacity and coping capacity as primary issues in mainstreaming for many LGUs. Studying the case of Quezon City, Raza developed the "Risk-Sensitive Comprehensive Land Use and Development Planning" Model to serve as a tool for LGUs in making CCA and DRR integral parts of development planners' and decision makers' agenda. Cuevas in 2015 and [10] likewise identified institutional capacity, information capacity, and resource capacity issues as barriers to mainstreaming CCA and DRR by looking at the experience of mainstreaming CCA into the CLUP of Albay. Institutional capacity issues, including fragmented national laws and regulations, overlapping policy requirements, and lack of guidelines for mainstreaming, are the primary issues. Aside from institutional issues, the availability and accessibility of information were also among the significant challenges.

Some literature provided tools to help cities and municipalities mainstream CCA and DRR. Raza [9] aims to improve the adaptive and coping capacities of LGUs by developing risk-sensitive Comprehensive Land Use and Development Planning Model. The model helps mainstream scientific-based risk assessment into local government planners' and decision makers' agenda and serves as a tool to support the achievement of the 2020 targets of the Sendai Framework, Paris Agreement, and Sustainable Development Goals.

Another is [11] study on using probabilistic or multi-scenario hazard maps to contextualize adaptive measures to climate change. The study provided examples that can be included in the local land use and development plans of several super typhoon Haiyan-affected areas.

4 Statement of the Problem

It has been generally recognized that CCA and DRR offer opportunities to achieve SDGs, and that SDGs highlight the importance of risk-sensitive development planning. This study thus asks to contribute to CCA and DRR. Conversely, lack of CCA and DRR can obstruct the attainment of SDGs. However, there is a need to exemplify these, particularly at the local level. To resolve above problem this study asks the following question:

- How do mainstreaming CCA and DRR in local development planning affect the achievement of SDGs?
- What are the policies in place to mainstream CCA and DRR in local planning?
- What should be the recommendations to achieve the Local level SDGs?

5 Objectives of the Study

The main objective of this study is to examine the implications of mainstreaming CCA and DRR on SDGs at the city/municipal level development plans in the Quezon City, Philippines. To achieve the main objective, this research needs to achieve the following specific objectives:

- To identify the current state of mainstreaming CCA and DRR in Quezon City development planning in terms of SDGs.
- To determine the existing policies and strategies in place to address CCA and DRR in Quezon City.
- To find out the effectiveness of the existing policies and strategies in addressing CCA and DRR in Quezon City CDP.
- To provide recommendations for improving the mainstreaming of CCA and DRR in local development planning towards achieving the SDGs in Quezon City.

6 Significance of the Study

The findings and recommendations of this study are expected to have implications for the achievement of SDGs related to climate action and resilient cities. This study will be significantly helpful to LGUs that are prone to climate and/or non-climate related hazards and with vulnerable populations such as fisher folk, farmers, women and children, among others. The mainstreaming of CCA and DRR in local development planning is crucial in ensuring the safety and well-being of communities, especially in vulnerable areas like Quezon City. The local government needs to prioritize these issues and take concrete actions to address them in line with the SDGs of the city.

7 Conceptual Framework Towards CCA, DRR, and SDGs Nexus

The study developed a Cyclical Conceptual Framework towards CCA, DRR, and SDGs nexus. It is to analyse the various factors contributing to Quezon City's vulnerability to the effects of climate change and natural disasters. It further provides the logic on how the mainstreaming of CCA and DRR can be Continuous Improvement Process (CIP) in attaining the local SDGs. This research is limited to operationalizing activities up to "formulation of recommendations" of the framework, thus the activities attached through dotted lines in the conceptual framework need to be dealt in the upcoming researches. See Fig. 2 for more details; certainly, it is self-explanatory framework.

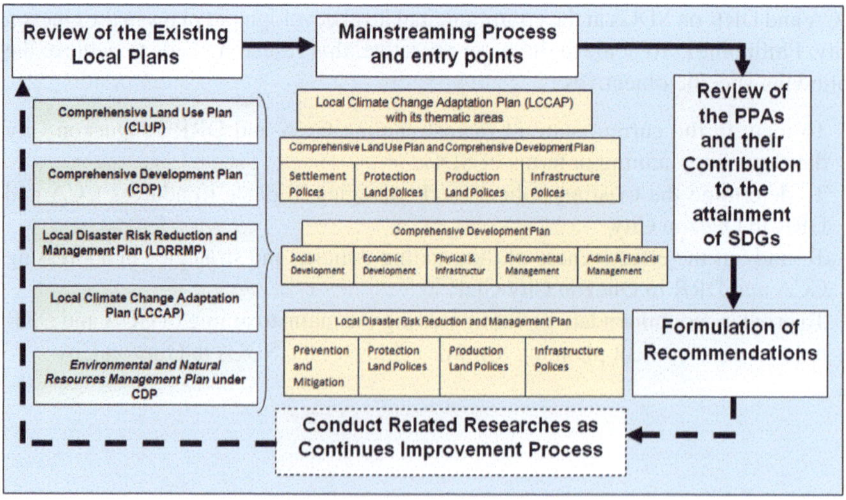

Fig. 2 Conceptual framework towards CCA, DRR, and SDGs nexus

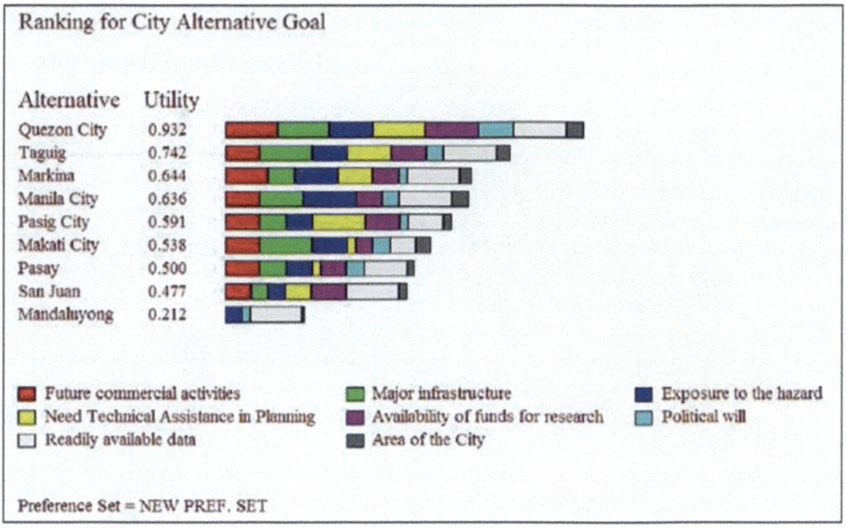

Fig. 3 Ranking results of the candidate cities (*Source* Courtesy of Raza [9])

8 Methodology

The study used a qualitative research design, specifically a literature review of existing studies and policies related to mainstreaming climate change adaptation and disaster risk reduction in Quezon City. This includes, but not limited to, the Comprehensive Land Use Plan (CLUP), Comprehensive Development Plan (CDP), and the Local Disaster Risk Reduction and Management Plan (LDRRMP). It will examine if CCA/DRR has been mainstreamed in the plan; determine the process of mainstreaming; assess the priority programs, projects, and activities/strategies (PPAs) are consistent with the result of risk assessment; and analyse how the PPAs and other actions in the above-mentioned plans contribute to the advancement of each of the 17 Sustainable Development Goals. Further, to justify the selection of Quezon City for piloting this research was done by considering nine (9) candidate cities as shown in Fig. 4. The factors considered in ranking the cities with the most potential for being a pilot LGU are shown in terms of different colour boxes at the lower part of Fig. 3 [9].

9 Results and Discussion

9.1 Selection of the Pilot Local Government Unit

Figure 3 shows Quezon City stood first on the rank. See location of the city in Fig. 4.

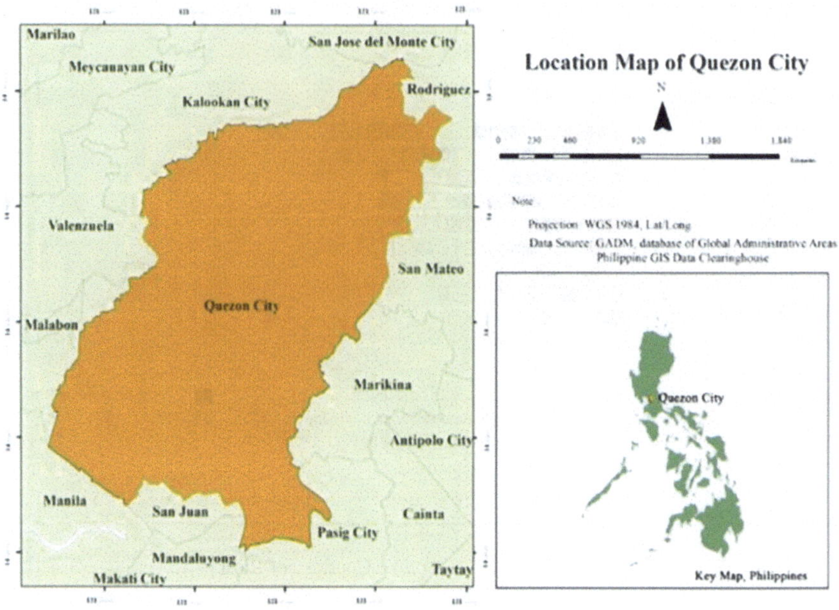

Fig. 4 Location of the Quezon city selected as pilot area (*Source* Courtesy of Raza [9])

9.2 *Analysis of the Plans*

All the cities and Municipalities in the Philippines are mandated to prepare their respective Comprehensive Land Use Plan (CLUP), Comprehensive Development Plan (CDP), Local Disaster Risk Reduction and Management Plan (LDRRMP), and Local Climate Change Adaptation Plan (LCCAP) are among others. Basically, it is the Philippines Constitution give highest priority towards the measures that protect and enhance the right of the people. Further, the 1991 Local Government Code of the Philippines mandates Local Governments to preparation their respective comprehensive land use and development plans. Further, Local Government Code also mandate that each LGU shall have a comprehensive multi-sectoral development plan i.e., Environment, social, economic, infrastructure, and institutional sectors. Thus, analysis of these plans can easily portray if the city is aligned with achieving the sustainable development goals through implementing above mentioned mandates. One of the main determinants to select the Quezon City as a pilot is due to the available updated mandated plans.

It is to be noted that the Quezon City Government (QCG) having an updated 1. Comprehensive Land Use Plan (CLUP) 2011–2025; 2. Comprehensive Development Plan (CDP) 2021–2025; 3. Local Disaster Risk Reduction and Management Plan (LDRRMP) 2021–2027; and 4. Environmental and Natural Resources Management Plan; and 5. Enhanced Local Climate Change Adaptation Plan (LCCAP) 2021–2050. Find below is the analysis of each plan.

9.2.1 Quezon City's Disaster Risk Reduction and Management Plan (2021–2027)

Quezon City's Local Disaster Risk Reduction and Management Plan (QC DRRMP) serves as the strategic roadmap for achieving disaster resilience aligns with the National DRRM Framework (NDRRMF) and the priorities of the NDRRMP. It offers comprehensive details on the local government's assessment of hazards, exposure, vulnerabilities, capacities, and outlines the goals, objectives, outcomes, action plans, and monitoring and evaluation mechanisms for the four thematic DRRM areas: disaster prevention and mitigation, disaster preparedness, disaster response, and disaster rehabilitation and recovery [5].

The need for an updated DRRMP is a response to the ongoing threats from impending disasters, may it be geologic, climate-change, and pandemic-related. It also seeks to provide the necessary elements to understand, communicate, and manage the various hazards and risks confronting the city by optimizing its existing capacities to minimize risks and safeguard its assets and communities. and eventually serve as the basis to institutionalize and implement various DRRM protocols, policies, procedures, functions, and resources for funding.

The plan has successfully incorporated local and national laws, including international frameworks such as the Sendai Framework for Disaster Risk Reduction, the Paris Agreement, the Sustainable Development Goals, and PDRRM Act of 2010 (R.A.NO. 10121) [4], among others. It has also been certified by both ISO 31000—Risk Management and ISO 9001:2015. Moreover, the plan has also incorporated the result of Climate and Disaster Risk Assessment (CDRA). A section of the plan provides the risk profile which includes a discussion of its natural and geographic conditions, hazards, vulnerabilities, and risk assessment. Given its key demographic characteristics, the city is considered to be the second biggest LGU in the country and with the number of its people and assets, it is important to take a proactive stand in ensuring that the damages, losses, and impacts of hazards will be minimized. While there is a discussion on the hazards, there are no hazard maps provided in the plan as well as the exposure database which is taken from CDRA that will provide a clear basis on the levels of sensitivity, adaptive capacity, and exposure of the different exposed units.

The situational analysis assessed the city's strengths, weaknesses, opportunities, and threats in terms of institutional capacity on DRRM. Strengths-Weaknesses Opportunities-Threats Analysis (SWOT) was used in various stages of the planning cycle. The QC TWG used the tool to develop strategies to improve its capacity and competencies in addressing its DRRM gaps and build on existing strengths. Key findings were observed from the areas on disaster prevention and mitigation, disaster preparedness, disaster response, and disaster recovery and rehabilitation. In terms of goals and objectives, the plan identified priority goals focused on addressing current and future risks of communities, strengthening risk governance, providing timely and anticipatory response actions, and recovering from disaster losses. Thereafter, programs and projects were identified and ranked based on their level of urgency.

9.2.2 Comprehensive Land Use Plan (2021–2025)

The CLUP is the overall development guide of the LGU in the management and use of its land. It assesses the development patterns and proposals for the changes or expansion of the different land uses in support of its vision and aligned with the city's functional role and development thrusts, particularly on the following policy areas: settlements, protection, production, and infrastructure.

Mainstreaming of CCA-DRR in the CLUP is vital in the identification of suitable areas for the different land uses such as residential, commercial, institutional, and industrial, among others. The plan also determines priority programs and projects based on its current and projected needs. Moreover, the CLUP is a tool for the enactment of measures such as appropriate land use regulations and zoning policies through its implementation tool—the Zoning Ordinance. Hence, it is an effective document in the implementation of measures for climate change adaptation and disaster risk reduction.

The Risk Sensitive CLUP provides climate and hazard information which serves as an input in the assessment of vulnerabilities and risks of its elements. The identified hazards are as follows: flood, earthquake, and fire. The plan did not conduct a separate risk assessment, but utilized the results of the Hazards, Vulnerability and Risk Assessment (HVRA) Report and DRRMP. A summary of sectoral consequences and interventions to address the issues is presented in the plan. The identified CCA-DRR interventions include the establishment of parks and open spaces, implementation of QC Shelter Program and Flood Control Program, identification and establishment or relocation and evacuation sites, and increased capacity development and information campaign.

Among its proactive climate mitigation strategies is the establishment of a Green Lung Network. The city acknowledges the vital role of open spaces in maintaining the ecological balance and addressing climate-related risks. The QC Parks Development and Administration Department (PDAD) reports a total of 285 parks within the city [12]. Although, some of these parks are yet to be developed due to the encroachment of illegal structures and informal settlers, plans for their development are already underway. The city also adheres to the strict implementation of the Water Code easement requirement and establishes a 3-m linear strip on both sides of its extensive network of rivers and creeks. This will add up to a total of 114 hectares of parks and open spaces. The city also incorporates a watershed-based approach in its development. The identified neighbourhood and community parks and linear parks will be connected to the La Mesa Watershed.

9.2.3 Comprehensive Development Plan (2021–2025)

The CDP of Quezon City has successfully integrated the result of climate and disaster risk assessment. Similar to the CLUP, detailed information on the process was not presented. The relevant climate and hazard information was taken from the comprehensive assessment report of HVRA and DRRMP. However, the plan has successfully identified mitigation and adaptation strategies for the different development sectors. Climate and hazard-related issues were also presented such as the presence of multiple geologic and hydrometeorological hazards, poorly-assessed structural integrity of public infrastructures, different levels of understanding of CCA-DRR-related concepts at the Barangay level, inadequate capacity development and information campaign, resettlement programs not aligned with the risk result, and poor enforcement of disaster-resilient structures in the barangays.

In pursuit of sustainable economic growth and promotion of the general welfare of the people through a CCA-DRR-centered planning process, the LGU has identified goals and supporting objectives for the attainment of a resilient city. One of its strategies is to strengthen its measures in the institutionalization of policies and mechanisms for climate change mitigation and disaster risk reduction. For the prevention and mitigation program, the city will employ measures through the installation of monitoring systems, establishment of permanent evacuation centers, development of drainage master plan, and establishment of communication infrastructure within its barangays. Under the disaster preparedness program, the strategies identified are the capacity development of the local communities, rigorous information campaign, and creation of a local planning body to closely monitor the implementation of LDRRMP and LCCAP such as the Environmental Sustainability Department. To address and lessen climate-related impacts, the city aims to strengthen its Climate Change Mitigation Program through the identification of alternative energy sources, installation of biodigester systems for public markets, establishment of an air-quality monitoring system, and reduction of emission with the implementation of Green Transport System.

Compared to the land use plan, the proposed programs and projects are more detailed in the CDP. However, both plans offer a comprehensive analysis of its existing situation and provide adequate interventions in addressing CCA-DRR issues. Proposals may vary due to the different implementation tools. Since the CLUP is geared towards physical planning, its interventions are spatially-focused. On the other hand, the CDP has included PPAs culled from the LDRRMP and LCCAP, which will be bridged towards investment planning.

9.2.4 Enhance Local Climate Change Action Plan (2021–2025) Mainstreaming in CDP in Term of Environmental and Natural Resources Management Plan

The effectiveness of the policies on mainstreaming CCA and DRR is apparent in the CDP, CCA- and DRR-related plans of Quezon City are primarily reflected in the city's Environmental and Natural Resources Management Plan under the QC-CDP. In the CDP 2021–2025, the stated goal of the environment and natural resources sector is to "create a clean, green, resilient, and sustainable environment that is conducive to healthy living and attracts not only the local residents but also visitors from outside the city". There are four objectives and corresponding strategies and programs, projects and activities to achieve each of these objectives see Fig. 5 for more details.

Several of the programs for the ENRMP are included in the "Top 51 Priority Programs for Inclusion in the QC LDIP 2021–2023". See Fig. 6 for more details.

10 Conclusion and Recommendations

The mainstreaming of CCA and DRR in local development planning is crucial in ensuring the safety and well-being of communities and achieving the local level SDGs, especially in vulnerable areas like Quezon City. The LGU needs to prioritize these issues and take concrete actions to address them in line with the SDG of the city. In addition, while Quezon City has existing policies and strategies to address CCA and DRR, there is always a big challenge ahead to build back better through mainstreaming CCA and DRR in local development and land use/Physical planning process. Based on the aforementioned findings, the researcher mentions the following recommendations:

- The City must strengthen its coordination mechanisms, increase public awareness and participation, and enhance its capacity to implement and monitor CCA and DRR programs that have been mainstreamed in the development plan.
- Develop and implement education and awareness campaigns to increase understanding and engagement with CCA and DRR tools and policies and its operationalization in achieving the local SDGs,
- Improve coordination and collaboration among local government agencies, community organizations, and other stakeholders to enhance the effectiveness of climate change and disaster risk reduction efforts,
- Develop and implement capacity-building programs to enhance the skills and knowledge of local government officials and staff related to CCA and DRR projects inclusion in ranking list of Projects, especially the projects that are cross cutting with development projects,
- Advocate for increased funding and resources for CCA and DRR activities, including through partnerships with other organizations and government agencies;
- Develop guidelines and tools to support the integration of CCA and DRR into local development planning processes towards achieving SDGs targets;
- Engage LGUs in the development of comprehensive guidelines to integrate CCA and DRR into the Local Public Transportation Route Planning (LPTRP) process. This approach aims to ensure the creation of resilient and climate-sensitive transportation infrastructure; and
- Advocate for the inclusion of CCA and DRR in national and regional development policies and plans to support local efforts.

Implementing these recommendations has the potential to significantly enhance the integration of Climate Change Adaptation (CCA) and Disaster Risk Reduction

Objectives	Strategies	Programs
Objective 1: To create a clean environment that is conducive to urban living	Strategy 1: Reduce the volume of solid waste reaching the final disposal site to 50% of the amount generated	Comprehensive Waste Management Program
	Strategy 2: Develop ecological waste disposal systems	Comprehensive Waste Disposal Program
	Strategy 3: Ensure efficiency in solid waste collection	Efficient Waste Collection Program
	Strategy 4: Ensure effective implementation of COVID-19 recovery and rehabilitation	COVID-19 Recovery and Rehabilitation Program
Objective 2. To nurture and sustain a green Environment where there is ample and convivial space for living	Strategy 1: Develop an integrated system and hierarchy of parks and open spaces to make Quezon City the greenest city and most disaster-resilient community in Metro Manila; and to transform and promote the city's parks as attractive public spaces for recreation, social interaction, education, cultural and historical promotion and environmental protection	Parks and Open Space Development Easement Development Program Parks Repair and support green development in the city
	Strategy 2: Promote and support green development in the city	Comprehensive Greening and Beautification Program
	Strategy 3: Effective implementation of the COVID-19 recovery and rehabilitation plan	COVID-19 Recovery and Rehabilitation Plan
Objective 3: To build a resilient city	Strategy 1: Strengthen measures and institutionalize policies and mechanisms for disaster risk reduction and climate change mitigation and adaptation	Disaster Prevention and Mitigation Program Disaster Response Programs
	Strategy 2: Increase/strengthen capacities of local communities to anticipate, cope with and recover from the negative impacts of disasters brought about by natural and man-made hazards	Disaster Preparedness Program
	Strategy 3: Increase awareness and instill preparedness to avoid loss of life and damage to social, economic and environment assets before, during and after emergencies and disasters	Disaster Prevention and Mitigation Program Disaster Recovery and Rehabilitation Program
	Strategy 4: Effective Implementation of COVID-19 Recovery and Rehabilitation Plan	Disaster Recovery and Rehabilitation Program Enhancement of the Implementation Arm of DRRM and CCA
Objective 4: To promote sustainable urban development Without compromising the future condition	Strategy 1: Urban development with the use of environmentally friendly practice	Climate Change Mitigation Program
	Strategy 2: Enhance the quality of the city's air to the levels meeting EMB standards	Air Quality Management Program
	Strategy 3: Control/manage the utilization of groundwater and other water sources	Water Quality Management Program Water Conservation Program
	Strategy 4: Protect and preserve the city's wildlife, educating the public on its richness and engaging the participation of the community, civil society and habitat landowners and harnessing the expertise /services of the academe and scientists in conservation programs and efforts.	Biodiversity Management Program
	Strategy 5: Rehabilitate and restore rivers and creeks to the level/condition that can support aquatic life and non-contact sports and recreation and keep the community safe from danger and disasters	River-ways Management Program
	Strategy 6: Effective implementation of COVID-19 Recovery and Rehabilitation	COVID-19 Recovery and Rehabilitation Program

Fig. 5 Objectives, strategies, and corresponding programs to the goal of the environment and natural resources sector under Quezon City CDP (*Source* Modified after QCG [13])

Program	Ranking in the List
Flood Control and Drainage Development Program	1
Responsive Water Management Operations	7
Establishment of Permanent Evacuation Centers	8
Establishment and Installation of Effective Monitoring Systems	9
Establishment of Satellite Offices of QCDRRMO	19
Establishment of an Effective Command and Control Structure for all city-owned, city-issued, partnership resources in the city	20
City's Water Quality Management Program Formulation of Water Quality Management Plan	23
Networking and Web Page Services, On-Line Computer Based Information System	27
Parks and Open Space /Vacant Lots Development Program	33
Supply and Installation of a Rainwater Harvesting System in 10 City-Owned Facilities	34
Comprehensive Greening and Beautification Program	34

Fig. 6 Programs for the ENRMP that are included Top 51 Priority Programs for Inclusion in the QC LDIP 2021–2023 (*Source* Modified after QCG [13])

(DRR) measures into local development planning efforts in Quezon City. This integration is crucial for aligning the city's objectives with the Sustainable Development Goals (SDGs). By incorporating CCA and DRR considerations into local development plans, Quezon City can effectively address environmental and disaster-related challenges, contributing to the attainment of the SDGs. This approach ensures that resilience-building measures are systematically integrated into urban development strategies, thereby fostering sustainable and inclusive growth. Additionally, mainstreaming CCA and DRR in local development planning enhances the city's capacity to respond to climate-related risks and natural disasters, ultimately improving the overall resilience and well-being of its residents.

References

1. World Food Programme (2015) Philippines, more intense typhoons: what does a changing climate mean for food security in the Philippines? https://documents.wfp.org/stellent/groups/public/documents/ena/wfp277192.pdf?_ga=2.19607551.650714770.1516641344-1846756425.1515078447. Accessed 5 Aug 2023
2. Sering LL (2013) World Bank Blogs, Filipinos, how are you adapting to climate change? You ask, we answer. https://blogs.worldbank.org/eastasiapacific/filipinos-how-are-you-adapting-to-climate-change-you-ask-we-answer. Accessed 1 Aug 2023
3. National Disaster Risk Reduction Management Council (NDRRMC) (2022) Situational report for TC PAENG (2022), SitRep No. 30 for STS PAENG (2022), NDRRMC operations center. https://ndrrmc.gov.ph/attachments/article/4211/SitRep_No_30_for_STS_PAENG_2022_Update.pdf. Accessed 1 Aug 2023
4. Republic of the Philippines (RoP) Congress of the Philippines (2010) Republic Act 10121': Philippine disaster risk reduction and management act of 2010. Metro Manila: republic of the Philippines, congress of the Philippines. https://ndrrmc.gov.ph/attachments/article/45/Republic_Act_10121.pdf. Accessed 1 Aug 2023

5. Republic of the Philippines (RoP) Congress of the Philippines (2009) Republic Act 9729: climate change act of 2009. Metro Manila: republic of the Philippines, congress of the Philippines. https://leap.unep.org/countries/ph/national-legislation/climate-change-act-2009-republic-act-no-9729-2009#:~:text=An%20Act%20mainstreaming%20climate%20change,Commission%2C%20and%20for%20other%20purposes.&text=This%20Act%20enters%20into%20force,publication%20in%20the%20Official%20Gazette. Accessed 1 Aug 2023
6. Housing and Land Use Regulatory Board (HLURB) (2015) supplemental guidelines on mainstreaming CCA/DRR in the CLUP, housing and land use regulatory board, climate change commission, united nations development programme, and department of foreign affairs and trade-Australian-Aid. https://www.scribd.com/document/411981634/HLURB-CDRA-Supplemental-Guidelines-07272015. Accessed 5 Aug 2023
7. Florano ER (2015) Mainstreaming integrated climate change adaptation and disaster risk reduction in local development plans in the Philippines. In: Leal Filho W (eds) Handbook of climate change adaptation, Springer, Berlin, Heidelberg. https://doi.org/10.1007/978-3-642-38670-1_36
8. Gabriel AG, Santiago PNM, Casimiro RR (2021) Mainstreaming disaster risk reduction and climate change adaptation in comprehensive development planning of the cities in Nueva Ecija in the Philippines. Int J Disaster Risk Sci 12:367–380. https://doi.org/10.1007/s13753-021-00351-9
9. Raza T (2018) Localizing disaster risk reduction and climate change adaptation in planners' and decision makers' agenda: technical comprehensive model, Quezon city, Philippines. Procedia Eng 212:1311–1318. https://doi.org/10.1016/j.proeng.2018.01.169
10. Cuevas SC, Peterson A, Robinson C, Morrison T (2016) Institutional capacity for long-term climate change adaptation: evidence from land use planning in Albay, Philippines. Reg Environ Change 16:2045–2058. https://doi.org/10.1007/s10113-015-0909-8
11. Lagmay AMFA, Santiago J, Pulhin JM (2021) Mainstreaming CCA-DRVRM using probabilistic multi-scenario hazard maps for future resilience in Haiyan-affected areas. In: Pulhin JM, Inoue M, Shaw R (eds) Climate change, disaster risks, and human security. Disaster risk reduction, Springer, Singapore. https://doi.org/10.1007/978-981-15-8852-5_11
12. Quezon City Government (QCG) (2011) 2011–2025 Quezon city comprehensive land use plan, QC, Philippines. https://quezoncity.gov.ph/wpcontent/uploads/2020/12/CLUP-2011-2025-Final-Version.pdf
13. Quezon City Government (QCG) (2021) Quezon city disaster risk reduction and management plan 2021–2027, vol 1, Quezon city disaster risk reduction and management council, Quezon city government, Philippines
14. Cuevas SC (2015) Challenges in mainstreaming climate change adaptation into local land use planning: evidence from Albay, Philippines. Int J Clim Chang Impacts Responses 7(3):45–65. https://doi.org/10.18848/1835-7156/CGP/v07i03/37246
15. Republic of the Philippines (RoP) (2010) Republic Act No. 10121, Philippine disaster risk reduction and management act of 2010, LAWPHIL, Philippines law and jurisprudence database. https://lawphil.net/statutes/repacts/ra2010/ra_10121_2010.html. Accessed 5 Aug 2023
16. Office of the Mayor (OM), Quezon City (2019) Reconstituting the Quezon city disaster risk reduction and management council (QCDRRMC), and reaffirming with modifications the composition, organization, functions, and funding thereof, Quezon city Philippines
17. Quezon City Government (QCG) (2021) Quezon City enhanced local climate change action plan 2021–2050. Climate change and environmental sustainability department (CCESD), Quezon city government, Philippines

Identifying Challenges to the Optimal Utilization of Early Warning Systems for Flood Management in Metro Manila: Advancing SDGs 11 and 13

Tabassam Raza, Krishna Buenaventura, Maryam Mazouz, Carmelita R. E. U. Liwag, Emmanuel Yasin, Jean Baptist Muhoza, Darling Ximena Velasco Caicedo, Mark Anthony M. Morales, and Dina C. Magnaye

Abstract The Philippines faces recurrent flood hazards, causing a range of issues from minor inconveniences to severe economic and human losses, disrupting essential resources and sanitation systems. Metro Manila, home to 15 million residents and a significant contributor to the national economy, grapples with persistent flooding that results in loss of life, property damage, and citywide traffic congestion. To address these challenges, the Philippine government collaborated with the World Bank initiated the development of the Metro Manila Flood Management Master Plan, which received approval from the National Economic and Development Authority (NEDA) Board in 2012. This plan emphasizes enhancing flood forecasting, establishing early warning systems, and implementing community-based flood risk management strategies. The Department of Science and Technology-Philippine Atmospheric, Geophysical, Astronomical Services Administration (DOST-PAGASA) developed an automated enhanced Flood Early Warning System (FEWS) to address escalating flooding issues in the region. However, flooding remains a significant problem, resulting in human and economic losses. Previous research has highlighted obstacles faced by early warning systems, such as insufficient infrastructure and financial constraints, yet a comprehensive assessment of these challenges in Metro Manila is lacking. This study aims to identify and analyze these barriers, contributing to enhanced flood management and early warning systems. Aligned with Sustainable Development Goals, particularly Goal 11: Sustainable Cities and Communities and Goal 13: Climate Action, the study uses a desk study methodology, analyzing articles, reports, and policy documents from secondary

T. Raza (✉) · K. Buenaventura · M. Mazouz · C. R. E. U. Liwag · E. Yasin · J. B. Muhoza · D. X. Velasco Caicedo · M. A. M. Morales · D. C. Magnaye
School of Urban and Regional Planning, University of the Philippines, Quezon City, Philippines
e-mail: tabassam.raza@planades.org

T. Raza
Planning and Development Resaerch Foundation Inc., UP SURP Building, Quezon City, Philippines

I. Pal et al. (eds.), *Proceedings of the 3rd International Symposium on Disaster Resilience and Sustainable Development—Volume 1*, Lecture Notes in Civil Engineering 623, https://doi.org/10.1007/978-981-96-2139-2_15

sources, with potential data bias due to reliance on secondary data. This research is significant as it can inform evidence-based policies to bolster community resilience and adaptability to flood disasters in Metro Manila.

Keywords Flooding · Early warning system · Sustainable development goals · Community based · Climate action

1 Introduction

Given the extensive land area in the country that is exposed to various hazards, flooding can have devastating consequences. Particularly vulnerable to these devastating consequences of floods is Metro Manila (Fig. 1), the most populous metropolitan region in the country, with more than 13 million inhabitants and a population density of 21,765 people per square kilometre [31]. As a vital economic center, the metropolis experiences disruptions in various sectors, including transport, commerce, and industry, which can have cascading effects on the national economy. In addition, damage to infrastructure, businesses, and houses generates substantial financial losses and hampers the region's overall development and productivity [2].

It is be noted that in the past twenty years, Metro Manila has experienced at least three major flood disasters. These include Typhoon Ondoy in 2009 and the monsoon rains known locally as "Habagat" in 2012 and 2016. The most significant event occurred in September 2009, when Typhoon Ondoy struck southwest Luzon in the Philippines, causing widespread flooding [34]. During the onslaught of the typhoon, the existing flood mitigation infrastructure proved to be inadequate. The Philippine Atmospheric, Geophysical and Astronomical Services Administration (PAGASA) serves as the main authority for information regarding storm intensity and flood potential in Metro Manila. PAGASA's flood forecasting system comprises two key components: (a) basin flood forecasting and (b) a flood forecasting and warning system for dam operation (FFWSDO). The basin flood forecasting monitors only four river basins in Luzon, excluding Metro Manila. Meanwhile, the FFWSDO oversees four major dams in Luzon but does not cover the primary flood-prone areas of Metro Manila. Notably, the dams monitored by PAGASA did not significantly contribute to the flooding that occurred in September 2009 [13].

To address this critical problem, the Philippine government, with technical and financial assistance from international aid organizations, has undertaken a comprehensive set of measures to mitigate its devastating effects. A crucial aspect of these efforts is the implementation of a Flood Management Master Plan, which outlines a series of priority actions to establish sustainable flood management practices [43]. Among them, the automated Flood Early Warning System (FEWS) developed by the Department of Science and Technology-Philippine Atmospheric, Geophysical, and Astronomical Services Administration (DOST-PAGASA) is recognized as a key element of disaster preparedness. Despite the concerted efforts of the Philippine government and international organizations to enhance flood management and

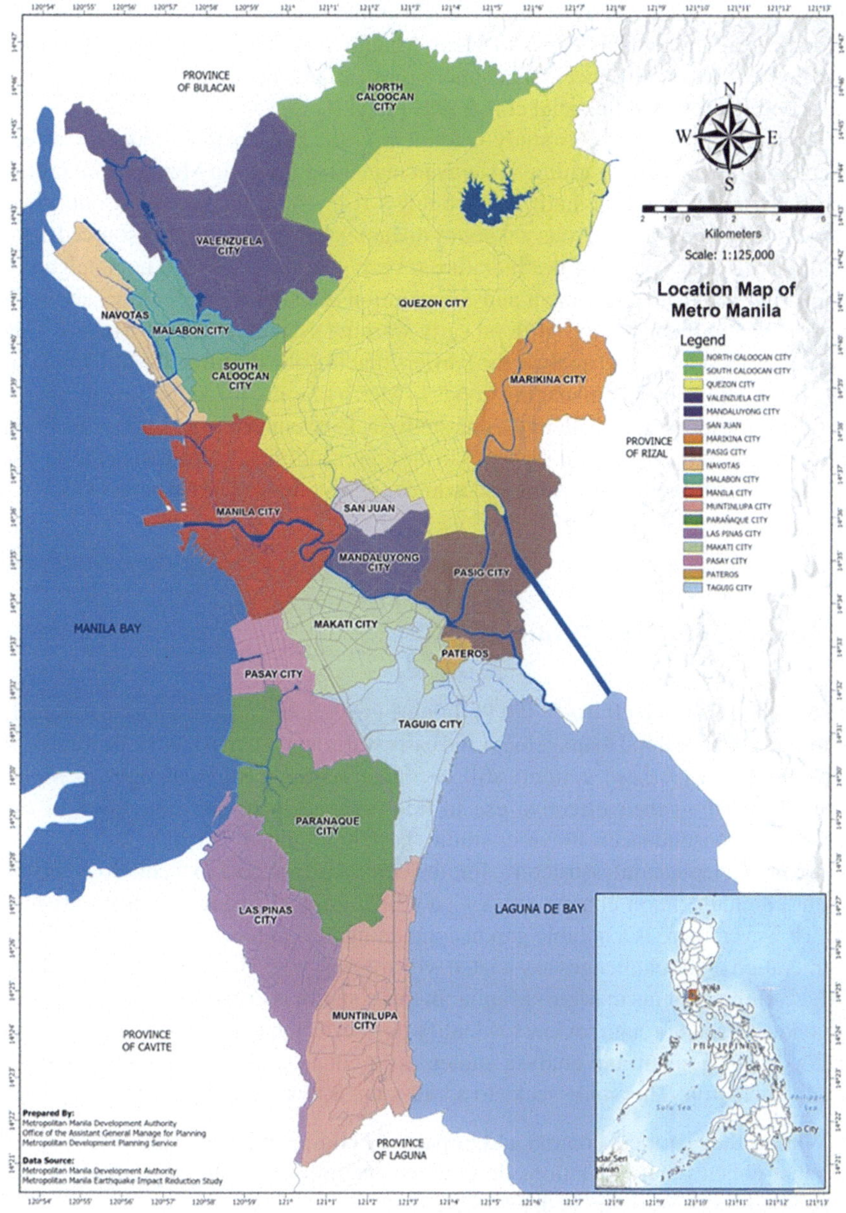

Fig. 1 Location map of the greater metropolitan [25] (*Source*)

early warning systems (EWS), the practical implementation of these systems still encounters significant challenges in Metro Manila, including inadequate infrastructure, limited community access to technology, insufficient communication channels, poor coordination, and financial constraints [32, 37].

The main objective of this study is to identify the barriers that hinder the effective utilization of Early Warning Systems in managing floods in Metro Manila. This research employed a desk study methodology. It involved the review of published articles, books, reports, and relevant policy documents from secondarily sourced data. A comprehensive review of the literature reveals that there exists a notable gap in conducting a holistic assessment and enumeration of the challenges associated with the practical implementation of flood early warning systems in Metro Manila. This study aims to identify and analyse the barriers that impede the effective utilization of early warning systems in managing floods in Metro Manila. By suggesting measures to address the identified challenges, this study seeks to contribute to improving flood management and early warning systems in the region, ultimately mitigating the devastating effects of flooding disasters and achieving Sustainable Development Goals 11 and 13.

2 Statement of the Problem

Despite the concerted efforts of the Philippine government and international organizations to enhance flood management and early warning systems (EWS), the practical implementation of these systems still encounters significant challenges in Metro Manila, impeding their effective use in flood management and mitigation of the devastating consequences. Previous studies have highlighted several key challenges, including inadequate infrastructure, limited community access to technology, insufficient communication channels, poor coordination, and financial constraints [32, 37]. However, there is a notable gap in conducting a comprehensive assessment and enumeration of the challenges associated with the real-world implementation of flood early warning systems in Metro Manila. In this sense, it is crucial to address this gap by conducting a thorough review to identify the barriers that hinder the improvement of flood management and analyse strategies to enhance the functionality of EWS within the region. This study seeks to address the following questions:

- What is the historical context and impact of flooding in Metro Manila?
- What policies and regulations do government entities employ regarding the use of flood Early Warning Systems (EWS)?
- What factors impede the effective utilization of flood EWS in Metro Manila?
- What recommendations can be proposed to enhance the utilization of flood EWS in flood management efforts in Metro Manila?

3 Objectives of the Study

The main objective of this research is to identify the barriers that hinder the effective utilization of Early Warning Systems in managing floods in Metro Manila. To achieve this objective this study, need to:

- Explore the history and impacts and flooding in Metro Manila,
- Examine the policies and regulations of governmental entities in the use of flood EWS,
- Investigate the prevailing factors that impede the effective utilization of flood EWS in Metro Manila, and
- Propose tenable recommendations to improve the use of flood EWS in flood management in Metro Manila.

4 Significance of the Study

This study holds significant importance as it aims to address the challenges hindering the effective utilization of early warning systems (EWS) in flood management in Metro Manila. By identifying these barriers, the study can inform policymakers, government entities, and local authorities about specific obstacles that need to be overcome to enhance the functionality and efficacy of EWS. Furthermore, exploring the link between EWS and Sustainable Development Goals (SDGs) 11 and 13 establishes a connection between flood management and broader sustainability and climate action objectives. This provides valuable insights for achieving SDGs, highlighting the role of EWS in promoting urban resilience, climate adaptation, and disaster risk reduction. Additionally, the examination of the role of policies, regulations, and local authorities in EWS implementation offers insights into institutional and governance aspects that influence EWS utilization. This can guide decision-makers in implementing necessary policy reforms, coordination mechanisms, and support structures. Lastly, the study's proposed recommendations aim to improve EWS utilization in flood management. These actionable suggestions, based on a comprehensive analysis of the challenges identified, can support evidence-based strategies and interventions to enhance EWS utilization, leading to more effective flood management practices.

5 Delimitation

This study acknowledges the constraints inherent in its scope and methodology. Primarily, it depends on secondary data sources, which could lead to data bias and restricted access to the latest information. Moreover, the study's emphasis on desk-based research might limit the depth of comprehension compared to primary data collection methods. Additionally, the complex nature of the factors affecting the effective utilization of early warning systems in flood management requires a thorough evaluation that might not be entirely captured within this study's confines. Nonetheless, this research endeavors to offer valuable insights and guide evidence-based policies and strategies to strengthen community resilience and adaptive capacity to flooding disasters in Metro Manila.

6 Review of Related Literature

6.1 Flooding

Flooding is a natural occurrence characterized by the inundation of typically dry land due to the elevation of water levels beyond the natural or artificial boundaries of a body of water [24]. This phenomenon is caused by a variety of factors, including heavy precipitation, rapid snowmelt, storm surges, and tropical cyclones [44]. The consequences of floods are far-reaching and can result in significant devastation, including loss of life, damage to personal property, and critical disruptions to public health infrastructure [44]. The destructive impact of floods is magnified by the immense force of moving water and the accumulation of debris during the receding phase [6].

Globally, the frequency and severity of floods are on the rise, particularly in low- and middle-income countries that lack sufficient preparedness measures and warning systems. Further, flood-related fatalities arise not only from drowning but also from physical injuries, diseases, disruptions to healthcare systems, and many more [44]. Furthermore, flooding has been associated with a myriad of disastrous consequences, ranging from minor inconveniences to significant economic and human losses. Even relatively small floods can give rise to widespread issues, while larger events can lead to extensive financial damages to infrastructure, agriculture, and industry, as well as loss of human life and forced displacement. Moreover, the aftermath of flooding events can trigger cascading effects such as shortages in essential resources and utilities, disruptions to sanitation systems, and potential contamination of water supplies, which can contribute to disease outbreaks [30].

6.2 Flooding and Early Warning System

According to [36], Early Warning Systems (EWSs) have gained global recognition as a key non-structural measure for disaster mitigation and preparedness. EWSs are comprehensive systems that include hazard monitoring, forecasting, prediction, risk assessment, communication, and preparedness activities. These systems allow individuals, communities, governments, businesses, and other stakeholders to take proactive measures to mitigate disaster risks before hazardous events transpire [41].

The concept of Early Warning Systems (EWSs) in flood disaster management involves tracking flood dynamics, simulating future scenarios, disseminating warnings, and implementing measures to mitigate impacts. These systems comprise four interconnected key elements. First, risk knowledge involves systematically collecting data and conducting disaster risk assessments. Second, monitoring and warning include detecting, analyzing, and forecasting hazards and their potential consequences. Third, dissemination and communication ensure timely, accurate, and actionable warnings, along with information on the likelihood and impact of hazards. Lastly, response capability focuses on preparedness at all levels to respond effectively to warnings. Coordination across various sectors and levels is essential for EWSs to function successfully. The failure of any single component or a lack of coordination among them can compromise the system's overall effectiveness [23, 28]. Flood-EWSs are vital in empowering communities at risk of flooding by delivering timely predictions about the scale, magnitude, duration, and potential damages of impending flood disasters at specific locations and times [38]. These predictions rely on data collected from sensors strategically placed in local water basins, such as rivers and lakes, as well as from flood defenses like dikes, dams, and embankments. Technological advancements have greatly enhanced the accuracy of flood forecasting, enabling predictions from days to mere hours before flooding events [12].

6.3 Flooding and Early Warning Systems (EWSs)

EWSs have gained global recognition as a key non-structural measure for disaster mitigation and preparedness. EWSs are comprehensive systems that include hazard monitoring, forecasting, prediction, risk assessment, communication, and preparedness activities [36]. These systems empower individuals, communities, governments, businesses, and other stakeholders to take timely actions to mitigate disaster risks before hazardous events occur [41].

In flood disaster management, EWSs involve monitoring flood dynamics, simulating future scenarios, issuing warnings, and implementing measures to mitigate impacts. These systems comprise four interconnected elements. First, risk knowledge entails systematically collecting data and conducting disaster risk assessments. Second, monitoring and warning involve detecting, analyzing, and forecasting hazards and their potential consequences. Third, dissemination and communication ensure timely, accurate, and actionable warnings, along with information about the likelihood and impact of hazards. Finally, response capability focuses on preparedness at all levels to respond effectively to warnings. Effective coordination across various sectors and levels is essential for EWSs to function successfully. The failure of any single component or a lack of coordination among them can compromise the entire system's effectiveness [23, 28].

Flood-EWSs play a crucial role in empowering flood-prone communities by providing timely predictions of the scale, magnitude, duration, and potential damages of forthcoming flood disasters at specific locations and times [38]. These predictions are primarily based on data collected from strategically placed sensors in local water basins such as rivers and lakes, or flood defenses like dikes, dams, and embankments. This is bolstered by the ever-evolving technological advancements which significantly improve the accuracy of flood forecasting, allowing for predictions ranging from days to hours before flooding events [12].

6.4 EWS and Sustainable Development Goals

As climate change intensifies, the frequency and severity of floods are expected to increase, making effective mitigation strategies crucial. Flood Early Warning Systems (EWSs) have become vital tools in minimizing flood damage. Both the Sendai Framework for Disaster Risk Reduction and the Sustainable Development Goals (SDGs) highlight the importance of expanding the availability of EWSs, a stance supported by the UN Office for Disaster Risk Reduction [32]. Specifically, SDG 11 (Sustainable Cities and Communities) and SDG 13 (Climate Action) stress the urgency of building resilient and sustainable cities and communities, as well as addressing climate change and its impacts. Targets 11.5, 11.b, and 13.1 focus on reducing the number of deaths and economic losses due to disasters, including water-related ones, and enhancing resilience and adaptive capacity to climate-related hazards and natural disasters globally [40].

Without a doubt, EWS play a crucial role in monitoring the implementation of the SDGs for DRR and climate change mitigation. EWS enable the delivery of clear and timely warnings that can help decision-makers assess the risks associated with impending floods [22]. EWS help achieve SDGs by disseminating critical information, empowering local people, and promoting proactive measures to address the problems brought on by climate change and to strengthen resilience capacity to natural hazards including floods. To achieve resilient and sustainable cities, governments, policymakers, and communities must continue to support the creation and implementation of strong EWS. However, the role of EWS may not be sufficient to reduce the likelihood and severity of floods, instead, efforts should be made to ensure their regular updating and to help the end-users understand their functioning [22].

6.5 Barriers to Flood Early Warning Systems

EWS serve a critical function in conveying scientific information to vulnerable communities facing the threat of flooding. However, the practical execution of EWSs is a highly intricate process influenced by factors such as scale, types of hazards, and the socio-economic and political circumstances in which they operate [11]. While the theoretical roles and functions of various components within EWSs are well-established, their effectiveness hinges upon all stakeholders' coordinated efforts. Any failure in a single component can undermine the overall performance of the system. Thus, collaboration and coordination among institutions and organizations, including governments, meteorological agencies, and local communities, are indispensable for effective risk monitoring and communication [9].

Numerous obstacles are known to impede the smooth flow of warning messages within EWSs, as extensively acknowledged in the scientific literature [14, 19, 18, 23, 39]. One major challenge lies in the uneven development of the four key elements of EWSs [28]. Other studies have indicated that inadequate infrastructure, limited community access to technology, insufficient communication channels, poor coordination, and financial constraints pose dire challenges to the effective functioning of early warning systems [32, 37].

While EWSs hold promise for mitigating flood risks, they are not devoid of challenges that may impede their effective implementation. These challenges encompass variations in scale, socio-political and geographical contexts, Therefore, efforts should be directed toward developing standardized and practical methodologies to enhance the efficiency of EWSs and achieve more effective flood risk reduction.

6.6 International Experiences and Flood EWS Best Practices

Worldwide, countries have been dealing with the damaging impacts of extreme weather events including floods. Governments and organizations have therefore increased their efforts to develop efficient flood warning systems that may mitigate the effects of floods. This section explores global experiences and best practices in flood warning systems, focusing on Japan, Germany, and England. We seek to shed light on the approaches that have been effective in reducing flood risks and protecting vulnerable populations by looking at the strategies and innovations used in these countries.

In order to provide early warnings, Japan's advanced Hydro-Meteorological and Flood warning system integrate meteorological information, river monitoring, and rainfall forecasting. The system accurately forecasts flood hazards by using real-time data and advanced modeling techniques. The main legislative frameworks dealing with flood protection include the River Act (established in 1896) and the Flood Fighting Act (established in 1949). The former stipulates that each local government office manages rivers, performs flood control operations, monitors hydrologic/hydraulic conditions, and conducts relevant river planning [20]. The latter mandates three flood-fighting actions namely (i) systematic flood forecasting and early warning in cooperation with meteorological sectors, (ii) the dissemination of flood hazard maps for specified rivers, and (iii) the promotion of "flood fighting brigades" [20]. The main river systems are under the direct management of the Ministry of Land, Infrastructure and Transport (MLIT), while prefectural or municipal governments are in charge of the remaining rivers [21]. The Japanese Meteorological Agency (JMA) collects weather data, including rainfall and snowfall, from observatories in Japan and abroad. They evaluate this data to forecast weather conditions and produce flood forecasts, informing municipal flood prevention management and other involved organizations [15] Japan's flood warning system comprises five levels, as detailed in Table 1. The primary objective of this system is to protect lives by clearly indicating varying degrees of danger, enabling individuals to prepare for evacuation and swiftly find safety during a natural disaster [8].

The effectiveness of the Japanese EWS relies on accurate real-time data integration, task separation, and effective collaboration among involved entities. The MLIT or prefectural government manages river water levels and flow volumes, the JMA oversees rainfall and snowmelt conditions, and municipal mayors issue evacuation orders based on warnings. The JMA also shares information with relevant weather, disaster management institutions, and broadcast organizations domestically and internationally [15]. The warning and alerts issued by JMA are transmitted through television, radio, sirens, mobile phone alerts, and dedicated websites.

The German flood early warning system is recognized for its efficiency and timely flood warnings, supporting flood management efforts. Forecasting, response, and warning duties are shared among federal, state, and local levels. The German national meteorological service (German Wetterdienst: DWD) handles weather forecasting and severe weather warnings at the federal level [37]. The DWD's extreme rainfall warnings are based on three levels of warning for specific intensity values. The hydro-meteorological data is produced by weather radars and rain gauge systems combined [17]. On the other hand, the responsibility for flood prediction and warning falls to the individual federal states [37]. Information about water levels, weather warnings, and flood warnings is published in different online portals such as the Inter-State Flood Portal, the waterway information system of the Federal Waterways and Shipping Administration, and the German Weather Service [45]. However, warning is primarily the responsibility of the local level. A variety of methods, including media operators and warning applications (like NINA and KATWARN) as well as sirens or loudspeaker announcements in some districts and municipalities are used to issue warnings to the population [37].

Table 1 Japan's flood warning system includes five levels

Warning level	Evacuation information	Meteorological information	Advised action
1		Early advisory	Be in a state of readiness
2		Heavy rain/flooding advisory	Verify evacuation routes
3	Evacuation preparation	Heavy rain/flooding warning	Evacuate elderly and other persons requiring assistance
4	Evacuation orders/ advisories	Landslide warning	Full evacuation
5	Disaster-level	Heavy rain emergency warning	Take optimal measures to save lives

Source [15]

Even if the German EWS is considered overall effective. The floods of 2021 raised several questions about its performance. Investigations revealed that not all people at risk received warnings through apps and other digital means. The responsiveness of the public authorities and the local population was also questioned.

England has established a robust Flood Early Warning System to protect vulnerable populations and mitigate flood impacts. This system, leveraging advanced technology, precise data analysis, and efficient communication, reflects England's commitment to disaster preparedness and resilience. The Environment Agency (EA) is responsible for constructing and maintaining flood defenses, coordinating multi-agency flood responses, providing accurate flood risk information, and issuing flood warnings. The EA monitors rainfall, river levels, and sea conditions 24/7 to forecast flood likelihood [7]. As a crucial component of its strategy for reducing flood risk, the EA has developed an advanced flood warning system. The three types of issued warnings are Flood Alert, Flood Warning, and Severe Flood Warning. Each warning type implies different measures to be taken by relevant institutions and individuals. The EA also commissioned design in action Global Fujitsu to create Flood Line Warnings Direct (FWD), which is a service designed to provide flood warnings and related information to residents and businesses in England that are at risk of flooding. The service was delivered in 2006 and was later on modernized in 2017. In addition to the FWD, warnings are also issued via other media such as the Environment Agency's official website, television and radio broadcasts, social media, and smartphone apps as well as via flood wardens, Sirens and loudhailers in some areas where these services are available [7].

An effective flood early warning system relies on several key factors to successfully detect and communicate potential flood events. These factors include robust monitoring and data gathering, advanced analysis and forecasting models, efficient communication and dissemination, institutional coordination, community preparedness and continuous evaluation and improvement. However, it's important to bear in mind that any early warning system can still face difficulties from extreme and unpredictable flood disasters as seen in Germany in 2021.

7 Methodology

Through the utilization of a qualitative research design and a comprehensive literature review, this study offers valuable insights to achieve the objectives of this research and provide the suitable solution to prevent if not totally resolve the research problem. The self-explanatory research paradigm is presented in the following Fig. 2.

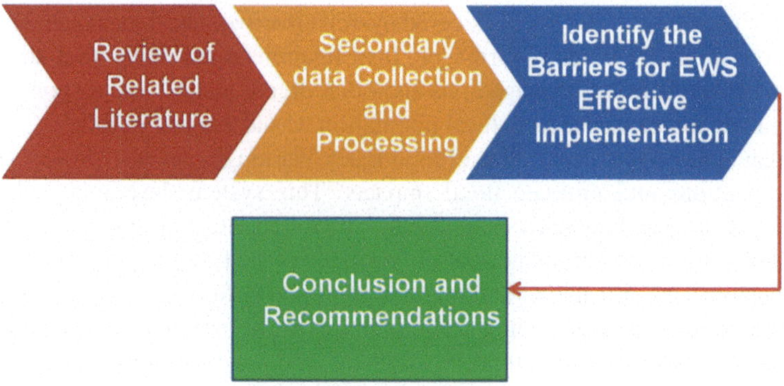

Fig. 2 Conceptual framework of the study

7.1 Data Collection

The main approach to gathering data was an extensive examination of literature encompassing policy documents, government reports, and academic articles [42] pertaining to the Flood Management Policies, Initiatives, automated Flood Early Warning System (FEWS), history and impact of flooding in Metro Manila, and barriers to effective EWS utilization in detail in Metro Manila. This approach was chosen due to the nature of the research topic, as it allowed for an extensive examination of existing literature and data related to early warning systems and flood management in Metro Manila.

The data collection process involved conducting a comprehensive search of electronic databases, such as academic journals, research databases, and government publications from credible repositories. Keywords and search terms related to early warning systems, flood management, barriers, and Metro Manila were used to identify pertinent sources. Additionally, relevant reports and policy documents from government agencies and international organizations were accessed to gather information on existing initiatives, policies, and strategies related to flood management and early warning systems. Critical scrutiny was employed to ensure the selection of high-quality and relevant sources. The selected sources underwent thorough reading and analysis to extract pertinent data and insights regarding the barriers to the effective utilization of early warning systems in Metro Manila.

7.2 Data Analysis Techniques

During this stage, the collected data is thoroughly analyzed and interpreted to discover significant patterns and insights [42]. Firstly, the collected data was carefully read and familiarized to comprehensively understand the information contained within the sources. Then, the data was coded by assigning labels or themes to relevant sections or segments that corresponded with the questions asked by the stipulated objectives. Similar labels were grouped together, forming broader themes or categories. This iterative process involved constant comparison and refinement of the identified themes as the analysis progressed. Once the identification, labelling, grouping, and thematic analysis were completed, the themes were elaborated upon and the identified barriers were synthesized and presented in a coherent manner. The findings were discussed and contextualized within the existing literature to provide a comprehensive understanding of the challenges faced in the utilization of early warning systems in Metro Manila.

8 Discussion and Findings

8.1 History and Impact of Flooding in Metro Manila

The metropolitan area, situated between Manila Bay and Laguna Lake, has evolved into an extensive urbanized drainage basin prone to frequent inundation from overflowing rivers and storm waters [2]. During the typhoon season, spanning from June to October when the Philippines receives 80 percent of its annual rainfall, Metro Manila experiences frequent and severe floods [26, 43]. Annually, Metro Manila encounters three to four significant flooding incidents primarily caused by typhoons, monsoon rains, and torrential downpours. Flood depths in the region vary from shallow inundations at gutter height, typically due to heavy rains leading to traffic disruptions, to over 5 m of flooding caused by storms or typhoons, resulting in extensive property damage and numerous fatalities [26].

Over the past two decades, Metro Manila has been devastated by at least three severe flood disasters, including Typhoon Ondoy in 2009 and two monsoon rains locally referred to as "Habagat" in 2012 and 2016. The impact of Typhoon Ondoy in September 2009 was particularly notable. Exceptionally heavy rainfall was recorded during this period, with 455.0 mm at the Quezon City Science Garden, 258.5 mm at the Port Area, and 331.8 mm in Tanay municipality, Rizal province [34]. The torrential downpour affected a total of 872,097 individuals across the metropolitan area, resulting in 241 fatalities, 394 injuries, and extensive damage to 65,521 buildings, including 12,563 completely destroyed [29]. Infrastructure damages, not including agricultural losses, amounted to 570,187,587 Philippine pesos. Moreover, the flooding caused significant disruptions, such as the closure of major city center routes like Epifanio de los Santos Avenue due to impassable floodwaters. Additionally, domestic and international flights at Manila Airport were cancelled [34]. Figure 3 depicts images from the aftermath of Typhoon Ondoy.

Due to climate change and human activities, floods are anticipated to become more frequent in this region. Despite heightened vulnerability across much of the metropolitan area, rapid development continues unchecked, resulting in residential buildings, industries, and commercial areas becoming increasingly susceptible to

Fig. 3 Images from Typhoon Ondoy Aftermath (*Source* Alchetron, web Mod, Norman, Jaime Goloyugo [1])

flood-related disasters [13, 26]. While flooding affects all parts of Metro Manila to varying extents, certain cities and municipalities are more vulnerable due to their geographical location and elevation above sea level. Areas situated to the east, southeast, and south of the capital, including Marikina, Pasig, Mandaluyong, Muntinlupa, and Parañaque, as well as those adjacent to Laguna de Bay, such as Taguig and Pateros, face particular vulnerability. Over the past fifty years. floods have become increasingly widespread and severe, shifting from low-lying coastal areas to suburban communities, newer urban developments, and along the shores of Laguna de Bay [2].

8.2 Flood Management Polices, Initiative, and Erly Warning System in Metro Manila

The scale and complexity of the issue dictate the responsibilities for flood management in Metro Manila. Localized flooding, associated with drainage basins spanning a few square kilometers, typically falls under the jurisdiction of city or municipal authorities. They address these issues by constructing gutters and storm drains to direct rainwater into natural or artificial river systems [2]. Until 2002, the Department of Public Works and Highways (DPWH) exclusively oversaw flood management activities in Metro Manila. Responsibilities were subsequently transferred to the Metro Manila Development Authority (MMDA), which now shares these duties alongside traffic and solid waste management [26].

Flood disasters in recent years have stimulated policy changes in flood risk management in urban settings, resulting in a global trend toward the adoption of the integrated flood risk management (IFRM) method [33]. Similarly, the Philippines began to take a proactive approach to flood disasters and risk reduction, resulting in the development of a Flood Management Master Plan in 2012 [5], and the creation of an integrated flood risk management (IFRM) strategy in 2013 for the greater Metro Manila region [26]. The IFRM is a relatively modern strategy that combines structural and non-structural elements and goes beyond flood protection by including preventative and preparedness measures [3]. By principle, IFRM mainly focuses on the concept of "*keeping people away*" rather than "*keeping water away*" [35].

He Metro Manila Flood Management Master Plan, developed with technical and financial assistance from the World Bank, outlined various measures to effectively manage future major flood events. These measures include mitigating flooding from river systems flowing through the metropolis by constructing a dam in the upper Marikina River catchment area to reduce peak river flows entering Metro Manila, eliminating long-term flooding in the Laguna de Bay floodplain, enhancing urban drainage systems by modernizing and increasing the number of pumping stations in Metro Manila, and improving solid waste management in villages near drainage systems [43]. The master plan also aimed to enhance flood forecasting and early warning systems, promote community-based flood risk management, relocate informal settlers encroaching on rivers, and establish institutional frameworks for improved flood control [4, 43].

Flood Early Warning Systems (EWS) that are effective are an essential component of disaster risk reduction and management (DRRM) and climate change adaptation methods. Floods Early Warning Systems (FEWS) have long been pushed in Metro Manila by the Department of Science and Technology (DOST) through the Philippine Atmospheric Geophysical and Astronomical Service Administration (PAGASA), which primarily handles weather and flood predictions [10]. Metro Manila's Early Warning System adheres to the four pillars of an effective "end-to-end" early warning system, which are as follows: risk knowledge, monitoring and warning, dissemination and communication, and response capacity (Fig. 4) [5].

According to DPWH and MMDA [5], when Metro Manila's early warning system detects a potentially dangerous rise in upstream water level as a consequence of heavy rain, the data is automatically transmitted to a local operations center, which then notifies the affected areas. Then, utilizing basic communication routes (such as Radios, TVs, Telephones, and household communications) a warning message based on established alert levels is distributed to the communities. Disaster response teams and authorities in the municipalities have also been educated and equipped to assist with the process, and the evacuation centers network offers refuge in the event of a disaster [5].

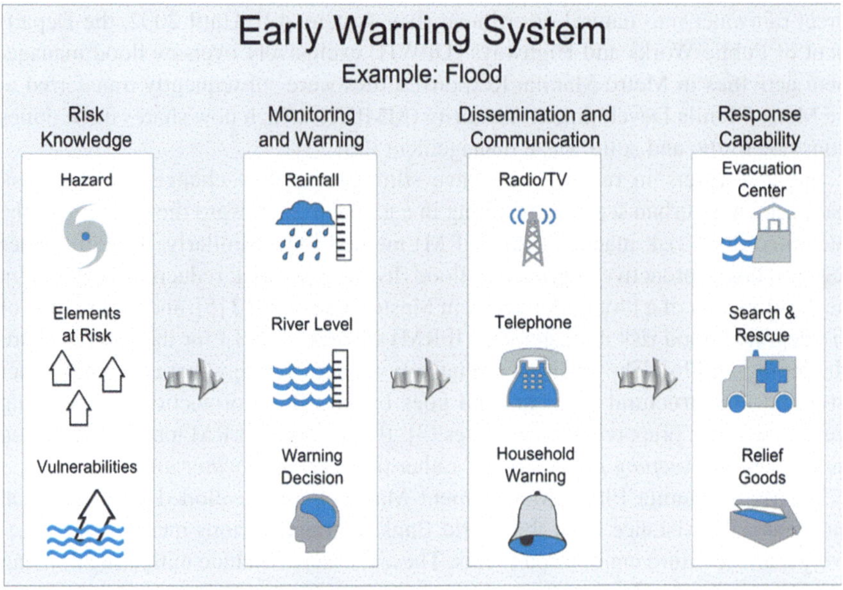

Fig. 4 Flood early warning system in metro Manila [5] (*Source*)

8.3 Barriers to Effective EWS Utilization in Metro Manila

The identified barriers include as follow:

8.3.1 Community Responsiveness and Negligence

According to 2006 reports from the United Nations for Disaster Risk Reduction (UNDRR) and the United Nations (UN), an Early Warning System (EWS) comprises four crucial elements: risk awareness, monitoring and warning, dissemination and communication, and response capacity. Numerous studies have highlighted the significance of community participation in ensuring the effectiveness of EWS design and operation. However, in the Philippines, there is still a predominant focus on individual elements, with an overemphasis on technology and equipment. This approach can result in the failure of the entire system, as the effectiveness of the technology relies on appropriate community responses to warnings. Previous disasters have demonstrated ineffective responses due to limited community integration and participation across all elements of the EWS. Some specific problems associated with this include the lack of adequate training for the community on protocols to follow during calamities and the lack of coordination between government officials and affected communities [16]. For instance, in Metro Manila, where flooding has been a recurring problem, many people have experienced flooding since childhood and have become complacent with flood warnings. This passive acceptance of flooding as a normal occurrence hinders the effectiveness of EWS. The mindset of complacency and non-proactive approach from community members impede them to take appropriate action during floods.

8.3.2 Poor Institutional Coordination

Despite advances in Metro Manila's Flood Early Warning System, solution implementation has been severely fragmented, with overlapping duties, insufficient coordination, and data sharing among stakeholders, and a lack of effect predictions at the local level. In Metro Manila, two government entities are in charge of flood control (Metro Manila Development Authority and the Department of Public Works and Highways), however, the division of labor is unclear [5]. The current array of diverse and independent agencies handling flood-related issues requires ongoing and laborious coordination and information exchange. However, there are no established policies governing the exchange of information and communication between the Metro Manila agencies responsible for flood management [5, 27].

8.3.3 Limited Budget and Lack of Flood Management Equipment

Flood control projects often receive less attention compared to other, less expensive, and revenue-generating urban infrastructure projects, leading to a shortage of effective flood forecasting technology. For example, the Metro Manila Development Authority's (MMDA) Effective Flood Control Operation and Warning System (EFCOS) was designed to mitigate flooding in Metro Manila. However, due to insufficient funding and data processing systems, it only monitors water levels and rainfall depths without providing real-time flood forecasting analysis [13]. Indeed, as of 2018, there were no real-time forecasts and updates on water level and rainfall depths in Metro Manila, flood forecasting systems from government agencies relied on storms' intensity as a warning mechanism, and they still used human tabulation and data processing. There is still limited and poor hydrometeorological information and flood data, and that results in flood estimations and forecasts being uncertain [13].

8.3.4 Lack of PAGASA Hydrological Simulation

The effectiveness of flood early warning systems relies on the accurate prediction of flood events through hydrological simulation. In the case of Metro Manila, the flood warning system operated by PAGASA is deficient in incorporating hydrological simulation techniques [13]. Consequently, the issued flood warnings may not possess the necessary precision regarding the spatial distribution and magnitude of flooding. This is exemplified by the occurrence on the 25th of September

2009, where PAGASA issued a flood bulletin for the whole of Metro Manila during typhoon Ondoy on the basis of storm warning signals (i.e., the wind speed of the storm). However, Accurate prediction of the location and extent of flooding was not available. This notorious disaster proceeded to affect 872,097 people throughout the entire metropolitan area, causing 241 fatalities, 394 injuries, and damaging 65,521 buildings (of which 12,563 were completely destroyed) [34]. The absence of hydrological simulations undermines the system's ability to provide reliable and timely warnings, compromising its overall effectiveness in mitigating flood impacts, hence the data processing and flood simulation capability of PAGASA needs to be greatly bolstered Non-Function EFCOS Warning Infrastructure.

8.3.5 Non-Function EFCOS Warning Infrastructure

The inadequacy of the warning system infrastructure in Metro Manila, specifically the Effective Flood Control Operation and Warning System (EFCOS) installed along the Manggahan Floodway, has significant implications for flood preparedness and response in the region. The existing infrastructure, consisting of speakers and radios, is designed to be activated when the Rosario weir, a flood control structure, is about to release water. A message announcing the release of water is sent to all nine warning stations along the Manggahan Floodway. However, the non-operational state of EFCOS during Typhoon Ondoy highlights the system's ineffectiveness in providing timely and reliable warnings [13]. This hampers the dissemination of crucial information to vulnerable communities at risk of flooding. Timely warnings are essential for residents to take appropriate actions and prepare for impending flood events. Without warning infrastructure, the ability to alert communities and provide them with accurate information regarding flood risks, evacuation measures, and safety precautions is severely compromised. As a result, the inadequate warning system infrastructure undermines the preparedness and response capabilities of communities during flood events. Residents are left unaware or insufficiently informed about the imminent danger, making it difficult for them to make informed decisions and take necessary actions to protect themselves, their property, and their livelihoods. The absence of timely warnings increases the likelihood of casualties, property damage, and disruption to critical services and infrastructure.

8.3.6 Insufficient Finance to Procure X-band Radars

According to an interview conducted by Philstar Global in July, 2022 with Vicente Manalo, the Administrator of the Philippine Atmospheric Geophysical Astronomical Services Administration (PAGASA), it is evident that the technological capabilities available to PAGASA are insufficient to adequately address the demands placed upon the agency. Manalo highlights the significant gap between the number of major river basins in the country, which amounts to 18, and the operational rain gauges or X-band radars possessed by PAGASA, which is only 10. These radars play a critical role in monitoring rainfall patterns, a crucial factor in forecasting floods, and disseminating timely warnings to the public regarding the associated hazards. Consequently, it is imperative that each river basin be equipped with a minimum of two of these monitoring devices. However, the high cost associated with acquiring and maintaining these radars poses a considerable challenge. With an estimated price tag of approximately P50 million per radar, the financial burden becomes a significant limiting factor. It is worth noting that the establishment of a comprehensive flood forecasting and warning system should have been completed, as emphasized by Manalo.

8.4 Inadequate Public Awareness and Understanding of Flood Warning Systems

A noteworthy barrier to the effective utilization of flood early warning systems in metro Manila is the insufficient public awareness and understanding regarding their existence and operation. As documented in research carried out by Mateo and Oki in 2011. Their survey results reveal that a considerable proportion of the population remains oblivious to the presence of these warning systems. Despite the availability of flood bulletins and warnings on the internet, their dissemination is not sufficiently publicized, impeding their discoverability and accessibility. Furthermore, the disseminated information often suffers from ambiguities and lacks clarity in terms of its content and significance. This underscores a pressing need for a redress of this knowledge barrier through the propagation of education and awareness programs to sensitize the populace on these systems and enable them to know the meaning and significance of each message passed across.

9 Conclusions and Recommendation

To ensure an efficient implementation of flood early warning systems in Metro Manila, several key measures should be taken to mitigate the existing barriers. These measures aim to address challenges such as insufficient financial resources, inadequate simulation and warning infrastructures, inadequate public awareness, community engagement and insufficient institutional coordination. Strengthening stakeholder consultations, and contribute to the achievement of SDGs 11 and 13.

- The importance of flood early warning systems, potential risks, and necessary actions to take during flood events should all be made clear to residents through awareness campaigns, workshops, and training sessions.
- Clear and well-defined roles need to be identified for each of the government entities involved in flood control and management. This will ensure that roles are clear and do not conflict.

- Relevant authorities should Advocate for increased funding allocation for flood control and early warning systems in Metro Manila and engage with government officials to prioritize funding for flood management projects and guarantee the allocation of sufficient and sustainable financial resources.
- Among the most important pieces of equipment are the X-band radars. Sufficient funding needs to be allocated to PAGASA for the its acquisition and maintenance of these radars in order to bridge the gap between the number of river basins and the available radars.
- PAGASA's hydrological stimulation needs to be improved through the enhancement of data collection methods and hydrological models by deploying additional monitoring stations, implementing advanced sensing technologies and incorporating sophisticated predictive algorithms.
- PAGASA also needs to foster collaboration with local agencies and communities that could help in collecting localized data and sharing knowledge. On the other hand, resources need to be allocated to regularly maintain and repair the EFCOS warning infrastructure including the conducting of inspections to identify any faults or malfunctions and ensure timely repairs.
- Substantial investment should be provided in terms of technological aspect of flood management. Insufficient technological resources hinder the ability of PAGASA to provide accurate and timely flood forecasts and warnings, which are crucial for safeguarding public safety and mitigating the impacts of flooding.
- Comprehensive public awareness campaigns could be launched in order to educate the population about the existence, purpose, and operation of flood warning systems. To reach a broad audience, these campaigns might make use of a variety of communication platforms, including radio, television, social media, and community outreach programs. The disseminated information needs to be easily accessible and understandable for the general public. Additionally, the incorporation of flood awareness and early warning system education into school curricula at various levels needs to be fostered. These mentioned public awareness campaigns need to be frequently evaluated using surveys, feedback systems, and focus groups in order to spot areas for improvement.
- By addressing these obstacles and implementing the recommended measures, Metro Manila can enhance its flood management capabilities, mitigate the impacts of floods, protect lives and property, and contribute to the achievement of SDGs 11 and 13. To achieve this, it is essential for all stakeholders to work together in order to ensure a sustainable and resilient future for the city and its inhabitants.

References

1. Alchetron N, Web M, Jaime G (2009) Roof top photo, bridge collapse photo and bus caught in flood. https://Alchetron.Com/Typhoon-Ketsana. https://Www.Typhoonondoy.Org/Category/Pictures/. https://Normannorman.Com/2009/09/28/Typhoon-Ondoys-Fury
2. Bankoff G (2003) Constructing vulnerability: the historical, natural and social generation of flooding in metropolitan Manila. Disaster 27(3):224–238. https://doi.org/10.1111/1467-7717.00230

3. Buchecker M, Ogasa DM, Maidl E (2016) How well do the wider public accept integrated flood risk management? An empirical study in two Swiss Alpine valleys. Environ Sci Policy 55:309–317. https://doi.org/10.1016/j.envsci.2015.07.021
4. CTI Engineering International CL, DPWH, JICA (2004) The study on flood control project implementation system for principal rivers in the Philippines
5. DPWH and MMDA (2017) Environmental and social management framework (ESMF): metro Manila flood management project
6. Doswell CA (2003) Flooding. In: CA Doswell III, University of Oklahoma, Norman, OK, USA, p 769–776. https://www.studocu.com/ph/document/zamboanga-state-college-of-marine-sciences-and-technology/stem-work-immersion/flooding-wew/39345041
7. EA (2012) Flood warnings a guide to the environment agency's flood warning code, a user guide for emergency responders version 1.0. In: Environment agency, horizon house, deanery road, Bristol BS1 5AH, vol 1. https://www.metoffice.gov.uk/binaries/content/assets/metofficegovuk/pdf/business/public-sector/hazard-manager/user_guide__flood_warning_codes.pdf
8. Fan J, Huang G (2020) Evaluation of flood risk management in Japan through a recent case. Sustainability 12(13):5357. https://doi.org/10.3390/su12135357
9. Foster HD (1980) Disaster planning. Springer, New York. https://doi.org/10.1007/978-1-4612-6093-6
10. GIZ (2009) Local flood early warning systems: experiences from the Philippines. In: Gesellschaft für Internationale Zusammenarbeit. https://reliefweb.int/report/philippines/local-flood-early-warning-systems-experiences-philippines
11. Garcia C, Fearnley CJ (2012) Evaluating critical links in early warning systems for natural hazards. Environ Hazards 11(2):123–137
12. Georgaka D, Mparmparousi M, Vitos M (2012) Early warning systems. Hosp Chron 7(1):37–43. https://doi.org/10.2015/Hc.V7i1%20Sup.482
13. Gilbuena R, Kawamura A, Medina R (2013) Gap analysis of the flood management system in metro Manila, Philippines: a case study of the aftermath of Typhoon Ondoy. IAHS AISH Publ 357:32–40
14. Glantz MH (2003) Early warning systems: do's and don'ts. Report of workshop, Shanghai, China, pp 0–76
15. GoJ: Government of Japan (2006) Early warning sub-committee of the inter-ministerial committee on international cooperation for disaster reduction government of Japan. In: Japan's natural disaster early warning systems and international cooperative efforts. https://www.bousai.go.jp/kokusai/kyoryoku/pdf/soukikeikai.pdf
16. Gonzales JN, Arnaldo AAP, Marquez-Lim WE, Ongkingco CMA, Pineda MVG (2003) Information Technology Department, De La Salle University-Manila
17. Hofmann J, Schüttrumpf H (2019) Risk-based early warning system for Pluvial flash floods: approaches and foundations. Geosciences 9(3):127. https://doi.org/10.3390/geosciences9030127
18. ISDR: International Strategy for Disaster Reduction (2006) Developing early warning systems : a checklist. In: Third international conference on early warning: from concept to action, pp 1–3
19. Ibrahim M, Kruczkiewicz A (2016) Learning from experience: a review of early warning systems, moving towards early action, World Vision. https://www.wvi.org/sites/default/files/WV_EWEA_Doc_FINAL_Web.pdf
20. Ikeda T, Yoshitani J (2006) Japan's strategic contributions to hydro-meteorological disaster mitigation in the world: planning to establish the UNESCO–PWRI Centre. Hydrol Process 20(6):1251–1261. https://doi.org/10.1002/hyp.6088
21. Ikeda T, Yoshitani J, Terakawa A (2005) Flood management under climatic variability and its future perspective in Japan. Water Sci Technol 51(5):133–140. https://doi.org/10.2166/wst.2005.0125
22. Kimuli JB, Di B, Zhang R, Wu S, Li J, Yin W (2021) A multisource trend analysis of floods in Asia-Pacific 1990–2018: implications for climate change in sustainable development goals. Int J Disaster Risk Reduct 59

23. Kundzewicz ZW (2013) Floods: lessons about early warning systems. In: Late lessons from early warnings: science, precaution, innovation
24. Luino F, Bobrowsky P, Marker B (2016) Encyclopedia of engineering geology. encyclopedia of earth sciences series, Springer. https://doi.org/10.1007/978-3-319-12127-7_126-1
25. MMDA: Metro Manila Development Authority (2022) Location map of greater metropolitan, data source: metro Manila development authority and metropolitan manila earthquake impact reduction study, Office of the Assistant General Manager for Planning, Metro Manila, Philippines
26. Mercado JMR, Kawamura A, Amaguchi H (2020) Interrelationships of the barriers to integrated flood risk management adaptation in metro Manila, Philippines. Int J Disaster Risk Reduct 49:101683. https://doi.org/10.1016/j.ijdrr.2020.101683
27. Mercado JM, Kawamura A, Amaguchi H (2018) Interrelationship of flood risk management barriers in Matro Manila, Vol, 74, No.5. https://www.jstage.jst.go.jp/article/jscejer/74/5/74_I_285/_pdf/-char/ja. (Accessed on December 10, 2024)
28. Molinari D, Ballio F, Berni N, Pandolfo C (2013) Towards more effective flood early warning systems-the Italian experience
29. NDCC: National Disaster Coordinating Council (2009) National disaster coordinating council situation report no.1–final report
30. PAGASA (2024) Floods. In: Philippine atmospheric, geophysical and astronomical services administration, department of science and technology. https://www.pagasa.dost.gov.ph/learning-tools/floods.
31. PSA (2020) Highlights of the population density of the Philippines 2020 census of population and housing. https://psa.gov.ph/content/highlights-population-density-philippines-2020-census-population-and-housing-2020-cph
32. Perera D, Seidou O, Agnihotri J, Rasmy M, Smakhtin V, Coulibaly P, Mehmood H (2019) Flood early warning systems: a review of benefits, challenges and prospects. UNU- INWEH, Hamilton
33. Samuels P, Klijn F, Dijkman J (2006) An analysis of the current practice of policies on river flood risk management in different countries. Irrig Drain 55(S1):S141–S150. https://doi.org/10.1002/ird.257
34. Sato T, Nakasu T (2011) 2009 Typhoon Ondoy flood disasters in metro Manila. https://doi.org/10.13140/RG.2.1.2817.5121
35. Scott M, White I, Kuhlicke C, Steinführer A, Sultana P, Thompson P, Minnery J, O'Neill E, Cooper J, Adamson M, Russell E (2013) Living with flood risk/the more we know, the more we know we don't know: reflections on a decade of planning, flood risk management and false precision/searching for resilience or building social capacities for flood risks?/participatory floodplain management: lessons from Bangladesh/planning and retrofitting for floods: insights from Australia/neighbourhood design considerations in flood risk management/flood risk management—challenges to the effective implementation of a paradigm shift. Plan Theory Pract 14(1):103–140. https://doi.org/10.1080/14649357.2012.761904
36. Sukhwani V, Gyamfi BA, Zhang R, AlHinai AM, Shaw R (2019) Understanding the barriers restraining effective operation of flood early warning systems. Int J Disaster Risk Manag 1(2):1–17. https://doi.org/10.18485/ijdrm.2019.1.2.1
37. Thieken AH, Bubeck P, Heidenreich A, von Keyserlingk J, Dillenardt L, Otto A (2022) Performance of the flood warning system in Germany in July 2021–insights from affected residents. EGUsphere, pp 1–26
38. Travis RW (2013) Design of a severe climate change early warning system. Weather Clim Extrem 2:31–38. ISSN 2212–0947
39. UNEP: United Nation Environment Programme (2012) Early warning systems: a state-of-the-art analysis and future directions. In: Division of early warning and assessment (DEWA), United nations environment programme (UNEP). https://na.unep.net/geas/getUNEPPageWithArticleIDScript.php?article_id=89
40. UNEP: United Nation Environment Programme (2016) GOAL 11: sustainable cities and communities. https://Www.Unep.Org/Explore-Topics/Sustainable-Development-Goals/Why-Do-Sustainable-Development-Goals-Matter/Goal-11

41. UNISDR (2009) Terminology on disaster risk reduction. In: United nations international strategy for disaster reduction (UNISDR), pp. 0–35. https://www.unisdr.org/files/7817_UNISDRTerminologyEnglish.pdf
42. Valdez AS, Raza T, Farolan MI, Mendoza CI, Perez LQ, Peralta JF, Valencia RI, Lim HAMP (2023) Primary and secondary data collection to conduct researches, write thesis and dissertation amidst COVID-19 pandemic: a guidepost, pp 269–288. https://doi.org/10.1007/978-981-19-6297-4_20
43. WB: World Bank (2016) Metro Manila flood management. In: flood-management. https://www.worldbank.org/en/country/philippines/brief/project-highlights-metro-manila
44. WHO: World Health Organization (2022) Floods. https://www.who.int/health-topics/floods#tab=tab_1. www.wvi.org/sites/default/files/WV_EWEA_Doc_FINAL_Web.pdf
45. umweltbundesamt.de (2022) August 10, Flood warning

Private Sector Engagement for Community Resilience and Sustainable Development

Kilian T. Murphy and Edwin M. Salonga

Abstract Key global frameworks and guiding strategies, such as the 2030 Agenda for Sustainable Development and the Sendai Framework for Disaster Risk Reduction 2015–2030, underscore the pivotal role of the private sector in fostering community resilience and sustainable development. While the importance of the private sector is acknowledged, there still remains a crucial need to explore the inherent 'business case' that motivates commercial enterprises to actively participate in disaster risk management (DRM) and contribute to humanitarian efforts in view of strengthening community resilience and sustainable development. This paper offers a conceptual exploration of the changing dynamics within the private sector's engagement in promoting community resilience and sustainable development through an analysis of the evolving landscape of private sector involvement in DRM. It explores how businesses are increasingly acknowledging the need to protect their operations from disasters while also supporting development initiatives that assist communities affected by such events. In addition, this paper critically examines the evolving relationship between private sector engagement, community resilience, and sustainable development. By analyzing this relationship, the study aims to provide evidence-based insights, shedding light on the mutually beneficial value addition that occurs when businesses actively participate in DRM initiatives. In doing so, it contributes to the ongoing discourse on how private sector involvement can bolster both community resilience and sustainable development goals, fostering a holistic approach to disaster risk reduction and management.

Keywords Private sector · Disaster risk management · Sustainable development · Multi-stakeholder

K. T. Murphy (✉) · E. M. Salonga
Asian Disaster Preparedness Center, Bangkok, Thailand
e-mail: kilian.murphy@adpc.net

I. Pal et al. (eds.), *Proceedings of the 3rd International Symposium on Disaster Resilience and Sustainable Development—Volume 1*, Lecture Notes in Civil Engineering 623, https://doi.org/10.1007/978-981-96-2139-2_16

1 Introduction

The occurrence of disaster events has shown a significant increase in frequency, from 4,212 incidents during the period of 1980 to 1999 to 7,348 incidents between 2000 and 2019 (Centre for Research on the Epidemiology of Disasters and United Nations Office for Disaster Risk Reduction, 2020). This rise in disaster events has had substantial economic consequences, with global economic losses totaling approximately US$1.63 trillion in the earlier period and escalating to approximately US$2.97 trillion in the later period (ibid).

The escalating occurrence of disasters, coupled with the effects of climate change, has had far-reaching implications for entire communities, nations, and regions. The economic impacts of disasters on the private sector are particularly significant, as businesses and industries face direct losses, supply chain disruptions, and long-term economic instability. Consequently, there has been an increasing recognition of the need to involve a wide range of stakeholders for prevention, mitigation, preparedness, response, and recovery for comprehensive disaster risk management (DRM) efforts. Traditionally, governments, development partners, and civil society organizations (CSOs) have been at the forefront of building resilience and mitigating the impacts of disasters. However, the private sector, which had been largely overlooked in DRM and humanitarian work, is now gaining prominence due to the intensifying disaster impacts.

This section aims to delve into the growing involvement of the private sector in DRM toward community resilience and explore the role played by international organizations and development partners in promoting and facilitating such engagement in alignment with the principles of sustainable development. The focus of the analysis includes examples from various regions, including Asia, which has emerged as a pioneer in this field of interest. By examining these examples, insights can be gained into the evolving role of the private sector and the collaborative efforts being undertaken to enhance disaster resilience, thereby contributing to the broader agenda of sustainable development.

The private sector, encompassing businesses and industries, possesses unique capabilities and resources that can contribute significantly to community resilience through DRM initiatives. Commentators have aimed to clarify the definition of the "private sector" in relation to the public sector. The private sector refers to entities in the economy that are privately owned, with many of them, particularly corporations, primarily focused on generating profits [22]. On the other hand, entities owned by the government are considered part of the public sector (ibid). It has been emphasized that it is crucial to disaggregate the "private sector" as it encompasses a wide range of entities, from small-scale micro enterprises to large multinational corporations operating globally. Treating it as a uniform or homogeneous entity has been noted as a limitation [11]. The involvement of businesses at different scales can help improve preparedness, strengthen infrastructure, and support local communities during and after disasters. Moreover, companies have expertise in risk assessment, crisis management, and resource mobilization, which are valuable assets in disaster response and recovery.

The escalating occurrence of disasters and their economic impacts have highlighted the need for a multi-stakeholder approach to DRM. The private sector, once considered a non-actor in this field, is now being recognized for its potential contributions toward resilience building [7]. International organizations and development partners are essential in encouraging and supporting private sector involvement in DRM, aligning with sustainable development principles that highlight the need to balance economic, social, and environmental factors. These collaborative efforts are crucial for enhancing disaster resilience and achieving more effective disaster risk management worldwide.

2 Private Sector Contributions to Disaster Risk Management

Recently, there has been notable growth in businesses participating in humanitarian efforts and community resilience. The private sector has become more involved in pre-disaster preparations, relief operations, which include cash donations, in-kind contributions, and the establishment of essential infrastructure [10, 13]. The insurance industry has also played a role in reducing the impact of disasters, assisting in post-disaster recovery through claim payments [33].

Acting as suppliers, donors, and operational partners in humanitarian endeavors, the private sector has been recognized as a key actor capable of driving transformative change within the humanitarian sector, particularly in the realm of disaster risk management. Leveraging their expertise, resources, and influence, businesses extend their efforts beyond immediate relief provisions to contribute to sustainable peace and development [41]. Given the mounting complexity of emergencies, which necessitate coordinated action among diverse stakeholders, the participation of the private sector has become increasingly crucial. The evolving focus of humanitarian actors has shifted from short-term outcomes towards a recognition of the need for better alignment between development, humanitarian, and peace-building endeavors to address the root causes of disasters [38].

The private sector's involvement in humanitarian assistance and community resilience is not limited to financial support. Between 2011 and 2015, private sector funding for humanitarian aid amounted to USD $6.6 billion, highlighting the potential for commercial enterprises to assist traditional stakeholders such as governments, development partners, NGOs, and CSOs in their crisis response and management efforts Development [12].

Private sector engagement in DRM has evolved beyond mere financial contributions towards establishing partnership approaches at various levels of business operations. Such collaborations facilitate collective action among stakeholders from different sectors, particularly those focused on reducing the impacts of disasters on vulnerable populations [9, 32]. Development partners have documented the capacity of businesses to enhance collaborative arrangements by providing technical expertise, efficient organizational skills, and innovative approaches to DRM practices [40]. The private sector's abundant resources, expertise, and technology make it a strong candidate for collaborative efforts aimed at disaster risk reduction [43].

Moreover, the private sector is deeply involved in disaster reduction and enhancing resilience, utilizing both philanthropic endeavors and commercial activities that leverage their core business expertise and supply of goods and services [33]. This active involvement in disaster management is demonstrated through various means: providing direct assistance to communities, implementing disaster preparedness measures within businesses, innovating products using business technology and expertise, collaborating with NGOs, governments, and international organizations, and creating private foundations, NGOs, and trusts dedicated to addressing disaster-related challenges.

The private sector's increasing involvement in DRM has transformed its role from being solely a financial contributor to becoming a proactive partner in strengthening community resilience toward sustainable development. Beyond financial support, businesses now act as suppliers, donors, and operational partners in humanitarian endeavors, recognizing the importance of long-term sustainable outcomes. The private sector brings valuable resources, technical expertise, and innovative approaches to DRM practices. Their involvement goes beyond philanthropic purposes and encompasses commercial activities that leverage their core business capabilities. This evolving engagement highlights the potential of businesses to play a transformative role in resilience building efforts.

3 Evolution of Private Sector Engagement

Historically, the private sector's engagement in humanitarian action and resilience building was limited, with businesses perceiving disasters as the sole responsibility of governments. Corporate social responsibility initiatives (CSR) related to disaster reduction were often seen as superficial and promotional, primarily benefiting businesses. Many of the initiatives documented lacked industry-wide approaches, leading to criticism of businesses being reactive rather than proactive in view of their engagement of disaster management [14, 33].

However, the private sector's viewpoint on DRM has evolved significantly. Businesses now acknowledge the significance of not only minimizing the potential impacts of disasters on their operations but also investing in reducing vulnerabilities in the communities where their employees live. There is growing encouragement for the private sector to incorporate DRM into their business models, integrating enterprise risk management and business continuity strategies [36]. To facilitate greater private sector engagement in resilience building down to the community level, it is crucial to address key issues related to legislation, incentives, and stakeholder engagement, which are considered mechanisms for effective private sector involvement (Izumi et al. 2016).

Despite the growing involvement of the private sector, concerns remain regarding the holistic nature and long-term sustainability of its disaster risk reduction interventions. Some companies still prioritize short-term business gains in their efforts, while others may exploit fragile ecosystems in pursuit of income generation [35]. However, it is worth noting that there is an increasing trend of collaboration between governments and the private sector, where businesses are expanding their response and recovery responsibilities beyond self-preservationist activities.

Recognizing the adverse effects of climate change and disasters on their operations, businesses are now seen as potential agents of change in building resilience, not only for themselves but also for the communities in which they operate [1]. Despite commercial competition, there is a growing realization that collective efforts throughout all phases of disaster management can benefit both businesses and communities. As more private sector entities join the global drive for disaster risk reduction and resilience, profound changes are expected. The private sector is shifting its focus from self-interest-driven disaster risk reduction efforts toward resilience-building initiatives that promote sustainable development and benefit both businesses and communities [21].

4 The Asia Region as a Pioneer in Engaging the Private Sector

Private sector engagement in disaster risk reduction and humanitarian action has gained significant attention and momentum at the global and regional levels, particularly in the Asia Pacific region [14, 40]. Global and regional DRM policy frameworks highlight the importance of a multi-stakeholder approach that includes businesses [25]. Development partners, such as the United Nations and the Asian Disaster Preparedness Centre (ADPC), have played a vital role in mobilizing the private sector through awareness-raising efforts and sensitization campaigns among business communities in disaster-prone countries of Asia [35].

Regional organizations in Asia, including the Association of Southeast Asian Nations (ASEAN), the South Asian Association for Regional Cooperation (SAARC), and the Asia–Pacific Economic Cooperation (APEC), have actively advocated for enhanced private sector involvement in disaster risk reduction and resilience building toward sustainable development. These organizations facilitate cooperation between corporations and national governments, recognizing the value of private sector engagement [6]. Business associations and chambers of commerce across Asia have formalized their role in disaster management by actively engaging in private sector initiatives coordinated by global and regional development partners such as the United Nations Alliance for Disaster Resilient Societies (ARISE) network, the

Connecting Business Initiative (CBI), and the Asian Preparedness Partnership (APP) platform established by the ADPC (Abe et al. 2019) [25]. These platforms serve as valuable spaces for stakeholders from both for-profit and nonprofit sectors to interact, exchange knowledge and resources, and overcome common challenges to private sector engagement in humanitarian action and disaster management [10, 27].

The Philippines stands out as a leader in engaging the private sector in DRM in Asia, showcasing the significant role that businesses can play in disaster resilience efforts. The country has enacted relevant legislation and policy frameworks that recognize the importance of private sector involvement in DRM [40]. In the context of disaster relief, the private sector in the Philippines has actively participated in organizing relief operations, delivering donations, and providing support to the affected population. Large enterprises, in particular, often prefer to channel their resources through corporate foundations and establish direct partnerships with government agencies and local communities [18]. There has been a notable shift in the private sector's approach, with increased investment in disaster resilience and proactive preparedness for disruptive events. The Philippines Disaster Resilience Foundation (PDRF) exemplifies private sector-led coordination efforts in disaster management, advocating a holistic approach to DRM and prioritizing proactive disaster preparedness and recovery over reactive recovery processes [42]. PDRF is credited with launching the world's first privately operated and self-sustaining Emergency Operations Centre (EOC). This innovative facility serves as a critical hub for coordinating relief and response efforts during major disasters, while also providing training courses on disaster preparedness (ibid). The EOC leverages data from local and international sources to effectively monitor hydro-meteorological hazards and pandemics, ensuring continuous surveillance of areas affected by disasters. In 2020, the EOC was operationalized during a 'triple whammy' emergency scenario, the overlapping Super Typhoon Goni and Vamco response during the COVID-19 pandemic [9]. The PDRF's endeavors are frequently cited as a model for enhancing private sector engagement in humanitarian action and DRM in the region [42].

In Indonesia, the country's National Disaster Management Agency (BNPB) has promoted the adoption of a 'Collaborative-Pentahelix' model to enhance coordination and joint planning in disaster management. This model emphasizes the active involvement and collaboration of various stakeholders, including government entities, media organizations, community-based groups, private sector actors, and academic institutions. At the provincial level, Disaster Risk Reduction Forums have been established under the oversight of the Regional Disaster Management Agencies (BPBD) to operationalize this collaborative approach. In Jakarta, business networks and individual business champions were engaged in consultations in order to include inputs from the private sector for earthquake contingency planning documents for the Daerah Khusus Ibukota (Special Capital Region) of Jakarta [26].

In Japan, the government has enacted policies to encourage private sector enterprises to enhance their preparedness to disasters through business continuity planning [20]. Following the Great East Japan Earthquake in 2011, the Cabinet Office and Ministry of Economy, Trade and Industry (METI) set a long-term goal of comprehensive Business Continuity plan (BCP) coverage for companies across the country.

The Japan Revitalization Strategy of 2012 set a goal for all large companies and half of SMEs to have their own Business Continuity Plans (BCPs) by the end of 2020. By 2017, approximately 60% of large companies and over 30% of mid-sized companies had established BCPs. When including companies in the process of developing BCPs, over 80% of large enterprises and nearly 50% of mid-sized enterprises were involved in BCP implementation (ibid). Also, in Japan, private sector companies have supported the implementation of revised evacuation procedures during the COVID-19 pandemic. Under the guidance and coordination of the central government, private hotels and accommodation industry associations made available over a thousand facilities to serve as public emergency evacuation shelters [37].

The Philippines, Indonesia and Japan exemplify countries where private sector involvement has been prominent, showcasing the value of businesses as key stakeholders in disaster resilience and management. However, challenges remain, and ongoing efforts are needed to ensure effective private sector engagement and to address issues related to coordination, legislation, incentives, and long-term sustainability.

5 Private Sector Engagement in Disaster Risk Management in Other Regions

Private sector engagement in humanitarian action and resilience building is not limited to the Asia Pacific region; it has gained attention and momentum in other regions as well. While the Asia Pacific region has been a pioneer in promoting private sector involvement, other regions have also recognized the value of business engagement in disaster resilience and management.

In North America, the private sector has played a significant role in disaster risk reduction and response. In the context of the USA, commentators have observed a growing trend of private sector entities playing a larger role in crisis and disaster management. This trend is linked to the broader concept of neoliberalization, which involves a shift from the state or public sector to non-governmental organizations, including commercial organizations, taking on responsibilities traditionally handled by the government [15]. In Mexico, since 2006 the government has leveraged private sector financing for the country's Natural Disaster Contingency Fund, becoming the first government to issue a parametric catastrophe bond to transfer Mexico's earthquake risk to the international capital markets [4]. In Latin America and the Caribbean, private sector engagement in disaster risk reduction has been gaining momentum. Regional organizations, such as the Caribbean Disaster Emergency Management Agency (CDEMA) and the Inter-American Development Bank (IDB), have recognized the importance of involving businesses in disaster management initiatives [8]. Countries in Central America, namely Costa Rica, Guatemala, Honduras, Nicaragua, Panama and El Salvador have included provisions for the role and involvement of the private sector in their national DRM policy frameworks [16].

Private sector entities have contributed to disaster response and recovery efforts by providing financial resources, technical expertise, and logistical support.

Meanwhile in Africa, the private sector's role in disaster risk reduction is also evolving. Various initiatives and partnerships have been established to enhance private sector engagement in disaster management and resilience building. For example, the African Risk Capacity (ARC) is a specialized agency that collaborates with governments, businesses, and other stakeholders to provide financial protection against climate-related disasters [19]. Private sector organizations in Africa have contributed to disaster response and recovery efforts through financial contributions, technical expertise, and the provision of essential goods and services (van Niekerk et al. 2015). Companies from prominent economies like South Africa, Nigeria, Egypt, Angola, and Algeria have demonstrated engagement in CSR initiatives and Public–Private Partnerships (PPPs) that actively contribute to disaster risk reduction (ibid). In the case of South Africa, a private insurance company designed their CSR strategies to tackle socio-economic challenges in collaboration with a youth development organization thereby contributing to mitigating the factors that render communities susceptible to disasters [3]. Efforts have also emerged from the private sector within the region, aiming to enhance pre-disaster preparedness. A notable initiative is the Business-led Emergency Operations Center (BEOC) located in Malagasy, Madagascar. The establishment of this center, led by the Private Sector Humanitarian Platform (PSHP) in collaboration with key partners including the National Office for Disaster Risk Management (BNGRC), the Directorate General of Meteorology, the United Nations Office for the Coordination of Humanitarian Affairs (OCHA), and UNDP, serves a dual purpose (ibid). It seeks to bolster the disaster response capabilities of Malagasy businesses in the face of natural hazards like droughts and cyclones while fostering effective collaboration among private sector entities, humanitarian partners, and governmental bodies, all coordinated under the Government's leadership.

In Europe, the engagement of the private sector in disaster management or civil protection has been more limited [2]. Some EU Member States and NGOs surveyed by DG-ECHO have been skeptical about prioritizing collaboration with the private sector, except when such engagement aims to enhance humanitarian response and as long as business partners comply with humanitarian principles and standards [24]. However, recognizing the potential benefits of private sector involvement, efforts have been made to encourage and facilitate partnerships between the business sector and disaster management agencies [17]. As the main agency responsible for humanitarian emergency and disaster response on behalf of the European Union (EU), the Directorate-General for European Civil Protection and Civil Aid Operations (DG-ECHO) several approaches to enhance private sector engagement in the field have been identified including partnerships for resource mobilization implementation, innovation, system coordination and advocacy, However, the actual outcome and tangible results of such collaborations are yet to be fully realized (ibid).

6 Private Sector Engagement Toward Attaining Sustainable Development

Since the inception of the Sustainable Development Goals (SDGs) in 2015, there has been a notable emphasis on engaging the private sector in their achievement [30]. The UN Global Compact (UN, 2012: 46) states that: 'the implementation of sustainable development will depend on active engagement of both the public and private sectors…. We support national regulatory and policy frameworks that enable business and industry to advance sustainable development initiatives considering the importance of corporate social responsibility. We call on the private sector to engage in responsible business practices, such as those promoted by the UN Global Compact". As governments take the lead in crafting policies to support the achievement of the SDGs, the private sector is seen to help realize social and economic progress [39]. It is recognized that while businesses are motivated by profit, they also aspire to improve their local communities and safeguard the environment [28]. In addition, it is noted that the private sector has already been implementing voluntary initiatives as its contribution to attaining the SDGs by 2030 [29].

Sustainable development and disaster risk reduction are interconnected, with each influencing the other. Disasters incur significant costs to societies through financial losses, infrastructure damage, and loss of life, often setting back development progress for years [34]. Environmental degradation and unsustainable development practices worsen disaster risk and its impacts. Additionally, climate change further increases risk and uncertainty, compounding the challenges faced in achieving sustainable development goals. In this context, the private sector also has the capacity to support societal climate resilience toward sustainable development. Aside from its provision of finance, the private sector can also be adaptors and solution providers [23]. It is argued that "the private sector is an indispensable partner in challenging complex issues in the current interdependent and interconnected world" [31].

Moreover, private sector initiatives in DRM have the potential to catalyze broader sustainable development outcomes, including poverty alleviation and climate action [30]. By investing in local communities, supporting livelihoods, and fostering inclusive growth, businesses can contribute to social resilience and equitable development. Leveraging technology, innovation, and cross-sectoral partnerships, the private sector can drive sustainable solutions to complex challenges, from climate change adaptation to disaster recovery and reconstruction.

7 Challenges and Barriers to Private Sector Engagement in DRM and Humanitarian Action

This section highlights several noteworthy examples of the private sector's proactive involvement in DRM to enhance community resilience. It showcases commercial enterprises implementing their initiatives and successful collaborations between traditional stakeholders in the field and businesses. However, alongside these achievements, significant challenges and barriers persist in effectively engaging the private sector.

First, the private sector is still considered as an "under-appreciated actor" in disaster management and resilience building efforts, particularly in many Asian countries [5]. This is despite the fact that the Sendai Framework emphasizes the importance of private sector collaboration with other stakeholders, including governments, civil society organizations (CSOs), and academia. It is essential to view the private sector as a complementary partner rather than a panacea to address resource limitations and technical expertise challenges in disaster management. Multi-stakeholder coordination should be mainstreamed to enhance cooperation among governments, NGOs, and the private sector. Regional organizations and international development partners play a crucial role in facilitating connections between these stakeholders and identifying areas of partnership and cooperation.

Second, businesses must incorporate disaster risks into their management practices, a critical task particularly challenging for micro and small enterprises due to their resource constraints. Given their exposure and vulnerability, micro and small businesses face difficulties in engaging in humanitarian action and resilience building initiatives. Failing to integrate disaster risks into their practices leaves businesses focused on immediate operational challenges during disruptions, restricting their ability to contribute effectively to development efforts for their partners and local communities.

Third, while the value of private sector engagement in DRM is increasingly recognized, there is still a lack of consensus regarding the definitive business case or motivation for corporate sector involvement. Many still perceive DRM as primarily the responsibility of government agencies, development partners, NGOs, and CSOs. Businesses, being profit-oriented, require a clear understanding of the value of their engagement in DRM initiatives at various levels, including global, regional, national, and local. Establishing a coherent and definitive business case is essential, addressing the need to safeguard operations during disruptive events and contributing to the resilience of communities and societies [40].

The private sector's growing role in DRM toward community resilience and sustainable development challenges traditional assumptions about which organizations can make meaningful contributions in this field. Business leaders and development partners have a crucial role to play in advocating for the value of businesses alongside traditional DRM stakeholders. Their collective efforts can help foster a better understanding of the benefits of private sector engagement and strengthen the overall effectiveness of DRM initiatives.

Private sector engagement in DRM holds great potential for addressing the complex challenges of disaster management and resilience building. While notable examples of successful collaborations exist, barriers to effective engagement persist. Integrating disaster risks into business practices and establishing a coherent business case are crucial steps to ensure meaningful private sector involvement. By recognizing the value that businesses bring and fostering collaboration between stakeholders, the full potential of private sector engagement in DRM can be harnessed to enhance overall disaster resilience and contribute to sustainable development.

8 Conclusion

The increasing frequency and severity of disasters, exacerbated by climate change, highlight the critical need for multi-stakeholder engagement in DRM. Private sector engagement is essential for achieving SDGs. Frameworks like the UN Global Compact emphasize responsible business practices and strategic partnerships. Businesses, driven by profit motives and a commitment to community and environmental well-being, are increasingly aligning their operations with the SDGs.

The interconnectedness of DRM and sustainable development is clear. Disasters hinder development, while unsustainable practices increase disaster risks. The private sector, with its resources, expertise, and innovation, can significantly contribute to climate resilience and sustainable development. This involvement safeguards business interests and supports community resilience and inclusive growth.

However, challenges persist, particularly for small and micro-enterprises that face resource constraints and lack a clear business case for DRM. Many still view DRM as primarily the responsibility of governments and NGOs. Addressing these challenges requires viewing the private sector as a complementary partner in DRM, alongside governments, NGOs, and international organizations.

Successful examples from the Philippines, Indonesia, and Japan demonstrate the potential for businesses to contribute to disaster resilience. By leveraging resources, expertise, and innovation, businesses can enhance community resilience and support sustainable development. Promoting knowledge exchange, developing supportive frameworks, and documenting successful collaborations are essential for harnessing the private sector's potential in an increasingly disaster-prone world.

Acknowledgements The authors acknowledge the support and guidance from the Asian Disaster Preparedness Center (ADPC) in the development of this paper.

References

1. Adhikari B, Safaee Chalkasra LS (2021) Mobilizing private sector investment for climate action: enhancing ambition and scaling up implementation. J Sustain Financ Invest pp 1–18
2. Boin A, Rhinard M, Visuri P (2014) Civil security and the European union: a survey of European civil security systems and the role of the EU in building shared crisis management capacities. https://www.diva-portal.org/smash/get/diva2:774421/FULLTEXT01.pdf. Accessed 19 Jul 2022

3. Bopape M, Nemakonde LD, Fourie K (2021) Private companies and community collaboration: towards building disaster resilience in Diepsloot, Johannesburg, South Africa. Jàmbá J Disaster Risk Stud 13(1)
4. Cardenas V, Hochrainer R, Mechler R, Pflug G, Linneroothbayer J (2007) Sovereign financial disaster risk management: the case of Mexico. Environ Hazards 7(1):40–53. https://doi.org/10.1016/j.envhaz.2007.04.005
5. Chatterjee R (2021) Private sector: under-appreciated actors in disaster response and recovery. In: Cook ADB, Gong L (eds) Humanitarianism in the Asia-Pacific. Springer, pp 15. https://doi.org/10.1007/978-981-33-4874-5_15
6. Chatterjee R, Shaw R (2015) Role of regional organisations for enhancing private sector involvement in disaster risk reduction in developing Asia. In: Izumi T, Shaw R (eds) Disaster management and private sectors, Springer, pp 4. https://doi.org/10.1007/978-4-431-55414-1_4
7. Clark-Ginsberg A (2020) Disaster risk reduction is not 'everyone's business': evidence from three countries. Int J Disaster Risk Reduct 43:101375
8. Collymore J (2011) Disaster management in the Caribbean: perspectives on institutional capacity reform and development. Environ Hazards 10(1):6–22
9. Connecting Business initiative (2021) Responding to multiple disasters: a case study on the Philippine disaster resilience foundation's emergency operations center and response to super Typhoon Goni and Vamco during the COVID-19 pandemic. https://www.connectingbusiness.org/sites/default/files/downloads/publication/PDRF%20Case%20Study_%20Responding%20to%20Multiple%20Disasters%20%28FINAL%20ao%2030June2021%29.pdf. Accessed 19 Jul 2022
10. Cozzolino A (2021) Platforms enhancing the engagement of the private sector in humanitarian relief operations. Sustainability 13(3024):1–18. https://doi.org/10.3390/su13063024
11. Crick F, Gannon KE, Diop M, Sow M (2018) Enabling private sector adaptation to climate change in sub-Saharan Africa. Wiley Interdiscip Rev Clim Chang 9(2):e505
12. Development Initiatives (2017) Global humanitarian assistance report 2017. https://reliefweb.int/report/world/global-humanitarian-assistance-report-2017. Accessed 19 Jul 2022
13. Fuchs A, Ohler H (2021) Does private aid follow the flag? An empirical analysis of humanitarian assistance. World Econy 44:671–705. https://doi.org/10.1111/twec.13021
14. Global Facility for Disaster Reduction and Recovery (2020) Private sector participation in disaster recovery and mitigation. https://www.gfdrr.org/en/publication/private-sector-participation-recovery. Accessed 19 Jul 2022
15. Gotham KF (2012) Disaster, inc.: privatization and post-Katrina rebuilding in New Orleans. Perspect Polit 10(3):633–646
16. Hori T (2015) Elements to enhance private sector engagement in disaster risk reduction in central America. In: Izumi T, Shaw R (eds) Disaster management and private sectors: challenges and potentials, Springer, pp 105–122
17. Hoxtell WADE, Norz M, Teicke K (2015) Business engagement in humanitarian response and disaster risk management. Global public policy institute, no. 1–74. https://www.gppi.net/media/Hoxtell_et_al_2015_Biz_Engagement_Humanitarian_Repsponse.pdf. Accessed 19 Jul 2022
18. International Labour Organisation (2015) Needs assessment for the private sector in the Philippines: disaster preparedness, response and recovery. https://www.ilo.org/asia/publications/WCMS_368162/lang--en/index.htm. Accessed 19 Jul 2022
19. Iyahen E, Syroka J (2018) Managing risks from climate change on the African continent: the African risk capacity (arc) as an innovative risk financing mechanism. In: Resilience, Elsevier, pp 243–252
20. Izumi T, Shaw R (2015) Overview and introduction of the private sector's role in disaster management. In: Izumi T, Shaw R (eds) Disaster management and private sectors: challenges and potentials, Springer, pp 1–10
21. Johnson DAK, Abe Y (2015) Global overview on the role of the private sector in disaster risk reduction: Scopes, challenges, and potentials. In: Izumi T, Shaw R (eds) Disaster management

and private sectors: challenges and potentials, Springer, pp 1–10. https://doi.org/10.1007/978-4-431-55414-1_2
22. Lienert I (2009) Where does the public sector end and the private sector begin? IMF working Papers (122):A001. https://doi.org/10.5089/9781451872699.001.A001. Accessed 19 Jul 2022
23. Lu X (2022) Accelerating private sector engagement in adaptation in Asia and the Pacific. ADB sustainable development working paper series, no. 84. https://www.adb.org/sites/default/files/publication/838831/sdwp-084-private-sector-adaptation-asia-pacific.pdf. Accessed 19 Jul 2022
24. Luchner J (2013) The Union's humanitarian aid: fit for purpose? Summary of responses to the stakeholder consultation. European union agency for Asylum. Malta. https://policycommons.net/artifacts/3364414/the-unions-humanitarian-aid/4163069/. Accessed 19 Jul 2022
25. Medel K, Kousar R, Masood TA (2020) Collaboration-Resilience framework for disaster management supply networks: a case study of the Philippines. J Humanit Logist Supply Chain Manag 10:509–553. https://doi.org/10.1108/JHLSCM-09-2019-0066
26. Miyamoto International (2021) Success story Jakarta disaster contingency planning and private sector participation. https://miyamotointernational.com/wp-content/uploads/USAID-ADVANCE-Story-2-PSE-in-DRM_Apr-2021_TO-2019_09b-FINAL.pdf. Accessed 20 Jul 2022
27. Oglesby R, Burke J (2012) Platforms for private sector—humanitarian collaboration. Deloitte. http://www2.deloitte.com/global/en/pages/about-deloitte/articles/private-humanitarian-sector-collaboration-platforms.html. Accessed 19 Jul 2022
28. Pérez L, Hunt V, Samandari H, Nuttall R, Biniek K (2022) Does ESG really matter—and why? McKinsey sustainability. https://www.mckinsey.com/~/media/mckinsey/business%20functions/sustainability/our%20insights/does%20esg%20really%20matter%20and%20why/does-esg-really-matter-and-why-vf.pdf. Accessed 19 Jul 2022
29. Rashed AH, Shah A (2021) The role of private sector in the implementation of sustainable development goals. Environ Dev Sustain 23:2931–2948. https://doi.org/10.1007/s10668-020-00718-w
30. Scheyvens R, Banks G, Hughes E (2016) The private sector and the SDGs: the need to move beyond 'business as usual'
31. Shimoda Y (2023) Toward the creation of effective partnerships with the private sector for sustainable development. In: Urata S, Kuroda K, Tonegawa Y (eds) Sustainable development disciplines for humanity: breaking down the 5Ps—people, planet, prosperity, peace, and partnerships, Springer, pp 149–166. https://doi.org/10.1007/978-981-19-4859-6
32. Taylor R, Amankwah F, Nalamada P (eds) (2016) Engaging the private sector and developing partnerships to advance health and the sustainable development goals, National Academies Press, Washington, D.C. https://doi.org/10.17226/23617
33. Twigg J (2002) Corporate social responsibility and disaster reduction: a global overview. Disaster Prev Manag 11(3):223–224. https://doi.org/10.1108/dpm.2002.11.3.223.7
34. Uitto JI, Shaw R (eds) (2016) Sustainable development and disaster risk reduction: introduction. In: Uitto J, Shaw R (eds) Sustainable development and disaster risk reduction, Springer
35. UNDRR (2019) Private sector. united nations office for the disaster risk reduction. http://www.undrr.org/implementing-sendai-framework/partners-and-stakeholders/partnerships-private-sector. Accessed 19 Jul 2022
36. UNDRR (2020a) Reducing risk and building resilience of SMEs to disasters, United nations office for the disaster risk reduction. https://www.undrr.org/media/48692/download. Accessed 19 Jul 2022
37. UNDRR (2020b) Private sector experience in Japan: supporting disaster. http://www.undrr.org/news/private-sector-experience-japan-supporting-disaster-preparedness-evacuations-under-covid-19. Accessed 19 Jul 2022
38. UNDRR (2021) Scaling up disaster risk reduction in humanitarian action 2.0: recommendations for the humanitarian programme cycle, United nations office for the disaster risk reduction. https://www.undrr.org/publication/scaling-disaster-risk-reduction-humanitarian-action, Accessed 19 Jul 2022

39. UNESCAP (2022) Advancing public-private partnerships and private sector networks to accelerate climate action and the implementation of the 2030 Agenda for sustainable development. Economic and social commission for Asia and the Pacific. https://www.unescap.org/sites/default/d8files/knowledge-products/2022_ESCAP_Knowledge_Product_Public-private-partnerships-private-sector-networks-climate-action-2030-Agenda.pdf. Accessed 19 Jul 2022
40. UNESCAP, ADPC, Ready Asia Pacific (2015) Resilient business for resilient nations and Communities. United nations economic and social commission for Asia and the Pacific, Asian disaster preparedness center & ready Asia Pacific. https://www.unescap.org/resources/resilient-business-resilient-nations-and-communities-0#:~:text=Resilient%20Business%20for%20Resilient%20Nations%20and%20Communities%20shows%20how%20the,country's%20overall%20resilience%20to%20disasters. Accessed 19 Jul 2022
41. UNOCHA (2017) The business case: a study of private sector engagement in humanitarian action. United nations office for the coordination of humanitarian affairs. https://www.unocha.org/sites/unocha/files/PSS-BusinessCase-FINAL_0.pdf. Accessed 19 Jul 2022
42. UNOCHA, DHL (2016) Combining capabilities: how public private partnerships are making a difference in humanitarian action. United nations office for the coordination of humanitarian affairs & DHL. https://eurodad.org/files/pdf/1546956-history-repppeated-how-public-private-partnerships-are-failing-.pdf. Accessed 19 Jul 2022
43. Izumi T, Shaw R, Djalante R, Ishiwatari M, Komino T (2019) Disaster risk reduction and innovations. Prog Disaster Sci 2:100033. https://doi.org/10.1016/j.pdisas.2019.100033
44. Chatterjee R, Cook ADB, Gong L (2021) Humanitarianism in the Asia-Pacific Engaging the Debate in Policy and Practice Private Sector: Under-Appreciated Actors in Disaster Response and Recovery Springer, Singapore, pp 85–89
45. Windiani W (2021) Pentahelix collaboration approach in disaster management: case study on disaster risk reduction forum-east. Java IPTEK J Proc Series 0(7):71. https://doi.org/10.12962/j23546026.y2020i7.9540

Land Readjustment Approach as an Urban Development Tool Towards Disaster Resilience: A Study of Japan's Experience and Its Applicability in the Philippines

Arlene D. R. Santiago and Tabassam Raza

Abstract Urbanization is a prevailing phenomenon worldwide. Similarly, consequences of disasters resulting from hazards are observed on a global scale. Japan exhibits both occurrences as an archipelago situated along the Pacific Ring of Fire. Japan faces a heightened risk of geologic hazards, including earthquakes, volcanic eruptions, and resulting tsunamis. Additionally, the country is susceptible to hydrometeorological hazards. The escalating global temperatures have contributed to the intensification of typhoons originating in the Pacific, leading to more frequent and severe impacts on Japan. In the midst of these challenges, Japan has undergone a process of extensive urbanization, with an urbanization rate of 91.7% as of 2021. Notably, land readjustment has played an essential role in Japan's urban development. Despite these vulnerabilities, Japan has distinguished itself with exceptional disaster management strategies that have garnered recognition from numerous nations. This research constitutes an exploratory investigation into the interplay between land readjustment practices and disaster resilience, drawing insights from case studies in Japan. This research employs a qualitative methodology, focusing primarily on an in-depth investigation through Key Informant Interviews (KII) and review of relevant literature. The core discussions within this study include the following themes: A review of the implementation of land readjustment in Japan; an enquiry on the role of land readjustment as disaster risk management tool within the Japanese context; and an exploration of potential lessons that the Philippines can draw from Japan's experiences with land readjustment, considering the shared characteristics in terms of geography, topography, and government structure. It provides valuable insights into the realm of policy, as well as strategies for its both developing and implementing. Further, policy formulation can be recommended for those readjustments that are not applicable due to lack of available policies in the Philippine context.

A. D. R. Santiago (✉) · T. Raza
University of the Philippines School of Urban and Regional Planning, Quezon City, Philippines
e-mail: adsantiago@up.edu.ph

T. Raza
Planning and Development Research Foundation, Inc, Quezon City, Philippines

I. Pal et al. (eds.), *Proceedings of the 3rd International Symposium on Disaster Resilience and Sustainable Development—Volume 1*, Lecture Notes in Civil Engineering 623, https://doi.org/10.1007/978-981-96-2139-2_17

Keywords Land readjustment · Spatial planning · Japan · Resilience · Disaster management

1 Introduction

Rapid urbanization across continents and disasters brought about by hazards are two major occurrences observable in the world today. Nations, along with international communities, continue venture with measures on how the incessantly growing population can prosper in a dynamic environment. These changes in the environment have attributed to anthropogenic, geologic, or hydro-meteorological nature. Consequently, a pressing challenge amidst this situation is how to attain resilience and sustainable development. Urbanization demands housing, transportation, infrastructure, social services, and employment requirements, among others. On the other hand, a continuously changing environment requires adaptation and resilience of its inhabitants to thrive. This study embarks on an exploratory journey to understand the dynamic relationship between land readjustment practices and disaster resilience, using Japan as a case study with examples of land readjustment (LR) done in different cities. Japan's urbanization journey can be traced back to centuries past. The Location Map of case study (Fig. 1) shows arrows pointing cities that are considered as examples of successful LR practices. The major criteria to select these cities is the post-disaster reconstruction sites land readjustment. Further, through a qualitative approach, the researchers investigate deeply into this relationship by conducting Key Informant Interviews (KII) and thoroughly reviewing pertinent literature. These methods offer rich, refined insights into how LR can enhance urban resilience in the face of natural disasters. The core questions this research aims to answer are: What is the role of land readjustment as a disaster risk management tool within the Japanese context? And what potential lessons can the Philippines draw from Japan's experiences with land readjustment for disaster risk reduction?

Japan's geography, topography, and government structure share similarities with many countries, making its urbanization process relevant for study. Sorensen [37] describes Japan's transformation from a rural country to an urbanized industrial state with distinctive urbanization patterns rooted in its pre-modern urban system, traditions, and material culture. Land readjustment (LR), known as "The Mother of City Planning" or Toshi Keikaku no Haha, has been pivotal in this transformation [36]. Approximately one-third of Japan's urban areas have developed through LR [27, 36].

LR, or kukaku-seiri, involves pooling ownership of scattered agricultural plots, constructing roads and infrastructure, and subdividing the land into urban plots [36]. It increases land value through development and shares profits with land rights holders and stakeholders [24]. The Land Readjustment Law (LRL) No. 119, established on May 20, 1954, regulates LR implementation and is divided into seven chapters [38].

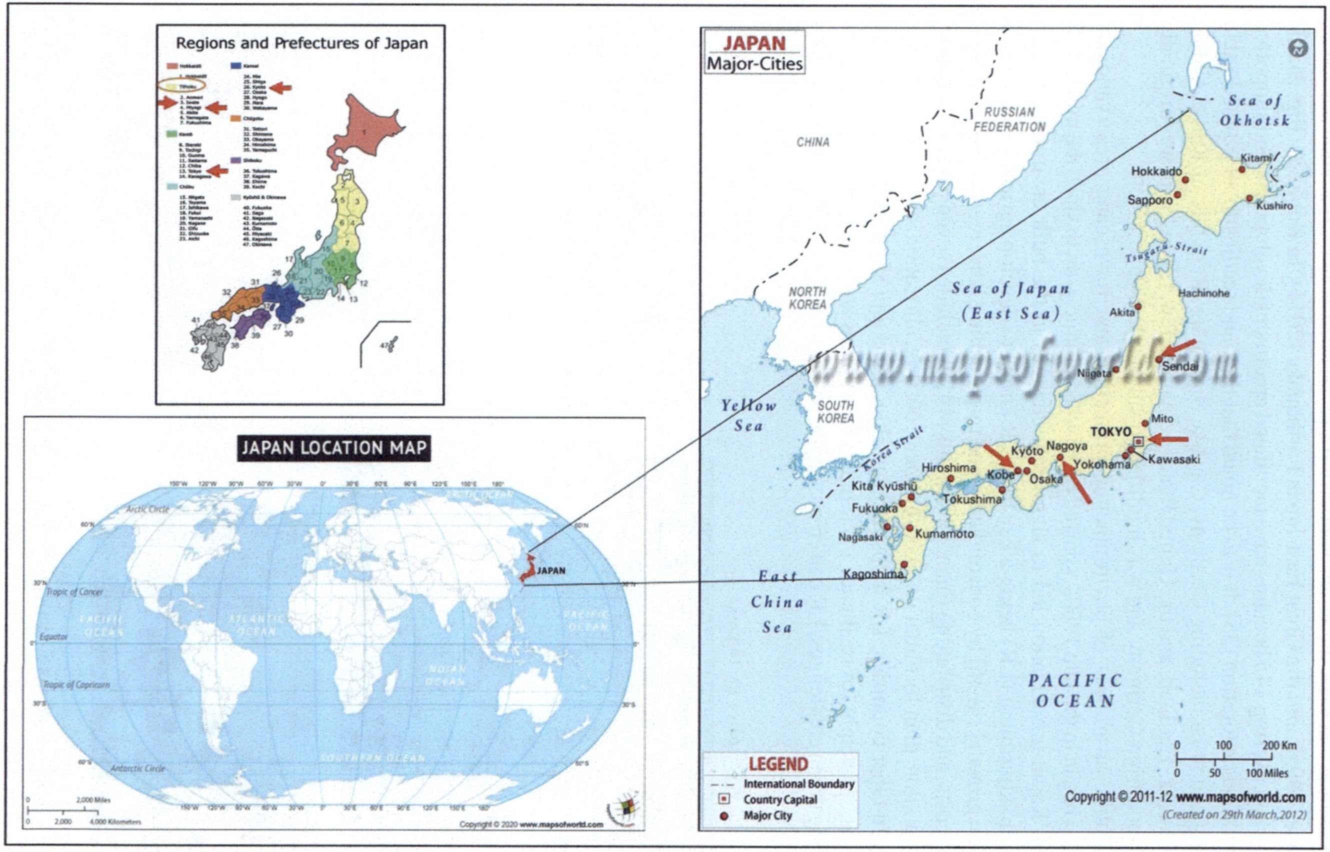

Fig. 1 Location map of Japan with red arrows showing the case studies cities (Modified after mapsoftworld [23])

LR has also been crucial for urban redevelopment and post-disaster reconstruction, applied after WWII, the Great Kanto Earthquake (1923), the Great Hanshin Earthquake (1995), and the Great East Japan Earthquake (2011). In Kobe (see Fig. 2), LR projects covering 145 hectares have facilitated apartment development and supported livelihood restoration [25]. It is to be noted that the Land readjustment is practiced in many countries including Germany, Canada, Singapore, Republic of Korea, China, Indonesia, Thailand, and the United States, to name a few. LR in Japan was originally adopted from Germany and was modified to suit the Japanese setting.

Public awareness of environmental protection and townscape preservation in Japan began in the late 1960s, spurred by rapid economic growth since the mid-1950s [12]. Rapid urbanization led to significant changes in the built environment, with land readjustment (LR) being a key tool. LR aimed to develop agricultural lands for urban growth [14, 36] by consolidating and re-parcelling land, recovering infrastructure costs, and distributing financial benefits [14]. It includes seven phases: initiation, declaration, plan formulation, measurement, costing, allocation, and implementation [8]. In other countries, LR helps manage slum areas and supports the "Build Back Better" principle for resilience [47]. In the Philippines and other developing nations facing urban challenges, LR can create inclusive, safe, resilient, and sustainable cities

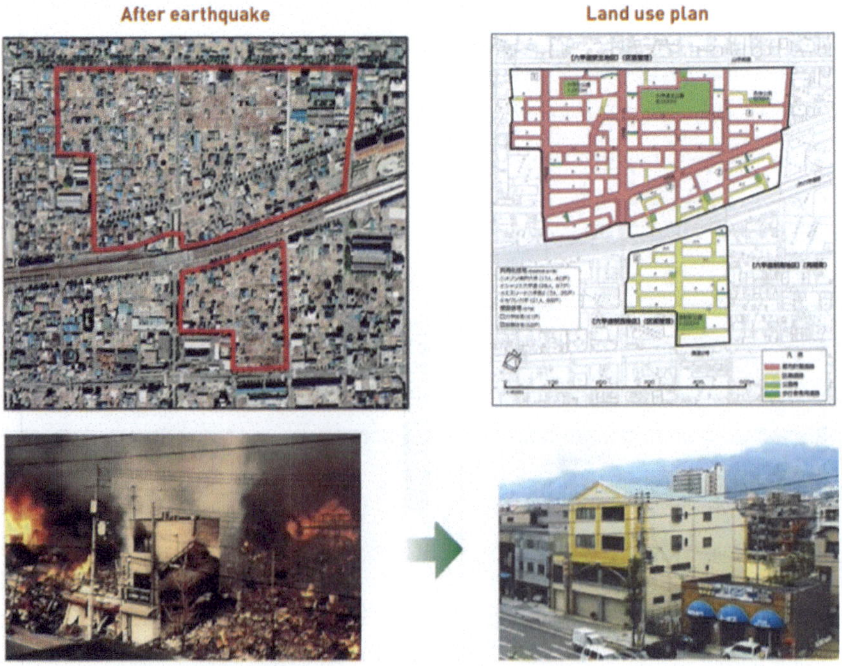

Fig. 2 Rokkomichi north station LR project and Rokkomichi western station LR Project, Kobe City. (*Source* [25])

[15]. This research provides policy insights to enhance LR practices and address gaps in the Philippine context.

1.1 Statement of the Problem

Urbanization is a widespread global trend, paralleled by the far-reaching impacts of disasters triggered by various hazards. Japan, situated in the Pacific Ring of Fire, epitomizes this dual reality, the target of several natural disasters. The nation faces heightened risks from geological events like earthquakes, volcanic eruptions, and ensuing tsunamis, as well as hydro-meteorological hazards. The rising global temperatures further exacerbate this situation, leading to more frequent and severe typhoons originating in the Pacific, which significantly affect Japan. Amidst these challenges, Japan has undergone extensive urbanization, reaching a rate of 91.7% as of 2021. These environmental changes stem from anthropogenic, geological, and hydro-meteorological factors. Consequently, the critical issue facing Japan is how to achieve resilience and sustainable development amidst these intertwined challenges. Over the years, Land Readjustment (LR) has also been used in Japan as a primary tool for urban redevelopment and post-disaster reconstruction.

1.2 Objectives

- Review the implementation of land readjustment in Japan
- Find valuable insights of Land Readjustment policies, as well as strategies for Japan's development and implementation.
- Investigate the interplay between land readjustment practices and disaster resilience, drawing insights from case studies in Japan.
- Find out the potential lessons that the Philippines can draw from Japan's experiences with land readjustment.
- Explore applicability of Land Readjustment in the Philippines context.

2 Literature Review

2.1 Japan's Natural and Physical Environment

Japan frequently experiences fatalities and property loss due to various hazards, including earthquakes and volcanoes. Sitting on the "Ring of Fire," Japan has about 10% of the world's active volcanoes, with 110 active volcanoes (Fig. 3) [6]. Its exposure to natural threats has led Japan to develop robust systems and mechanisms for

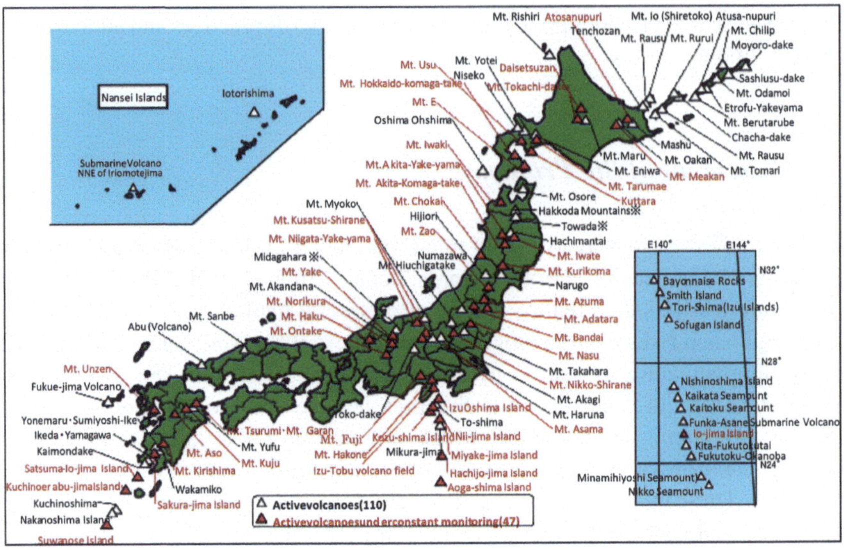

Fig. 3 Distribution of active volcanoes in Japan. (*Source* Cabinet office Japan, White Paper on Disaster Management)

hazard response. The Japan Meteorological Authority (JMA) monitors 47 volcanoes 24/7 for eruption warnings. These physical attributes necessitate effective disaster risk reduction (DRR) efforts. The UN defines DRR as actions to reduce disaster risks and impacts through systematic efforts [44]. Japan's Disaster Countermeasures Act of 1962 supports its DRR plans. Despite natural hazards, Japan continues to develop its urban landscape, emphasizing preparedness and vulnerability reduction.

2.2 Japan's Governance Structure and Urban Planning System

Japan has a parliamentary form of government. It is also a constitutional monarchy; however, the Imperial Throne has no sovereign power. The emperor performs state matters and ceremonial functions as a symbol of Japan and its people while the Prime Minister derives its authority from its 1947 Constitution [41].

Japan has a national government and a two-tiered local government structure comprising 47 prefectures and municipalities within those prefectures. As of January 1, 2015, Japan had 790 cities, 745 towns, and 183 villages, along with special local entities like wards, cooperatives, and public property districts [42]. The urban planning system in Japan operates at three levels: national, prefecture, and city (Fig. 4) [25]. Prefecture and city-level urban planning are crucial for urban management,

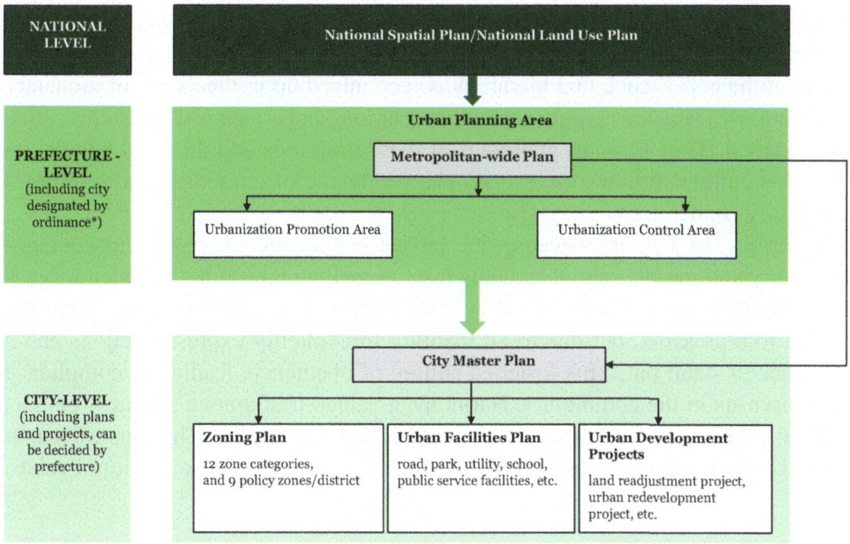

Fig. 4 Urban planning system in Japan

while the central government dominates Tokyo's planning [11, 19]. National governments are essential in disaster risk reduction and management [16], providing strong leadership and oversight for local governments [33]. Decision-making authority is given to local governments [40], supported by Japan's Local Autonomy Law [39]. Japan's rule of law balances societal interests, ensuring laws serve the common good [31].

2.3 Community Participation in Japan

Enhancing engagement of local populations [35] as well as international actors, is significant in Japan's local arena. With respect to DRRM, community participation in Japan [2] has received attention which extends beyond the local setting. Consequently, assistance extended during disaster response facilitated the formal structure of volunteerism in Japan. When disaster strikes, not only the national government, LGUs, and local community are mobilized in response operations, the international community likewise enters the scene to deliver humanitarian aid.

Japan has enjoyed a tradition of collaborative action as demonstrated in their local communities. Not only during the need to protect one's community (such as during disasters) do local communities come together, but the roots of this collective action or consensus can be traced back centuries ago [43]. Japanese people come together to observe various festivals or matsuri several times a year. Many festivals are celebrated in relation to agricultural activities, some are influenced by the Shinto religion which is largely recognized in Japan. These matsuris also function as a way to connect the community as a collective body—the role of temples and their belief system at the core—such that hierarchy is recognized up to the extent of monetary contributions. There are neighbourhood associations that organize these events called the chonaikai. They serve as guardians of these traditions and the local population [43]. This cultural framework also provides a degree of consensus and reinforces community connections.

According to key informants, the Japanese educational system discourages elementary and middle school students from "speaking their mind" (Yoneno-Reyes 2023). Students are expected to follow school authorities without voicing opinions, not due to repression, but due to an inability to explicitly express opinions and a reluctance to stand out. This fosters a culture of obedience, leading to compliance and consensus in the community. Harmony, a valued trait among Japanese people, stems from the need for peace in a confined island nation [30]. Children are taught to avoid inconveniencing others, highlighting consideration as a key cultural value.

2.4 Overview of the Physical and Political Environment of the Philippines

Some similarities in the physical attributes of Japan and the Philippines are observable. Geographically, both lie along the Pacific Ring of Fire and are in constant risks of tremors and volcanic eruptions. See the distribution of active faults and trenches in the Philippines in Fig. 5.

Moreover, the Philippines is also an archipelago composed of island groups that experience torrential rains and tropical cyclones throughout the year. As such, earthquakes, volcanic eruptions, flooding, landslide, among others, are but some hazards that at times result in exponential negative impacts onto affected population of both nations. Bankoff [3] in 2002 asserts that the Philippines have had long exposure to disasters caused by natural hazards and such experience has affected the "evolution and adaptation of the societies in the archipelago". These frequency and magnitude suggest significant cultural implications among peoples and societies. Given these conditions, the Philippines has set considerably high priority in disaster risk reduction and management which is embedded in policies both in the national and local arena.

Politically, both Japan and the Philippines enjoy a democratic form of government where the citizens have the right to choose their governing officials. The government structure in the Philippines is divided into national government and local government units—composed of provincial, city/municipal, barangay levels, respectively [21]. Participation of constituents in decision making is also practiced through consultations initiated by either the government or the civil society. Ironically, these similarities do not seem to transpire in the overall governance system of the two states.

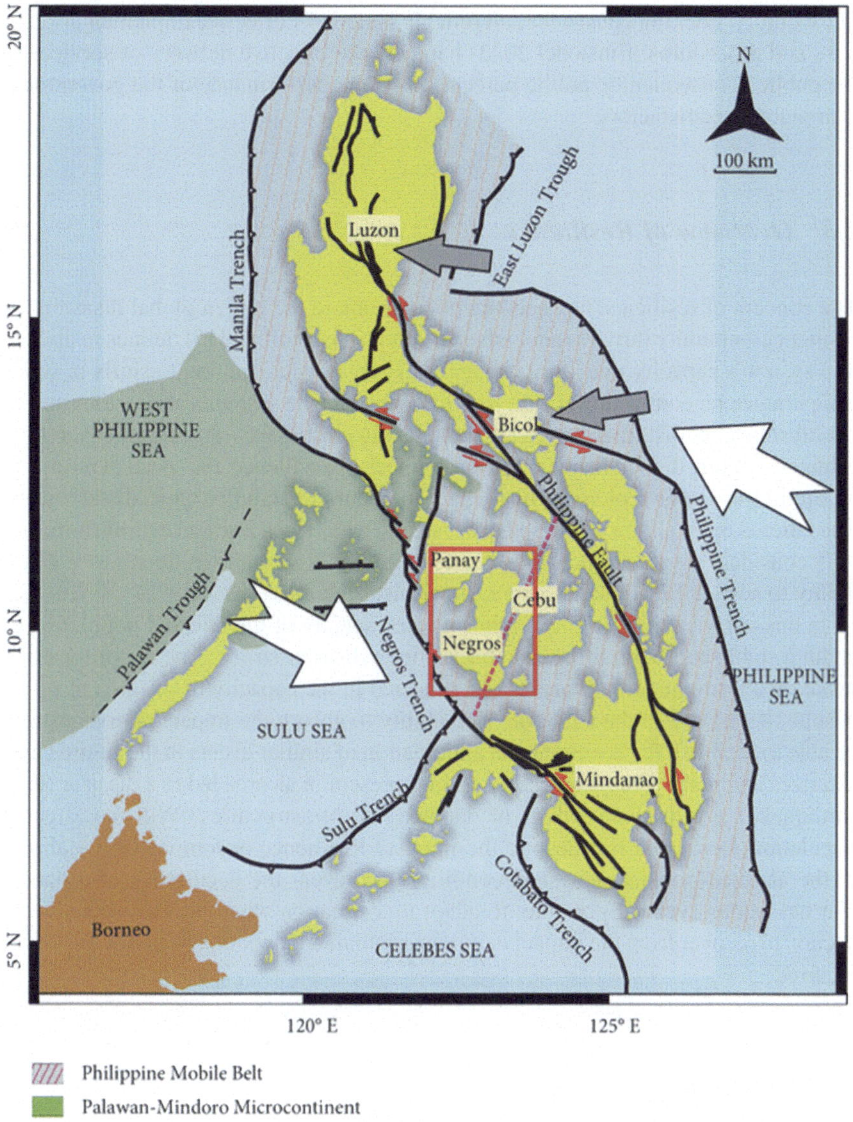

Fig. 5 Distribution of active and inactive and potentially active volcanoes in the Philippines (*Source* [28])

While Japan may have gained a reputation of an "efficient" government, the Philippines struggles to uphold a positive image in terms of governance. Despite the legal measures available, the execution and implementation of policies and regulations appear to fall short of public approval. Some argue that many regulatory policies and administrative procedures in the Philippines are deemed "recommendatory" instead of executory. Lacking consistent and reliable system for effective implementation of laws and procedures (Einsiedel 2023) hampers the effective delivery of services to the public. Consequently, public perception on the performance of the government is arguably unsatisfactory.

2.5 *Overview of Resilience*

The concept of resilience has become popular due to the rise in global disasters. A resilient community thrives under adverse conditions. Holling [13] defines resilience as a system's capacity to endure changes, absorbing and adapting to shifts in states and parameters, contrasting it with stability, which is the system's ability to return to equilibrium after disturbances. Resilience enables entities to withstand and overcome changes. Holling distinguishes between engineering resilience, the speed of returning to equilibrium, and ecological resilience, maintaining stability amid disturbances. The latter is more apt for cities and ecosystems, which lack a single equilibrium. Nel [27] considers resilience non-normative, emphasizing urban resilience as a city's ability to sustain functionality and adapt to disturbances.

In this study, resilience is examined as the capacity of the affected population to attain a stable condition or bounce back after a disaster struck. The conditions that make up a resilient community can be reflected in the capacity of the area to thrive despite the gravity of the situation. The ability to absorb the impact/aftershock and be able to establish improvements in anticipation of similar events in the future characterizes resilience. For instance, if a hazard present in an area led to a disaster (e.g., earthquake), are there casualties or damages in infrastructures? Will the affected population recuperate fast despite the impact? Resilience encompasses the ability of the affected population/system/condition to prevent the occurrence of a particular hazardous event, if possible; to adapt to a situation when the hazardous event occurs; to recover from the impact or damages obtained; or compensate for the losses incurred.

3 Methodology

The first step involved conducting literature reviews to gain a more profound comprehension of the research subject. Subsequently, a concept note for the study was created, encompassing essential elements such as study background, problem statement, research objectives, theoretical framework, and conceptual framework, this methodology, anticipated outcomes, and the study's implications in Urban LR in Philippine context [46]. The study uses qualitative analysis to examine the relationship between land readjustment (LR) and disaster management in Japan, particularly focusing on resilience. A descriptive method was employed to observe and describe events. Reports from actual disaster operations served as the basis for understanding the relationship between variables. While direct case study comparisons were ideal, available reports and data sources were used to explore associations. Primary and secondary data were gathered from official documents, libraries, online sources, documentaries, video clips, and news articles. Key informant interviews (KIIs) with experts, including urban planners from Japan and the Philippines, a former government official, academics, a Japanese resident, and a Taiwanese urban planning expert, provided additional insights [20]. Interviews were conducted in person or online, with some respondents sharing knowledge via email. An interview guide was used. Factor analysis was not feasible due to insufficient sample size, but relationships from existing case studies were identified, providing assumptions for future testing. Each LR attribute was matched with a resilience dimension to evaluate the relationship.

3.1 Theoretical Framework

According to Sorensen [36], landowners are attracted to developing their lands through readjustment in the hope of increasing the profitability and marketability of their land after the process. Serote [34] explained how economic considerations affect the decisions of property owners towards preferred land utilization. Further, he describes how the highest and best use principle provides for the most profitable use to which the property is adapted. As such, value, or "present worth of future benefits arising from the ownership of real property" can be taken as a key consideration in understanding the adaptability of land readjustment. To give emphasis on the significance of developed property, the study uses the urban land value theory developed by economist Eugene F. Brigham. His study was conducted to examine a model of residential land values in Los Angeles, California in the 1960s. This model assumes that the "value (V) of a particular urban area site is functionally related to its accessibility to economic properties (P), to its amenities (A), to its topography (T), to its present and future use (U) and certain historical factors that affect its utilization (H)" [5] (Box 1).

Box 1 Urban Land Value by E.F. Brigham

The value (V) of a particular urban site is a function of its

P = accessibility to economic activities
A = existing amenities
T = topography
U = present and future use
H = historical factors that affect its utilization

Accessibility value—relative ease with which the site occupant can reach other activity areas in the city

Amenity value—subjectively perceived "pleasantness" of the environment in which the site is located

Topography—natural characteristics of the site like slope, elevation, soil structure. It affects both amenity value and development cost

Utilization—existing and desired use of the area where the site is located, as reflected in the current land use map and zoning map. Usually categorized into residential, commercial, industrial, institutional, agricultural, open space, etc.

Historical factors—the succession of the past land uses of the site and of the area. This indicates whether existing use is still appropriate or is no longer suitable.

(*Source* [34])

Following Sorensen's observation, the expected benefits of stakeholders to increase the marketability of its property after engaging in land readjustment process appeal to landowners to participate in the projects.

3.2 Conceptual Framework

The study draws from the "Build Back Better" (BBB) framework of the United Nations Office for Disaster Risk Reduction (UNISDR). The UN General Assembly in 2016 defines BBB as "the use of the recovery, rehabilitation, and reconstruction phases after a disaster to increase the resilience of nations and communities through integrating disaster risk reduction measures into the restoration of physical infrastructure and societal systems, and into the revitalization of livelihoods, economies, and the environment" [45]. The main principle of BBB is to make stakeholders less vulnerable to disasters and strengthen the resilience of nations and communities. Figure 6 suggests that land readjustment influences the resilience of the area. The framework shows that LR has the application for urban reconstruction after disaster impact at a particular site, disaster management measures are employed to address the situation. Part of the recovery and rehabilitation from disaster is to enable the site to be resilient in the event that similar occurrence takes place in the future. In Japan, disaster management strategies integrate land readjustment, particularly during the reconstruction phase in disaster-affected areas. This approach leads to significant improvements in the affected sites, such as the widening of city roads, the creation of community roads and open spaces, the construction of new buildings with earthquake-resistant and fireproof structures, and the installation of fire prevention water tanks. These enhancements equip the area with the capacity to "build back better," ensuring greater resilience and preparedness for future crises.

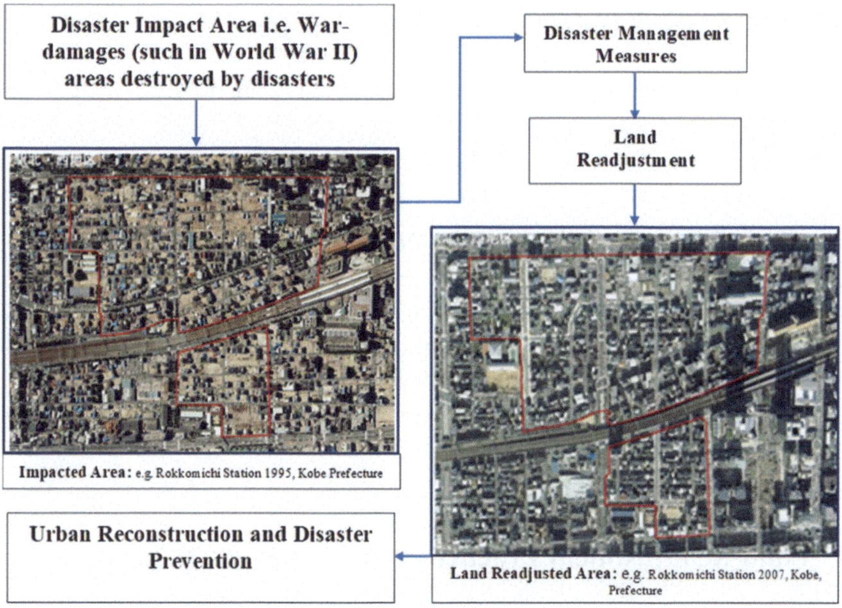

Fig. 6 Conceptual framework for risk sensitive land readjustment (*Source* Modified after Santiago [32] and Raza [29])

3.3 Data Collection

The main approach to gathering data was an extensive examination of literature encompassing policy documents, government reports, academic articles and information available in the library and online pertaining to the Land Readjustment as a mandatory technique practice in Japan since 1954 for post-disaster construction (Secondary Data). The data gathering and analysis for this research were also done using primary data collection through a key informant interview (KIIs) of select respondents or "experts".

3.4 Data Analysis

During this stage, the collected data is thoroughly analysed and interpreted the existing case studies, the researchers were able to draw relationships that provided assumptions for future testing. This process of drawing relationships involves identifying a set of attributes that characterize each LR feature and resilience dimension. Each LR attribute is matched with a resilience dimension to evaluate the relationship.

3.5 Presentation of Findings

In this section, outcomes and explanations derived from the data analysis are presented.

3.6 Conclusion and Recommendation

This section presents an overview of the study's purpose, the main findings, and the implications. Additionally, it offers valuable insights into the realm of policy, as well as strategies for development and implementation. Further, policy formulation is recommended for those readjustments that are not applicable due to lack of available policies in the Philippine context.

4 Land Readjustment Practices and Challenges in Japan

4.1 Urban Planning in Japan

The urban planning system in Japan has three tiers: national, prefectural, and city-level planning. Two zones outline any land use change in an area as dictated by the prefectural government: (1) urbanization promotion area (UPA); and (2) urbanization control area (UCA) (Fig. 4). The UPA strategically promotes urbanization based on the development policy of the prefecture. The UCA, on the other hand, aims at preserving agricultural areas and natural resources thus prohibits building or infrastructure development activities [25]. As in Serote [34], the UPA is the "area of development" while the UCA is the "area of no-development". Urban planning is the institutional framework that dictates LR in Japan. Land readjustment observes the urban planning map and adheres to the upper-level plan. The implementation plan of LR projects must follow designated urban facilities.

4.2 Overview of Land Readjustment Policies and Strategies in Japan

Japan's mountainous terrain limits arable land, necessitating intensive cultivation for rice production, leading to fragmented agricultural lands. Population growth has increased the need for higher yields and efficient land use. Post-World War II land reforms directed by the Allied Forces further fragmented these lands, capping individual holdings and resulting in many plots smaller than two hectares [36]. These

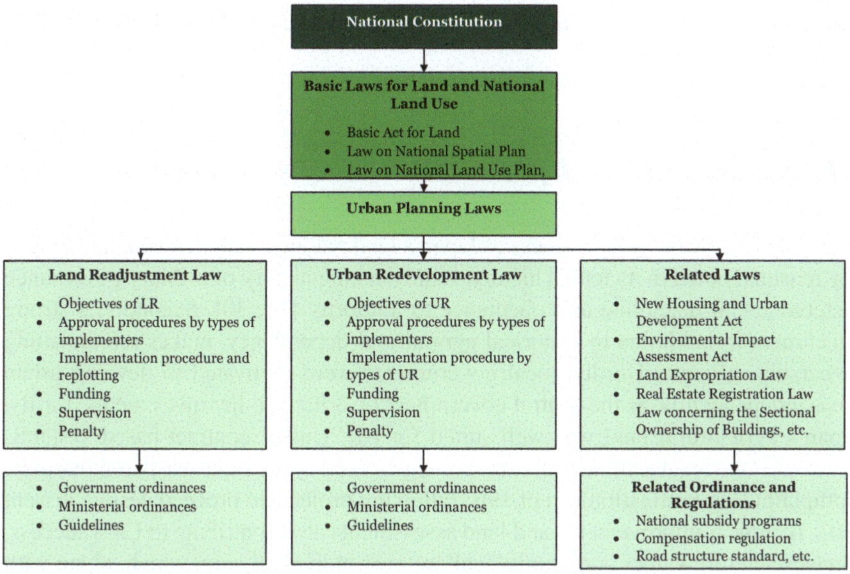

Fig. 7 Legal framework of LR and UR in Japan

reforms, hastily implemented without considering land consolidation, created scattered small farms, making it difficult for farmers to sustain full-time farming and forcing many into part-time farming. Despite this, farmers often retain their land, hindering urbanization, especially on urban fringes [36].

From the 1920s to 1950s, Japan refined land readjustment (LR). Initially governed by the Agricultural Land Consolidation (ALC) Law in the late nineteenth century, it was replaced by the Land Improvement Law in 1949. This transition temporarily undermined LR's legal authority until the LR Law was established in 1954. Rapid economic growth from the 1950s to 1990s necessitated large-scale LR. Post-Bubble Economy in the 1990s, LR policies shifted towards urban redevelopment, emphasizing city center renovations and areas around transit stations, and introduced new system implementers [25]. Disasters have continually shaped LR's functions and scope, as shown in Fig. 7 [25].

5 Case Studies: Adopting LR in Post-Disaster Reconstruction and Potential Lessons Drawn

Table 1 summarize the Case studies related to LR in post-disaster reconstruction. These case studies portray effective use of collective land replotting into a large-scale site which has supported apartment development, and land right exchanging with money has provided support for livelihood restoration of the disaster-affected people.

5.1 Success of LR in Japan and Potential Lessons Drawn

Hayashi [10] attributes the success of Japan's land readjustment (LR) method to four key reasons rooted in its feudal history. Firstly, a long history of military governance fostered social discipline and obedience to authority [10, 30]. Secondly, a strong attachment to land, due to historical agricultural dependency, makes relinquishing ownership difficult. Thirdly, local governments were motivated to develop urban areas independently, as the central government prioritized other investments. Lastly, Japan's agricultural land was well-suited for LR. Unlike contract-based systems, Japan's LR is a legal administrative measure governed by the Japanese Administrative Complaint Reinvestigation Act of 1962, allowing projects to proceed despite dissent [10]. Technical improvements and land assessments also contribute to LR's success. Japanese culture values community welfare over individual gain, which, along with a historical culture of obedience and democratic consultation practices, encourages engagement in LR projects for their perceived communal and individual benefits (Key Informants: Okamoto 2023, Einsiedel 2023) [10, 30] (Annex A).

6 Relationship of LR and Resilience: Towards Factor Analysis

The dimensions of resilience and features of LR were reviewed to assess the probable relationship of the two concepts. There are several discussions on resilience dimensions, indicators and capacities found in literature [1, 4, 9, 18] that encompass the physical, mental, social, and spiritual aspects. In this study, four dimensions of resilience were used to consider the relationship of land readjustment and resilience. These are avoidance, adaptation, recovery, and compensation. These dimensions are drawn from the array of resilience dimensions discussed in literature. Moreover, the dimensions identified are based on consultation with scholars and practitioners in urban planning in the course of this study (Annex A).

These dimensions and underlying attributes are defined in this study as follows:

Avoidance. The ability of the system (including people) to avert, evade, or prevent the occurrence of an event that may result in negative impacts.

- Will the event happen at a determined time?
- Will the event cause negative impacts on the system?
- Will the system be able to prevent the direct effects of the event?

Table 1 Case studies: adopting LR in post-disaster reconstruction

Item	City adopted LR in post-disaster reconstruction	Description regarding post disaster reconstruction
1	Tokyo	The Great Kanto Earthquake of 1923 was essential in shaping Japan's urban development policies. The disaster's devastation led to large-scale reconstruction projects, especially in Tokyo, which was severely affected. The ensuing fires further damaged infrastructure, prompting a reconstruction plan that included fire prevention zones and land readjustment districts. Nearly half of Tokyo was involved in land readjustment, managed by either the central government or the municipality. Only a small portion of the city remained without such adjustments.
2	Kobe	Kobe City experienced severe damage from the Great Hanshin Earthquake on January 17, 1995, resulting in 4,600 deaths and 15,000 injuries within the city. The disaster led to the collapse or partial destruction of approximately 123,000 buildings, with an additional 7,000 structures burned [25]. The Kobe City Government responded with a reconstruction project using land readjustment in the Shin-nagata Ekikita area, which was densely packed with wooden houses, small factories, and shoe workshops. The project aimed to create new housing areas with disaster-prevention features, as well as commercial and cultural facilities [25]. The first phase covered 42.6 hectares and began in 1996, followed by an additional 17 hectares in the Takatori area in 1997. Completed in 2011, the project cost 103 billion JPY (1.03 billion USD) in public funds. Improvements included wider city roads, community roads, open spaces, earthquake-resistant and fireproof buildings, and fire prevention water tanks. By 2011, the population in the project area increased from 7,587 in 1994 to 9,400 [25]. The development of underground power lines enhanced the cityscape, as planned by the Residential Town Planning Committee. Due to the small plot sizes, the land contribution ratio was reduced to 9 percent, with the government acquiring land to compensate for decreased land values, aiding residents in their livelihood rehabilitation

(continued)

Table 1 (continued)

Item	City adopted LR in post-disaster reconstruction	Description regarding post disaster reconstruction
3	Tohoku region	The magnitude of damage to lives and properties of the Great East Japan Earthquake that took place on 11 March 2011 left 18,456 recorded casualties and 400,438 houses/ building infrastructures destroyed. The government estimated the loss in direct damage to be 16–25 trillion Japanese Yen (JPY) The earthquake triggered the gigantic tsunami which caused the subsequent meltdown of the Fukushima Nuclear Power Plant. This series of unfortunate events prompted the government to prepare reconstruction plans suitable to each respective cities/towns which incorporated measures to improve disaster prevention functions. Large-scale site preparations were employed to "relocate" communities through land readjustment. Affected population were moved from lower to higher ground for safer and more comfortable environment. On the tenth year since the 3/11 disaster, Nippon Hoso Kyokai (NHK) documented that the residential areas in Minamisoma, Fukushima were converted to industrial zones so as not to risk people in the area [26].
4	Nagoya City	In Nagoya, land readjustment (LR) projects under the 1954 Land Readjustment Act have covered 173 areas, accounting for nearly one-third of the city. Including LR projects under the former law and farm LR projects (307 areas, 20,600 hectares), the total reaches nearly two-thirds of Nagoya. Kukaku-seiri projects have improved public facilities like roads and parks, controlled urban sprawl, and allowed for planned development and large-scale redevelopment without direct land purchases. LR projects in Nagoya City include the Ozone Area (northeast) and Tokushige Nanbu Area (southeast). The Ozone Area, a key commercial hub, saw traffic improvements and enhanced site utility through LR projects. Conversely, Tokushige Nanbu, primarily agricultural and forest land, was designated an urban promotion area. Landowners, accepting urbanization trends, agreed to convert their land to mainly low detached houses per city zoning plans. In the Ozone project, increased public facility land requirements led to no reserve land generation and decreased site values, which the project funds compensated. To manage this, the government pre-empted some project area land for public use, reducing the contribution ratio from 22.25% to 15.41% while maintaining site value balance [22].

(continued)

Table 1 (continued)

Item	City adopted LR in post-disaster reconstruction	Description regarding post disaster reconstruction
5	Other sites	Iwate Prefecture was severely affected by the 3/11 catastrophe. In Rikuzentakata City, rebuilding the new town involved creating "raised land." Before the old town was buried for readjustment, the community commemorated it with festivals and parades, blending mourning and celebration [17]. Older residents felt a mix of sorrow for the old town and hope for the new one. The population decreased from 24,246 in February 2011 to 18,618 in January 2021. A key informant stated that the LR system will be resilient against disasters. Organized road networks and drainage systems enhance resilience, and land readjustment can elevate flood-prone areas. The greatest benefit is integrating disaster countermeasures into the project.

Adaptation. The ability of the system (including people) to anticipate, prepare for, and/or adjust to the effects of a hazardous event. It includes the capacity of the system to mitigate the impacts of a hazardous event based on predetermined actions designed to address the anticipated hazard.

- Are there risk and/or vulnerability assessments conducted prior to project implementation?
- Are infrastructures installed observant of state regulatory standards?
- Is the target community receptive and/or cooperative of the actions taken?

Recovery. The ability of the affected system (including people) to function to its acceptable capacity despite the effect of the hazardous event. It incorporates programs, projects, and/or activities adopted to address the impact of the hazardous event.

- Is the affected system able to comply with the necessary conditions to function properly?
- Is the expected timeframe to return to acceptable condition achieved?
- Is the affected system able to support its needs to return to a stable condition?

Compensation. The restitution received by the affected system (including people) to cover the cost of losses incurred due to the changes brought about by an event. The payment may be a subsidy, portion of, or actual cost incurred by the affected system.

- Will the cost burden the affected system?
- Does the state subsidize/cover the cost of losses of affected system?
- Does accommodation of changes in cost implications occur when necessary?

On the other hand, features of Land Readjustment method were derived from the case studies and descriptions provided in the literature. Each feature encompasses specific attributes. Table 2 summarizes the identified characteristics covering features of LR and corresponding attributes based on the case studies reviewed.

From the above descriptions, the following assumptions and observations in the relationship of LR and resilience are deduced:

Avoidance. The effect of land readjustment to avoid occurrence of disaster in an area is basically unknown. The existence of hazards, particularly natural hazards, may be innate in an area, as in the case of Japan, and hence may be impossible to avoid. Even in low-risk areas, the dynamic environment can still pose unexpected turn of events and cause disturbance. Mitigation measures can be adopted to minimize the impact of disaster but not necessarily to eliminate the risk. From the literature consulted, there was no identified relationship that can link land readjustment to avoid adverse situations.

Adaptation. Land readjustment (LR) sites are master-planned with risk and hazard assessments. Land use conversion, infrastructure improvements, and facility upgrades enhance resilience through adaptive measures. LR projects require majority consensus (at least 2/3 of affected landowners) and cannot be imposed by the government. While the government can propose LR projects, private groups, associations, and cooperatives can also initiate them [10, 30].

Recovery. The principle of Build-Back-Better is embedded in post-disaster reconstruction process. In the case of Japan, LR has become an element of the post-disaster reconstruction measures. LR incorporates stringent standards imposed by laws, like the building code, on built infrastructures. Wide roads and streetscapes, open spaces, public structures, risk-mitigating facilities such as flood channels, levees, among others, are also integrated in post-reconstruction of affected areas in Japan. As such, there is a relationship in the recovery dimension of resilience and LR. In addition, LR projects are not contract-based projects, as explained in the previous section, in that delays in implementation can be prevented and expected timeframe of completion is relatively short. Another notable LR feature is that LR projects are self-financing. The cost of the project will not burden the public funds which can then be channelled to other priority projects.

Compensation. The damage and loss of properties aggravate the impact of disaster. Compensation of losses is part of the resilience dimensions. The government also extends this assistance to affected individuals in LR in cases when the victim will be aggravated by standard contribution of land portions if the LR project proceeds. There are situations wherein the government would pay for the cost of the decrease in land value of properties (as stipulated in LR Law) included in the LR project. In special cases, the government also buys properties and converts it to public lands before proceeding to the LR project. This balances the total value of public and private lands before and after the LR project. As such, the government compensates the landowners in case decrease of the land value occurs due to already high unit price of existing land. In disaster-affected areas, land acquisition by the government helps speedy resettlement of disaster-affected people.

The following illustrates the observed relationship between LR and resilience based on the case studies. The relationship of LR and resilience is considered

Table 2 Characteristics of LR features

Features	Attributes
• Has an overall master development plan of the project site in correspondence with higher level plans (e.g., City land use plan, zoning plan)	1. Feasibility study is conducted including risk and vulnerability assessments 2. It aligns with higher level plans of the LGU 3. Site plan on plot distribution/replotting is prepared
• Built infrastructure incorporates standards imposed by the Law (e.g., building code)	1. Strict implementation of standards provided by state regulations (fire, earthquake, flooding, etc.) on infrastructures 2. Inclusion of public facilities such as parks, open spaces which serve as fire buffers, roads and drainage systems 3. Widened roads, passageways, alleys, street and building frontage, etc., for more accessibility and network mobility
• Is self-financing	1. All landowners involved in LR project contributes a portion of its land property to serve as reserve land and to use for public utility. Cost of project implementation is taken from the sale of the reserve land 2. The government instituted mechanisms to handle the project management from inception to implementation. There are specific authorized entities to undertake the project and oversee the transfer of properties back to respective owners 3. There is considerably no cost or, if at all, minimal subsidies from the government are tapped on case-to-case basis to support the LR projects
• Is flexible to adapt to the locality's conditions and requirements including compensation for decreased value	1. Can be government initiated due to urban redevelopment/renewal projects, post-disaster reconstruction, or private sector-initiated proposals 2. Calculations on land value and ratio of land contribution may be modified to ensure that balance is achieved 3. The government may subsidize the project in cases when LR project will decrease the value of original properties included

(continued)

Table 2 (continued)

Features	Attributes
• Is not contract-based; timeframe is usually defined	1. A legal administrative measure covered by the Japanese Administrative Complaint Reinvestigation Act of 1962; cannot immediately be taken to Court 2. The project implementation can proceed whether dissent arises from one of the landowners 3. The timeframe of project implementation is usually followed under normal circumstances
• Is implemented by majority consensus and not government-imposed	1. The involved property owners need to have a 2/3 majority consensus before the project is approved 2. The 2/3 agreeing property owners must constitute at least 2/3 of the overall consolidated property under the project 3. LR project can be initiated by both government and private entities

following these conditions: (1) Prevention and recovery; (2) Nature and magnitude of impact; (3) State intervention (Table 3).

Prevention and recovery. As illustrated in the previous LR projects discussed, LR implementation in disaster management comes as an element of post-disaster reconstruction and/or redevelopment. The myriad other requirements imposed by the law (e.g., reconstruction plan, adherence to building code and other standards)

Table 3 Proposed relationship of LR and resilience

	Dimension of resilience			
	Avoidance	Adaptation	Recovery	Compensation
Feature of land readjustment				
• Has a master plan of the overall development of the project site	Unknown	✓		
• Built infrastructure incorporates standards imposed by the Law (e.g., building code)	Unknown	✓	✓	
• Is self-financing	Unknown	✓	✓	
• Allows flexibility to adapt to conditions (includes compensation)	Unknown	✓		✓
• Is not contract-based; timeframe is usually defined	Unknown		✓	
• Is implemented by majority consensus (not government-imposed)	Unknown	✓		

in such projects also affect the general resilience of the project itself. Better transportation network access, improved built infrastructure, and engineered facilities to withstand flooding or tremors are some of the elements incorporated in a reconstruction/redevelopment of an area to improve resilience. Land readjustment becomes a factor in overlaying the "framework" of the project to warrant a more equipped condition in case of future incidents.

Nature and magnitude of impact. Land readjustment was adopted in disaster-stricken Tokyo after the Great Kanto Earthquake and ensuing fire in 1923, leading to extensive reconstruction. World War II, however, left much of the capital in ruins again. Despite the complex urban designs post-1923 disaster, Japan could not escape the destruction from war. Air bombings decimated major cities, reducing a third of the country's wealth to ashes: 51% of Tokyo, 58% of Yokohama, 40% of Nagoya, and 99% of Toyama were destroyed. Hiroshima and Nagasaki suffered beyond measure. War-damaged cities totalled 115, with 63,153 hectares affected [7]. Japan's post-war reconstruction set high standards for public facilities, focusing on modernization. In 2019, Typhoon Hagibis caused widespread flooding and damage, overwhelming city systems, especially Tokyo, now below sea level and more vulnerable to intense typhoons. Resilience in readjusted areas can be tested by extraordinary situations, requiring continuous review, especially for earthquakes.

State intervention. Japan's disaster risk reduction investments can be viewed as a mitigating factor to evade even greater damage. For example, the city has built an underground discharge channel that extends more than 6-km long system of tunnels of floodwater facility. Levees built along the rivers were designed to withstand precipitations with a 200-year interval. Despite this, Hagibis water level (e.g., in Tama River) exceeds the 5.9 m threshold by the highest record of 6.24 m water level. Nevertheless, it was deemed that far worse impacts could have been experienced if Japan had not levelled up its engineering interventions in disaster prevention measures and flood control. Communication systems and notifications were also regularly issued to ensure the safety of residents. Institutions are engaged to promote disaster resilient communities. In this respect, land readjustment may not directly affect the resilience of the area unless the government has identified specific conditions to address disaster management to be embedded in implementing LR. Awareness of this condition may result in exponential benefits considering that minimal public funds are used for LR while investments in engineering technicalities can be given more funding.

7 Applicability of Land Readjustment in the Philippines

In 1985, the Ministry of Construction in Japan (now Ministry of Land, Infrastructure, Transport and Tourism or MILT) sponsored an international seminar on Kukaku-seiri held in the cities of Tokyo, Tsukuba, and Nagoya. This was in celebration of the 30th year of the Japanese Land Readjustment Law. It was a venue participated in by researchers, officials, practitioners, and international organizations from 18 countries.

Many country representatives presented on their experiences in land readjustment projects. The Philippines was among the presenters in the assembly.

The success of LR method in Japan opens the inquiry of its applicability in other countries such as the Philippines. In the past decades, Japan initiated a strong push for adaptation of LR method in developing countries in Asia. As mentioned previously, there was an attempt to introduce the LR method in an urban site in a Philippine city as initiated by the Philippine Government in partnership with the Japanese Government. Several challenges and difficulties were met in the process although the proponents were hopeful that the implementation of the process will yield high benefits if accomplished.

In the Philippines, developing land for infrastructure utility requirements and socialized housing is executed through government intervention and control. More often, the state needs to secure land by acquisition. Several obstacles can be identified such as limited financial resources, inefficient/inadequate methods of acquisition, taxation, management, and defective land registration system [8]. This condition is different from Japan wherein the government does not necessarily need to acquire the land before it is developed through land readjustment. Instead, land inputs are usually of private individual property owners participating in the LR projects.

7.1 Challenges in Implementing Urban Development in the Philippines

According to Einsiedel (2023) [7], the following major constraints in the implementation of urban development projects in the country are present:

1. Land acquisition—lacking lands for low- and middle-income housing;
2. Legal constraints—land registration system needs to be systematized; land titling needs to be streamlined; status of land titles unclear; reliable and well-organized land registration and land tenure system needed; lack of legal enforcement mechanism;
3. Financial or economic constraints—limited funds and scarcity of capital;
4. Limited private sector participation—decline of private sector participation in investments;
5. Limited/inadequate local government participation—usually attributed to shortage of skilled manpower and resources to assess and implement the process;
6. Lack/inefficient coordination/monitoring system—no strict or effective monitoring of actual developments as against the structure plan;
7. Absence of planning database—local development plans, cadastral maps, and detailed surveys are not always available and at most times neglected;
8. Socio-economic problems—mobilizing private landowners to participate; difficulty in securing cooperation and sustain support in implementation.

As such, the case study conducted in the Philippines did not reflect the expected outcome. The area considered for LR was occupied by informal settlers on government property and not on privately-owned lands. From the case study, Einsiedel emphasized that difficulty in acceptance, mobilization, and participation in LR projects can be attributed [8] to the following:

(a) Land is viewed as a commodity and a valuable asset which should not be risked;
(b) Landowners possess various options in managing and disposing their lands which could circumvent existing methods of acquiring or developing land;
(c) Landowners perceive government's efforts as more of a political drama, hence, not credible;
(d) In times of economic difficulties where capital is scarce and the value of peso is low, the highest priority is to earn an income. Participation either in the form of financial contribution or community undertaking (which may indirectly involve some monetary contribution) is the least of considerations.

Land is regarded as a prime ultimate possession in the Philippines that despite the expected manifold benefits of the community, the implementation of urban development projects is difficult to pursue. In addition, city and residential development is primarily executed by the private sector while the government takes charge in the improvement of infrastructure (i.e., roads), zoning, and regulations/subdivision control. In both cases, the only means to undertake the project is to purchase all the properties involved [8] (Einsiedel 2023) which entails large-capital public funds.

Despite these obstacles, Einsiedel (2023) deemed LR, given its attributes, as a promising and needed urban development system in the Philippines. To make this viable, however, there are hard decisions that are needed to take place (Key Informant: Morales 2023) (Annex A). Given that the biggest challenge to implement urban development in the country would be land acquisition and capital, both are addressed in LR process if handled properly.

The Philippines needs to improve the percentage of completed land titling and registration especially on land parcels earmarked for readjustment (Morales 2023). It is necessary to revisit development and regulatory policies in local government units to promote a conducive environment for readjustment. Morales (2023) emphasized the importance of creating and strengthening the LGU's urban renewal authority and designate a specific office that handles land readjustment along with the revitalization/rehabilitation/redevelopment to be implemented on the readjusted land. Among the areas that would greatly benefit in LR projects are slum areas for rehabilitation, urban expansion areas, and station areas for railways, among others (Einsiedel 2023).

8 Conclusion and Recommendations

Land readjustment is an urban development tool, benefits of which extends beyond the stakeholders involved. The role of institutions is crucial given that established legal framework is prerequisite for its effective implementation. Cultivating a culture of engagement, collaboration, consensus building, open-mindedness, and trust and support to the government vision and action is foreboding to achieve success in LR implementation. The origin of LR is not based on disaster management. As a planning tool, however, it can contribute to resiliency through its integration of disaster risk reduction measures such as risk mapping and vulnerability assessment.

As Japan is exposed to a wide range of hazards, readjusted areas remain at risk. Resilience of an area is influenced by infrastructure development, programs, and activities integrated onto the readjusted land. This should intertwine with an urban revitalization plan aimed at connecting land readjustment with spatial facilities to maximize development potential.

Resilience to hazards or external threats can be addressed by strong institutions, effective governance, efficient urban planning and reliable infrastructure, to name a few. Land readjustment that incorporates disaster management features can amplify the benefits to stakeholders. Further, the significance of community participation cannot be undermined. The collective effort of the community can make a difference in the success of any project at hand.

Japan's land readjustment integrates resilience from recovery to reconstruction, mitigating disaster impacts. Reconstruction strategies enhance disaster management and adaptation. Special considerations, like reduced land contribution and monetary support, bolster both physical and economic resilience for affected communities.

Japan's "culture of obedience" aids the successful implementation of Land Readjustment (LR) projects, unlike in the Philippines. Clear land tenure and effective communication with interest groups are essential for faster implementation. A strong legal framework and willingness to participate are also crucial.

In the Philippines, the historic struggle for land ownership makes Filipinos value their property highly. Government inefficiency, malpractice, and political motives breed distrust in government projects affecting property. Building trust and competence, and communicating a clear, effective system to stakeholders, is essential for cooperation and project success.

Development is not just about the physical environment but also the people. Technical approaches may promise excellent results, but understanding the local context is crucial. A cultural perspective can guide decision-makers, ensuring government actions prioritize the general welfare of the people.

Further assessment on whether readjusted areas are resilient in the facc of disaster needs empirical data for analysis. There is a need for more scientific evidences on the assumptions and claims suggested in this study. The study has provided some assumptions of relationships to be tested using actual data samples for future studies. This exploratory research, however, was too limited to encompass this level of assessment.

Acknowledgments My sincere gratitude goes to my mentors and advisers at the University of the Philippines: Professor Ernesto M. Serote, Dr. Maria Dulce F. Natividad, and Dr. Matthew C.M. Santamaria; to the Planning and Development Research Foundation, Inc. (PLANADES); to my resource persons, to the faculty and staff of UP Asian Center and UP School of Urban and Regional Planning; to friends and colleagues in the profession especially to Professor Kristine Aspiras and Jade Agapinan; and to my family. All glory to God!

Annex A

Profile of the Key Informants

Name	Position	Affiliation	Date of interview
Anota, Apolonio Jr. B. (JD/EnP)	National president	Philippine institute of environmental planners	16 May 2023
Chang, Kuei-Feng Daphne (PhD)	Professor and chairman	Department of real estate management, national Pingtung university, Taiwan	11 May 2023
Einsiedel, Nathaniel von (PhD/EnP)	College of fellows chairman and principal planner	Philippine institute of environmental planners comprehensive Environmental planning, Inc	16 May 2023
Morales, Mark Anthony M. (Dr. of Engineering/EnP)	Professor	University of the Philippines Diliman	15 May 2023
Okamoto, Naohisa (Dr. of Engineering)	Professor	University of Tsukuba, Japan	16 May 2023
Yoneno-Reyes, Michiyo (PhD)	Professor	University of Shizuoka, Japan	17 May 2023

References

1. Argiroff E (2018) Identifying valued community resilience indicators and dimensions for King County. Master's Thesis, University of Washington. https://digital.lib.washington.edu/researchworks/handle/1773/42215
2. Bajek R, Matsuda Y, Okada N (2008) Japan's Jishu-bosai-soshiki community activities: analysis of its role in participatory community disaster risk management. Nat Hazards 44:281–292. https://doi.org/10.1007/s11069-007-9107-4
3. Bankoff G (2002) Cultures of disaster: society and natural hazard in the Philippines. Routledge, London. https://doi.org/10.4324/9780203221891
4. Bergamini N, Blasiak R, Eyzaguirre PB, Ichikawa K, Mijatovic D, Nakao F, Subramanian SM (2013) Indicators of resilience in socio-ecological production landscapes (SEPLs). United Nations University—Institute of Advanced Studies (UNU-IAS)
5. Brigham EF (1964) A model of residential land values. https://www.semanticscholar.org/paper/A-model-of-residential-land-values-Brigham/a11e22f9e50109b756c067d782301db9140ae7cf
6. Cabinet Office Japan (2015) White paper, disaster management in Japan 2015, Government of Japan. http://www.bousai.go.jp/kaigirep/hakusho/pdf/WP2015_DM_Full_Version.pdf
7. De Souza FF, Ochi T, Hosono A (2018) Land readjustment: solving urban problems through innovative approach. JICA Research Institute. https://www.jica.go.jp/Resource/jica-ri/publication/booksandreports/l75nbg00000ltoak-att/LAND_READJUSTMENT_web.pdf
8. Einsiedel NV (1985) A case study on the applicability of the land readjustment system in the Philippines. In: International seminar on kukaku-seiri 1985, Japan Ministry of Construction, pp 428–475

9. Graveline MH, Germain D (2022) Disaster risk resilience: conceptual evolution, key issues, and opportunities. Int J Disaster Risk Sci 13:330–341. https://doi.org/10.1007/s13753-022-00419-0
10. Hayashi K (1982) Land readjustment in Nagoya. In: De Souza FF, Ochi T, Hosono A (2018) Land readjustment: solving urban problems through innovative approach. JICA Research Institute. https://www.jica.go.jp/Resource/jica-ri/publication/booksandreports/l75nbg00000ltoak-att/LAND_READJUSTMENT_web.pdf
11. Hein C, Pelletier P (eds) (2006) Cities, autonomy, and decentralization in Japan. Routledge, New York
12. Hohn U (1997) Townscape preservation in Japanese urban planning. Town Plan Rev 68(2):213–255. http://www.jstor.org/
13. Holling CS (1973) Resilience and stability of ecological systems. Annu Rev Ecol Syst 4(1):1–23
14. Home R (2007) Land readjustment as a method of development land assembly: a comparative overview. Town Plan Rev 78(4):459–483. http://www.jstor.org/
15. Hosono A. (2022) Addressing challenges of urbanization for quality of growth. In: SDGs, transformation, and quality growth. Sustainable development goals series. Springer, Singapore. https://doi.org/10.1007/978-981-16-9748-7_6
16. Ishiwatari M (2010) Japanese experiences of disaster reduction. Asian J Environ Disaster Manag 2(3):247–248. Research Publishing Services. http://rpsonline.com.sg/journals/101-ajedm/2010/0203/S1793924011000447.php
17. Japan Times (2017) Tsunami-hit Rikuzentakata rebuilding on raised ground, hoping to thrive anew. https://www.japantimes.co.jp/news/2017/03/07/national/tsunami-hit-rikuzentakata-rebuilding-high-ground-hoping-thrive-anew//
18. Jeans H, Castillo G, Thomas S (2017) The future is a choice: absorb, adapt, transform resilience capacities. Oxfam International, UK. https://oxfamilibrary.openrepository.com
19. Kataoka M (n.d.) Direct popular participation through referendum: grass-roots conservative elitism vs. populism in Japan. In: CS Ahn (ed) New development in local democracy and decentralization in East Asia: Korea in comparative perspective, Seoul National University Press
20. Kumar K (1989) Conducting key informant interviews in developing countries. Agency for international development, Program design and evaluation methodology report no. 13
21. Local Government Code of the Philippines (1991). https://officialgazette.gov.ph/
22. Mano M (1985) The case studies of the kukaku-seiri project in Japan—in case of Nagoya. In: International seminar on kukaku-seiri 1985, Tokyo, Japan
23. Mapsoftworld (2020) 2011–12. www.mapsoftworld.com. Accessed 6 Feb 2024
24. Matsui M, Deguchi C (2014) The characteristics of land readjustment systems in Japan, Thailand, and Mongolia and an evaluation of the applicability to developing countries. In: Proceedings of international symposium on city planning, Hanoi, Vietnam. ISCP2014. https://www.cpij.or.jp/com/iac/sympo/Proceedings2014/3-fullpaper.pdf
25. Matsui M (2019) World bank case study: land readjustment in Japan. https://documents1.worldbank.org/curated/zh/481571569562840686/pdf/Land-Readjustment-in-Japan-Case-Study.pdf
26. NHK (2022) Decommissioning Fukushima 2021: ten years on from the nuclear accident. https://www3.nhk.or.jp/nhkworld/en/tv/documentary/20220501/4001395/
27. Nel DH (2022) Morphologies of resilience: Towards an approach to the study of urban spatial resilience. Doctoral Dissertation, National University of Singapore. https://www.researchgate.net/publication/362070577_Morphologies_of_resilience_Towards_an_approach_to_the_study_of_urban_spatial_resilience/link/62e40f497782323cf1848030/download
28. Pastoriza LR, Holdsworth RE, McCaffrey KJW, Dempsey E (2018) Tectonic evolution of the Southern Negros geothermal field and implications for the development of fractured geothermal systems. Hindawi Geofluids 2018:6025038 (20 pages). https://doi.org/10.1155/2018/6025038, https://onlinelibrary.wiley.com/doi/epdf/10.1155/2018/6025038

29. Raza T (2018) Localizing disaster risk reduction and climate change adaptation in planners' and decision makers' agenda: Technical comprehensive model, Quezon city, Philippines. ProcediaEngineering, 212(2018):1311–1318. https://doi.org/10.1016/j.proeng.2018.01.169, https://www.sciencedirect.com/science/article/pii/S1877705818301978
30. Rhie Won-bok (2002) Korea unmasked: In search of the country, the society and the people. Paju: Gimm-Young.
31. SGI, Sustainable Governance Indicators (2019) About Japan's sustainable governance indicators. SGI Network. https://www.sgi-network.org/2019
32. Santiago ADR (2023) Land readjustment and disaster resilience in Japan: Exploring possible applications in the Philippines, Master's Thesis, University of the Philippines (unpublished)
33. Scott Z, Tarazona M (2011) Study on disaster risk reduction, decentralization and political economy. Global assessment report on disaster risk reduction. http://www.preventionweb.net/english/hyogo/gar/2011/en/bgdocs/Scott_&_Tarazona_2011.pdf
34. Serote EM (2023) Property patrimony and territory: Foundations of land use planning in the Philippines, Revised edn, University of the Philippines
35. Sheppard A, Tatham P, Fisher R, Gapp R (2013) Humanitarian logistic: enhancing the engagement of local populations. J Humltarian Logist Supply Chain Manag 3(1):22–36. http://www98.griffith.edu.au/dspace/bitstream/handle/10072/52673/86528_1.pdf;jsessionid=0D63BB39388817A3AD890D6482E9741C?sequence=1
36. Sorensen A (1999) Land readjustment, urban planning and urban sprawl in the Tokyo metropolitan area. Urban Studies 36(13):2333–2360
37. Sorensen A (2002) The making of urban Japan: cities and planning from Edo to the twenty-first century. Nissan Institute/Routledge Japanese Studies Series, London
38. Souza F (2018) Land readjustment: concepts and practice in Japan, law, environment science. https://api.semanticscholar.org/CorpusID:219178302, https://www.semanticscholar.org/paper/Land-Readjustment%3A-Concepts-and-Practice-in-Japan-Souza/921ce392721fc362ce7a6a21fdba53f3373d5699
39. Stevens A (2012) Local government in Japan: a variety of taxes funds Japan's prefectures and municipalities, city Mayors government. http://www.citymayors.com/government/japlocgov.html
40. Takao Y (1998) Participatory democracy in Japan's decentralization drive. Asian Surv 38(10). http://www.jstor.org/stable/2645645
41. The Constitution of Japan (1947). http://japan.kantei.go.jp/constitution_and_government_of_japan/constitution_e.html
42. Tokyo Metropolitan Government (2006–2018) (2018) About government structure in Japan. https://www.metro.tokyo.lg.jp/
43. ULI, Urban Land Institute (2021) In the eye of the storm: How centuries of disaster make Tokyo a case study in urban resilience. Urban Land Institute, Washington, D.C. https://knowledge.uli.org/-/media/files/research-reports/2021/uli-culturalresiliencyreport-tokyo_fin.pdf?rev=e7780918f73d4428bced9b879f5fa7c3
44. UNISDR: United Nations Office of Disaster Risk Reduction (2008) Climate change and disaster risk reduction. UNISDR briefing note 01. http://www.unisdr.org/eng/risk-reduction/climate-change/docs/Climate-Change-DRR.pdf
45. UNISDR (2017) Build back better. https://www.unisdr.org/files/53213_bbb.pdf
46. Valdez AS, Raza T, Farolan MI et al (2023) Primary and secondary data collection to conduct researches, write thesis and dissertation amidst COVID-19 pandemic: a guidepost. In: Proceedings of the 2nd international symposium on disaster resilience and sustainable development, Lecture Notes in Civil Engineering 294:269–288 Springer Nature, Singapore. https://doi.org/10.1007/978-981-19-6297-4_20
47. Wismansyah AR, Dewi MS, Rahmi, Syahzaeni T, Hastut W (2019) Dynamic system development: an approach for land re-adjustment in managing slum areas. https://doi.org/10.18502/kss.v3i21.4960, https://knepublishing.com/index.php/Kne-Social/article/view/4960/9931#content/contributor_reference_3. Accessed 10 Mar 2024

Role of Micro-Finance in Poverty Reduction: A Case Study from Neelkantha Municipality, Nepal

Shiva Raj Khanal, Romi Ghimire, and Pradeep Shrestha

Abstract The study analyzes the role of Microfinance in poverty reduction in Neelkantha Municipality, Nepal. This descriptive and analytical study was conducted with a well-structured questionnaire that was developed to collect data from 223 respondents using a purposive sampling technique. Simple descriptive statistical tools as well as inferential statistical tools, i.e., paired t-tests, have been used to analyze the role of Microfinance in poverty reduction. Data was analyzed using the Statistical Package for Social Science (SPSS) to generate mean and standard deviations, and a t-test was used to diagnose the collected data. According to the findings of this study, microfinance financial services boosted the annual average monthly income. Most consumers profited from microfinance in various areas, including food security, education, health care, loans, shelter, and diversification of income sources. As a result, the study contends that microfinance had an essential role in raising living standards and lowering poverty. This study is crucial for policymakers because it provides a more decisive direction for anti-poverty policies based on various dimensions, geographies, and demographic distributions.

1 Introduction

Nepal is classified as one of the most underdeveloped nations globally, with over 25% of its population residing below the poverty threshold. [3, 4]. The nation relies significantly on remittances, which constitute up to 30% of the Gross Domestic Product (GDP) [12]. Agriculture is the predominant sector of the economy, supporting the livelihoods of over two-thirds of the population, although contributing to less than one-third of the Gross Domestic Product (GDP). Nepal has a longstanding history of implementing microfinance programs for over three decades. [10]. During this

S. R. Khanal (✉) · P. Shrestha
Nilkantha Multiple Campus, Neelkantha Municipality, Dhading, Nepal
e-mail: shiva@actionnepal.org

S. R. Khanal · R. Ghimire
Action Nepal, Neelkantha Municipality, Dhading, Nepal

I. Pal et al. (eds.), *Proceedings of the 3rd International Symposium on Disaster Resilience and Sustainable Development—Volume 1*, Lecture Notes in Civil Engineering 623, https://doi.org/10.1007/978-981-96-2139-2_18

period, many Microfinance Institutions (MFIS) emerged through government, non-government, and donor initiatives. Microfinance plays a crucial role in developing nations such as Nepal, where improving livelihoods can be a key method for reducing poverty [6]. Such microfinance programs have materialized in different modalities in different regions and parts of the country [5]. Regular assessment and appraisal of programs are essential in selecting and developing new programs that effectively improve the lives of the intended beneficiaries. Conversely, Microfinance programs offer those living below the poverty line the chance to become self-employed. Over 70 percent of impoverished households in Nepal continue to rely on informal channels to fulfill their credit needs [12]. Thus, it is imperative for Microfinance programs to have the capacity to improve the quality of life for individuals and empower women who are involved in their program regions.

Microfinance institutions are the exclusive financial institutions that offer loans to their members without requiring collateral. Microfinance banking is a specialized approach designed specifically to alleviate poverty. World Bank [14] stated that microfinance, often known as microcredit, refers to a banking service that is specifically designed for individuals or groups who are unemployed or have little income and lack access to traditional financial services. The contemporary notion of microfinance originated from the endeavors of Professor Mohammad Yunus, who founded Grameen Bank, a distinctive financial institution catering to impoverished individuals. According to contemporary understanding, microfinance encompasses the provision of small loans, savings accounts, insurance, money transfers, healthcare, education, vocational training, and community engagement. It aims to cater to the financial and non-financial needs of individuals typically excluded from traditional financial institutions.

The main objective of this study is to identify the role of microfinance programs in poverty reduction in Neelkantha Municipality, Nepal. The study's specific objectives are to analyze the saving habits of microfinance clients of the municipality and to examine the loan use practice of clients of Neelkantha Municipality.

2 Study Area

The study was conducted at Neelkantha Municipality, Nepal. Annual Household Survey 2015–2016 reports show the municipality's total population is 71,131 [11]. This municipality holds 17% of the total population of the district. Among 14 wards, Ward No. 3 is more advanced with infrastructure development and developed as the central market of the municipality. It has a high population growth rate, which is 3.4 according to the census 2011. The municipality has 3 Government banks, 14 private banks, and 96 Finance/cooperative institutions. Similarly, there are currently 5 microfinance companies registered within the municipality, and they have launched different programs to uplift the lifestyle of the people of Nilkantha Municipality [11]. Most of the microfinance companies formulated saving and credit groups where they can provide small-scale loans that help people meet their short-term credit. It also generates self-employment opportunities by enhancing their skills.

3 Methodology

This study has employed a descriptive technique to analyze the association between saving behaviors, income, and loan utilization practice. The descriptive strategy was primarily employed to conceptualize the study objectives and research topic, while the analytical approach was utilized to analyze and interpret the quantitative data.

3.1 Population, Sample, and Sampling Design

The National Microfinance Company was selected for this study through judgmental sampling. Altogether, 1895 total population from Neelkantha Municipality are the registered clients until 2020 at National Microfinance Company. The sampled households were taken only from the loan users and regular savers, where only 505 were identified. In this research, these loan users and regular savers are the population of the study. The sample was selected through a simple random sampling method with a sampling intensity of 44.15% (n-223) considered for the questionnaire survey. The sample consisted of 223 clients from the different centers out of the 236 questionnaires distributed, as it was impossible to cover the whole population of Neelkantha Municipality. The data for before and after status of microfinance clients has been taken for a 5-year period, i.e., 2017–2022. The calculation of the sample size has adopted a similar approach as [8].

Calculation of Sample size of finite population:

Here;

N = population size 505.

e = sampling error/margin of error = 5% = 0.05.

$n = N/(1 + Ne^2)$.

$= 505/ [1 + 505 \times (0.05)^2]$.

= 223.20.

~ 223.

Calculation of sampling intensity.

X % of 505 = 223.

Sampling intensity = 44.15%

Response Rate = 94.67%

3.2 Nature and Source of Data

This study is mainly based on primary data. Hence, a quantitative approach was used for data analysis. Structured questionnaires were used to collect the primary data.

3.3 Instruments of Data Collection

This study is based on primary data. Primary data was collected through a structured questionnaire. The researcher self-facilitated the questionnaire for the uneducated clients and provided the questionnaire to the educated clients. Each client interacted with person to person to collect their response based on a structured questionnaire. The questionnaire was designed to collect information regarding their income, saving habits, and economic status in two periods before and after the participation of MFIs.

3.4 Methods of Analysis

The gathered data was revised, categorized, and organized in a suitable manner. The data was analyzed and displayed using various statistical techniques such as tables, averages, percentages, and bar graphs. The characteristics of the respondents, as well as the independent and dependent variables, were described using descriptive methods such as percentage, frequency counts, and mean. The paired t-test, an inferential statistical tool, was employed to assess the significance of the parametric test for sampling attributes.

3.4.1 Paired t-test

The paired sample t-test, also known as the dependent sample t-test, is a statistical method used to ascertain if the average difference between two sets of observations equals zero. A paired sample t-test involves the measurement of each subject or entity on two separate occasions, resulting in pairs of observations. Paired sample t-tests are commonly used in case–control studies or repeated-measures designs.

A paired T-test is a test that is based on the differences between the values of a single pair, one of which is deducted from the other. Using the paired t-test, we get two values from before and after involvement in MFIs. In this study, the data for before and after the status of microfinance clients was taken over a 5-year period, i.e., 2017–2022.

In the equation for a paired t-test, the discrepancy is denoted as d. The formula for the paired t-test is calculated by dividing the sum of the differences between each pair by the square root of n, multiplied by the sum of the squared differences minus the sum of the differences squared, all divided by n-1. The formula for the paired t-test is provided as follows:

$$t = \frac{\sum d}{\sqrt{\frac{n(\sum d2)-(\sum d)2}{n-1}}}$$

where,

d = difference of each pair of two samples given.
Σd = sum of the differences.
n = sample size.

4 Findings and Discussion

This section of the study attempts to analyze and evaluate the information received from the collected data from members. The analysis process and contents are determined per the set of objectives for the study. It consists of saving habits, loan use practices, and the economic status of the respondents.

4.1 Factor Influencing Joining Microfinance Institution

As per the data from the structured questionnaire, respondents of Neelkantha municipality were asked about their main aim in joining microfinance institutions. The major options were increasing saving habits, empowering economic status, receiving loan facilities, and all of the above. So, to determine the opinion of respondents, Table 1 illustrates the summary of the main aim of joining MFIs.

Table 1 Respondents' main aim of joining MFIs

Aim of joining MFIs	Frequency	Percent (%)
To increase saving habit	28	12.56
To empower economic status	132	59.19
To receive a loan facility	51	22.87
All of above	12	5.38
Total	223	100

Table 2 Levels of monthly saving by members of microfinance

Level of monthly saving (in Nepalese Currency)	Before		After		Frequency difference
	Frequency	Percent (%)	Frequency	Percent (%)	
1–100	75	34	30	13	– 45
101–200	55	25	37	17	– 18
201–300	46	21	29	13	– 17
301–400	27	12	22	10	– 5
401–500	18	8	61	27	+ 43
501–above	2	1	44	20	+ 42
	223	100	223	100	0

4.2 Saving Habits of Members of Microfinance

Saving is stored for future consumption. Nowadays, people are habitual in saving some of their earned money for future consumption. Considering Neelkantha Municipality, organize members in six kinds of regular monthly savings. They are explained in Table 2.

Table 2 shows the monthly savings levels before and after involvement in microfinance programs. Before joining MFIs, 34% of members were used to saving 1 rupee to 100 rupees per month, 25% of members were used to saving rupees 101 to rupees 200, 21% of members used to save rupees 201 to rupees 300, 12% members used to save rupees 301 to rupees 400, 8% members were used to save rupees 401 to rupees 500 and 1% members were used to save above rupees 500. Most respondents were used to save monthly rupees 1–100. There has been a dramatic reduction from 34 to 13% in the monthly savings of rupees 1–100, which was the major level of savings for the majority of the population previously. Regular public awareness about MFIs' benefits might be needed for easy access to loans, training, and facilities provided to the number of registered clients. Similarly, monthly saving levels of rupees 101 to rupees 200, rupees 201 to rupees 300 rupees and rupees 301 to rupees 400 also reduced from 25 to 17%, 21% to 13%, and 12% to 10%, respectively, after the introduction of MFIs.

4.3 Sectors and Amount of Loan

The loan amount and sectors of using loans show which sector the microfinance provides their loan and how much it gives their client. So, to reveal the sectors where respondents have utilized loan amounts, Table 3 represents sectors and the amount of loans used.

Table 3 Sectors and amount of loan used

Sectors of loan use	Frequency	Percent (%)	Loan amount (In '000')
Agriculture	16	7.2	1,571,300.45
Small entrepreneur	83	37.2	8,151,121.08
Livestock	47	21.1	4,615,695.07
Health expenses	17	7.6	1,669,506.73
Education	41	18.4	4,026,457.40
Food and clothing	19	8.5	1,865,919.28
Total	223	100.0	21,900,000.00

Most respondents used loans from MFIs to six different sectors, with the highest loan of 37.2% being used for small entrepreneurship. Similarly, the loan amount used for the livestock sector was 21.1% and the educational sector 18.4%. Due to the changing economic status of the country, everyone wants to uplift their living standard through microfinance loans. Most respondents used microfinance loans for small entrepreneurs due to the easy allocation of microfinance loans. Also, people could use the loan amount to pay for their children's higher studies and quality education. The loan amount was used for health expenses, helping them to benefit from well-equipped health service provider and their routine medications. Due to food insufficiency and lack of regular income sources, some of the respondents indicate that they also use a particular portion of microfinance loans in the food and clothing sectors.

4.4 Major Sources of Income

A Microfinance Institution (MFI) offers financial services to communities that lack the ability to provide collateral for the loans they receive yet possess the necessary skills and motivation to engage in economic activities that generate income and self-employment. At Neelkantha Municipality, MFIs have contributed to the many households to uplift their previous socio-economic status. Some indicators of poverty reduction are Food sufficiency, access to quality education, good residential status, access to quality health facilities, satisfactory income level, livestock, Sources of fuel, and land holding size before and after the microfinance intervention. The impact can be measured by assessing the deviations in such variables. MFIs provide different life skills training to support their livelihood and income-generating activities.

Sources of income refer to something that provides a regular supply of money where respondents are involved in different income-generating activities to operate their daily lives. Similarly, at Neelkantha Municipality, many households were involved in various occupations for their livelihoods before the microfinance intervention. According to their response, they could engage in different income-generating occupations after they got skill development training and in-kind support

Table 4 Comparison of sources of income of the clients

Sources	Before		After		Frequency difference
	Frequency	Percent (%)	Frequency	Percent (%)	
Agriculture	65	29	47	21	− 18
Small entrepreneurship	7	3	24	11	+ 17
Livestock farming	44	20	77	35	+ 33
Daily wage based income	89	40	41	18	− 48
Salaried	18	8	34	15	+ 16
Total	223	100	223	100	0

from MFIs. Table 4 illustrates the status of households with different major income sources before and after microfinance's introduction in the study area.

Table 4 shows the sources of income before and after the involvement in the microfinance program. 29% of respondents were engaged in agriculture, 3% in small entrepreneurship, 20% in livestock farming, 40% in daily wage-based income, and 8% in salaried occupations. However, right after the microfinance intervention, there was a dramatic reduction in daily wage-based occupation (18%), a major source for the majority of respondents previously. Regular public awareness about MFIs' benefits might be needed for easy access to loans, training, and facilities provided to the number of registered clients. Similarly, there was a slight reduction in the agriculture sector (21%). However, there has been a rapid increase in small-scale entrepreneurship, 3% to 11%, livestock farming, 20% to 35%, and salaried occupations, 8% to 15%. This is because members of MFIs started to invest the borrowed amount from MFIs in small-scale entrepreneurship and livestock farming, and they also invested in easily accessible loans for quality education purposes. Quality education led them to a regular salary-based occupation in government and other corporate jobs. Thus, the above table indicates that the MFI clients have started to diversify their income-generating activity and sources of income in order to have a quality life due to easy access to financial and other life skill training from MFI.

4.5 *Landholding Size*

Land refers to the physical assets that are considered a factor in production. This implies that the economic status of the households can be determined by the size of land owned by the people. Larger land holding by respective households is considered economically sound and vice versa. To understand the land holdings of different clients, we will categorize those land holding sizes into four groups. A comparison of the land size that the respondents had before and after getting the MFIs services

Table 5 Land holding size of the clients before and after getting MFI services

Average land holding size in Ropani (land measurement unit practiced in Nepal)	Before		After		Frequency difference
	Frequency	Percent (%)	Frequency	Percent (%)	
Less than 2 Ropani	45	20	34	15	− 11
2–4 Ropani	91	41	52	23	− 39
4–6 Ropani	67	30	58	26	− 9
More than 6 Ropani	20	9	79	35	+ 59
Total	223	100	223	100	0

could show the entire change. The summarized form of land holding size among the respondents before and after obtaining MFI services is presented in Table 5.

From Table 5, It can be said that previously, the majority of the clients with average land holding size was 2–4 Ropani (41%), followed by the 4–6 Ropani (30%), less than 2 Ropani (20%) and more than 6 Ropani (9%). In contrast, after the microfinance intervention in the study area, the majority of the respondent's land holding size was more than 6 Ropani (35%). In the numbers, 59 respondents came from other landholding categories and had the highest landholding size. Similarly, fewer clients hold less than 2 *ropani* of land, and this is steadily decreasing. The presented data can distinctly show that the land holding size of the clients increased after the establishment of the microfinance services in the study area. And have brought a positive change in the size of the landholding. Due to easy access to MFI loans to its members, they could diversify their borrowed loans into different income-generating activities, and their income was reinvested and multiplied by the income able to acquire more land for production activities.

4.6 Types of Livestock

Livestock refers to domesticated land animals raised for various purposes, including traction, meat, milk, eggs, skins, fibers, and feathers. Livestock substantially impact the rural livelihoods and economics of emerging nations. Households of Neelkantha Municipality are adapting different types of livestock as an economic asset before and after being connected to microfinance institutions. Table 6 illustrates the comparative description of livestock adapted by the respondents.

Table 6 shows the types of livestock that respondents had before and after being involved with the microfinance program. According to the data, the majority of households had cows (44%) as livestock. Additionally, 17% had buffaloes, 14% had goats, 10% had pigs, 8% had chicken, 5% had honey bees, and 2% had fish as livestock animals and birds before they had any services from microfinance. After their involvement in microfinance, they received different livelihood development training programs regarding beekeeping, fish farming, goat farming, and buffaloes

Table 6 Livestock of the clients before and after getting MFI services

Types of livestock	Before		After		Frequency difference
	Frequency	Percent (%)	Frequency	Percent (%)	
Buffaloes	37	17	51	23	+ 14
Cow	98	44	24	11	– 74
Goats	32	14	47	21	+ 15
Pigs	23	10	39	17	+ 16
Chicken	18	8	31	14	+ 13
Honey bees	11	5	22	10	+ 11
Fish	4	2	9	4	+ 5
Total	223	100	223	100	0

farming. They were attracted to the increased number of livestock in their households. Microfinance programs encourage them to keep such income-generating activities by providing easy access to loans. So, the results have changed after involving the MFIs. The majority of the respondents were attracted to having buffaloes (23%) as livestock because they can use buffaloes for milk production and, later on, meat purposes as well. The massive decrease in the number of cows as livestock was 44% to 11% because most respondents were attracted to goats, chicken, beekeeping, and fish farming due to their rapid income-generating activities.

4.7 Sources of Fuels

The fuel source refers to the energy used to cook food or perform household activities. Using different sources of fuel represents the different levels of economic status and shows how economically they are empowered. Since there are various types of fuels, some of them are categorized as LPG, biogas, fuel wood, and electricity. Here, we assume fuel wood as a traditional material used for cooking and others as modern sources. In the study area, households use different fuels before and after connecting to the microfinance institution. Table 7 shows the comparative type of cooking fuels used by HHs before and after MFI services.

Before the MFI program, 95% of HHs had wood as a fuel source; following it, only 3% had electricity. 4 out of the 223 respondents were found to use LPG as a source of fuel, and only 2 respondents used biogas. After microfinance interventions, 45%, 39%, 13%, and 2% HHs use wood, LPG, Electricity, and Biogas as fuel sources, respectively. Table 7 shows that after involving the microfinance program, the number of people using LPG, electricity, and biogas has shifted from wood fuel as a fuel source.

Table 7 Sources of fuels of the respondents before and after getting MFI services

Sources of fuel	Before		After		Frequency difference
	Frequency	Percent (%)	Frequency	Percent (%)	
LPG	4	2	88	39	+ 84
Bio gas	2	1	5	2	+ 3
Fuelwood	211	95	101	45	− 110
Electricity	6	3	29	13	+ 23
Total	223	100	223	100	0

4.8 Food Sufficiency

Food is an essential requirement for human existence. Food comprises nutrients needed for the development, restoration, and upkeep of bodily tissues and for controlling crucial physiological functions. The availability of sufficient nutrient foods for consumption reflects the landmark of the good economic status of people. People who can afford hygienic and nutritious food are considered standard and happy. The table below represents the food sufficiency level of respondents of Neelkantha Municipality.

Table 8 shows that the majority of the respondents had food sufficiency for 3 to 6 months in their households before getting MFI services. Following this, 36% had food sufficiency for 6–9 months, 16% for more than nine months, and only 4% for less than 3 months. In contrast, after getting MFI services, respondents were able to increase their land holding size and income level, so they started to afford quality and adequate food. There were zero out of a total 223 respondents who had food sufficiency for less than three months after involving microfinance services. The majority of the respondents were able to have food sufficiency for more than 9 months (75%) after getting microfinance services such as easy access to loans, agriculture & farming training programs etc.

Table 8 Food sufficiency of the respondents before and after getting MFI services

Food sufficiency	Before		After		Frequency difference
	Frequency	Percent (%)	Frequency	Percent (%)	
Less than 3 months	10	4	0	0	− 10
3–6 months	98	44	13	6	− 85
6–9 months	80	36	42	19	− 38
More than 9 months	35	16	168	75	+ 133
Total	223	100	223	100	0

Table 9 Health facility of the respondents before and after getting MFI services

Health facilities	Before		After		Frequency difference
	Frequency	Percent (%)	Frequency	Percent (%)	
Traditional (Dhami Jhakri-Jharfuk)	154	69	22	10	− 132
Small health unit	48	22	152	68	+ 104
Well equipped hospital	21	9	49	22	+ 28
Total	223	100	223	100	0

4.9 Health Facility

The health facility is another indicator of poverty reduction, which is directly proportional to the financial status of respondents. People who can afford quality health services are considered economically sound and have a quality life. In our study area, HHs have different sorts of health services before and after being involved in MFI. Table 9 shows the comparative illustration of health services they received before and after being connected to the microfinance program.

Table 9 reveals the data of different households of Neelkantha Municipality on what sort of health facilities they are using. Among 223 respondents, most of the households, i.e. 154 or 69%, had used the prescription of Traditional Dhami Jhakri, whereas 22% and 9% had health services from small health units and well-equipped hospitals, respectively, before being linked to MFI. Meanwhile, after receiving awareness and supportive programs from microfinance institutions, health facilities were tremendously changed. Most respondents, i.e., 68%, have a health safety prescription from a small health unit, while 22% have health services from well-equipped hospitals. And 22 households are found to have traditional dhami jhakri services. The number of respondents with traditional dhami jhakri services decreased from 154 to 22. This means there will be exponential positive changes in health facilities for the people of Neelkantha Municipality after the microfinance program. The data presented in table shows that before getting the MFIS service, almost half of the respondents had health service from the Traditional (Dhami chakra). After getting the MFIS service, the number of respondents receiving services from Traditional (Dhami chakra) is declining, and the number of services from small health units and well-equipped hospitals is increasing because of MFI's support and health awareness program.

4.10 Access to Quality Education

One of the key attributes that measures the impact of microfinance on poverty reduction of any individual or family is access to quality education. Accessing quality

Table 10 Access to quality education of the respondents before and after getting the MFI program

Access of education	Before		After		Frequency difference
	Frequency	Percent (%)	Frequency	Percent (%)	
Government school	219	98	135	61	− 84
Private boarding school	4	2	88	39	+ 84
Total	223	100	223	100	0

education is about people's financial status. Table 10 shows efforts to show that MFIs made a difference in access to education. In the case of Nepal, there are two kinds of schools, i.e., government and private schools. Government schools lack quality, and those with good financial conditions prefer to send their children to private schools. In private schools, the ones in the city also provide better quality services than those in rural areas.

The above table shows that 98% of household children are in government schools, and only 2% of 2223 respondents were in private boarding schools before getting the microfinance service. However, after getting the microfinance service, the number of children attending government schools for education decreased to 61% from 98%. There is also a hike in the rate of sending children to private boarding, which is 39%. This means that microfinance programs have assisted in changing qualitative education for children of Neelkantha Municipality.

4.11 Type of House

Also, the type of house reflects the worth of people's property. A good house requires spending at large, which means the respondent has a sound economic status. The type of the house has been classified under mud, RCC, thatched roof, and iron roof. Among such categories, thatched roofs to RCC-type houses can be more economically worthy. A comparative summary of the type of house before and after getting microfinance services is given in Table 11.

Table 11 indicates that most houses were made from Mud and Thatched roofs (91%), 5% were made from RCC structures, and 4% were made from mixed cement, sand, aggregate, stone, and mud. However, after getting MFI services and government grants to construct the new earthquake-friendly house after the 2015 earthquake, most houses were constructed as mixed-earthquake friendly (70%). Following this, the number of RCC houses also increased from 11 to 48. And houses made from mud-thatched roofs decreased from 91 to 8%. From this, the economic status of the microfinance clients has been empowered after getting the MFI services if we consider the variable type of house. Figure 6 shows the same. The presented data shows that most respondents have houses made from mud-thatched roofs. After

Table 11 Type of house of the respondents before and after getting MFI services

Type of house	Before		After		Frequency difference
	Frequency	Percent (%)	Frequency	Percent (%)	
Mud-Thatched roof	203	91	18	8	− 185
Mixed-Earthquake friendly	9	4	157	70	+ 148
RCC	11	5	48	22	+ 37
Total	223	100	223	100	0

getting the microfinance services, most respondents shifted from mud-thatched roof houses to mixed-earthquake-friendly houses and RCC houses. It shows that they change their shelter type with the increment of income.

4.12 Average Monthly Income

The economic status of an individual can be measured through his/her average monthly income level. If an individual has a higher income level, he/she will have a higher level of economic status, which helps to reduce poverty. As a result, respondents' financial and income levels can be defined, and they can be further categorized to determine the circumstances of poverty under which HHs fall. On the other hand, the average income level of respondents before and after indicates the impact of microfinance programs to uplift the clients' economic status. The average income status before and after were comparatively analyzed in the following paragraphs with the help of SPSS statistical tools (Tables 12 and 13).

A paired sample t-test was used to assess the microfinance program's significance on the clients' average monthly income. The result showed a significant increase in the average monthly income before (Mean = 9384.23 and SD = 7181.00) to after (Mean = 21,014.5 and SD = 7347.57), $t(223) = 26.3$, $p < 0.05$ (two-tailed). The

Table 12 Paired sample statistics for the average monthly income before and after getting microfinance service

Average monthly income		Mean	n	Std. deviation	Std. error mean
Pair 1	Average monthly income of clients before joining the MFI	9384.23	223	7181	4.8087
	Average monthly income of clients after joining the MFI	21014.5	223	7347.57	4.9203

Table 13 Paired sample test for the average monthly income before and after getting the MFI service

Average monthly income	Mean	SD	Std. error mean	95% confidence interval of the difference		T	Df	Sig (2-tailed)
				Lower	Upper			
Average monthly income of clients before joining the MFI Average monthly income of clients after joining the MFI	11,630.26	6602.9	442.16	10,758.89	12,501.64	26.3	222	0

mean increase in the test score was 11,630.26, with a 95% confidence interval ranging from 10,758.89 to 12,501.64.

Similarly, comparing through the paired sample t-test, the calculated P-value (0.000) is lower than the tabulated value at 5% confidence interval. Therefore, there is a significant difference between average monthly income before and after getting microfinance services. A significant change was brought in the HHs of the clients regarding annual average income due to the introduction of microfinance.

4.13 Satisfy with the Interest Charge of Microfinance

During the survey respondents' opinions regarding the interest charged by the MFI to their clients. 201, i.e., 90.1% out of 223 respondents, are satisfied with the interest charge, while only a few (9.9%) respond with an unsatisfactory opinion.

4.14 Views of Clients Towards Service Provided by Microfinance Institutions

The survey showed that 15 represents the interpretation given by households on services and facilities provided by MFIs after being involved in it. The majority of respondents, i.e., 218 or 97.8%, have positive responses on MFI services, while only 1 have negative feelings, and 4 out of 223 are neutral.

4.15 Attitude Toward Society After Involving the MFI Program

The survey also showed that 98.2% of the respondents have a positive attitude towards the MFI program after being involved in it. Very few, i.e., 2 out of 223 respondents, have a negative attitude towards society after involving the MFIs, and only 0.9% of respondents have neutral feelings.

5 Major Findings of the Study

The study showed that the majority of the clients aim to join microfinance institutions to empower their existing economic status. In addition, most clients improved their monthly saving habits after joining the MFIs, as MFIs conduct monthly meetings with their members to collect the deposits and provide loans from the collected deposits. Most respondents used loans from MFIs to six different sectors, with the highest loan of 37.2% being used for small entrepreneurship. Major household sources of income were agriculture before, while after the microfinance intervention, it has diversified to livestock farming and small-scale entrepreneurship. Previously, the majority of land holding size was 2–4 Ropani (35%), which increased after microfinance's involvement, i.e., the majority (41%) of clients now have more than 6 Ropani of land. Initially, about 95% of clients were fully dependent on fuel wood from the forest, and now 45% of them use fuel wood. LPG and electricity users have increased after their involvement in microfinance. Previously, only 16% of the population was able to fulfill their food requirement throughout the year from the crops they have cultivated. Still, now, it has increased to 75% due to the increment in land holding size and diversified income sources. The majority of people were asked traditional Dhami chakra if they had health-related issues; their thoughts have changed now, and they started to visit small health units and well-equipped hospitals due to the awareness program, education level, and increase in income through MFI. About 98% of the respondents got an education from government schools. Still, due to an increase in income and MFI services, they started to join and send their children to private schools, decreasing the percentage of going to government schools by 37%. Before, a majority (91%) of clients had mud-thracked roof houses, which has been reduced to 8% after the MFI intervention. Now, they are diverted to mixed-earthquake-friendly houses and RCC houses. Almost all people are satisfied with the interest charged by microfinance, and they perceive their services positively. Microfinance is one of the important tools to reduce the country's poverty level.

6 Conclusion

Microfinance provides a beneficial and efficient function in empowering clients in the study region. Training programs and a range of credit products enhance farming skills. Additionally, micro-enterprises have enabled members to transition from traditional agriculture to income-generating activities like crop cultivation and animal husbandry, resulting in increased profitability. The results consistently demonstrate the beneficial influence of MFIs on the economic effects of microfinance in the studied region. The rural population of Neelkantha Municipality experienced a major shift in their traditional agricultural activity towards income-generating enterprises, thanks to the financial help and skillful training provided by MFIS. The study's overarching finding is that the degree of women's empowerment inside the home is deemed sufficient.

The findings of this study differ from those of [7] since it determined that there were no sufficient economic consequences in terms of asset development and self-employment. Sohn and Ume [13] discovered a statistically significant influence of microfinance on various aspects such as education, savings behavior, health, household income, food consumption, ownership of household assets, and more. The findings indicate that advancing microfinance is an effective means to enhance welfare and alleviate poverty in Pakistan. The outcome of the findings aligns with the conclusions of my investigation. Shakya found that the microfinance program effectively raised income levels by facilitating the establishment of businesses. The acquired result aligns with the findings of this investigation. This study supports the findings of [2] regarding the impact of microfinance on addressing inadequate housing. The study reveals a significant rise in the number of clients with houses constructed using a combination of earthquake-resistant materials, rising from 4 to 70%. Additionally, 22% of the houses were built utilizing reinforced concrete construction. The study conducted by [1] found that the economic condition of borrowers has consistently improved. The results of this study are consistent with previous findings. Kasali et al. [9] determined that certain explanatory variables, such as microfinance loans, gender, living standard, health standard, marital status, and household size, were statistically significant in determining the effectiveness of microfinance loans in reducing poverty among beneficiaries in Nigeria. This finding suggests that similar factors may also be relevant in the context of Nepal.

References

1. Adhikari DB, Shrestha J (2013) Economic impact of microfinance in Nepal: a case study of the Manamaiju village development committee, Kathmandu. Econ J Dev Issues 15(16):1–2
2. Agrawal M (2016) Role of microfinance in empowering Indian women. In: 5th international conference on research in humanities, sociology and education 5:26–30
3. Atreya K, Rimal NS, Makai P, Baidya M, Karki J, Pohl G, Bhattarai S (2023) Dalit's livelihoods in Nepal: income sources and determinants. Environ Dev Sustain 25(11):12629–12657. https://doi.org/10.1007/s10668022025822

4. Bhattacharjee A, Anadón J, Lohman D, Doleck T, Lakhankar T, Shrestha B, Thapa P, Devkota D, Tiwari S, Jha A, Siwakoti M, Devkota N, Jha P, Krakauer N (2017) The impact of climate change on biodiversity in Nepal: current knowledge, lacunae, and opportunities. Climate 5(4):80. https://doi.org/10.3390/cli5040080
5. Dhungana BR (2023) Perceived impact of microfinance on livelihood improvement in Kaski district of Nepal. Interdiscip J Innov Nepal Acad 2(1):81–95. https://doi.org/10.3126/idjina.v2i1.55968
6. Dhungana G, Pal I, Bhattarai PC (2023) Indigenous knowledge of Chepang in disaster risk reduction. In: Pal I, Kolathayar S, Islam T, Mukhopadhyay A, Ahmed I (eds) Proceedings of the 2nd international symposium on disaster resilience and sustainable development: disaster risk science and technology, Springer Nature, Singapore 2:243–256. https://doi.org/10.1007/9789811962974_18
7. Dulal S (2010) Impact of microfinance program in socio-economic empowerment of women in Nepal (Master's Thesis), Shankar dev campus, Tribhuvan University, Kathmandu, Nepal
8. Joseph OO, Kingsley O (2016) Statistical approach to the link between internal service quality and employee job satisfaction: a case study. Am J Appl Math Stat 4(6):178–184. https://doi.org/10.12691/ajams463
9. Kasali TA, Ahmad SA, Ean LH (2015) The role of microfinance in poverty alleviation: empirical evidence from South-West Nigeria. Asian Soc Sci 11(21). https://doi.org/10.5539/ass.v11n21p183
10. Lamichhane BD, Lama PB (2023) Challenges and prospects of Nepalese microfinance program: a mini review. Nepal J Manag Res 3(1):41–49. https://doi.org/10.3126/njmgtres.v3i1.57914
11. Neelkantha Municipality (2020) Municipal profile, pp 1–25. https://www.neelakanthamun.gov.np/sites/neelakanthamun.gov.np/files/Neelakantha-Profile%28Male%2CFemale%2CEthnicity%20and%20Eductaion%29_0.pdf
12. Pant CR, Bashyal S, Pant KR (2023) Role of remittance in GDP growth of Nepal. AMC J (Dhangadhi) 4(1):45–51. https://doi.org/10.3126/amcjd.v4i1.58832
13. Sohn W, Ume L (2019) The impact of microfinance on poverty alleviation: the case of Pakistan. J Asian Dev 5(3):16–36. https://econpapers.repec.org/RePEc:mth:jad888:v:5:y:2019:i:3:p:1636
14. World Bank (2017) Microfinance and economic development. https://documents.worldbank.org/pt/publication/documents--reports/documentdetail/107171511360386561/microfinance-and-economic-development

The manufacturer's authorised representative in the EU is Springer Nature Customer Service Centre GmbH, Europaplatz 3, 69115 Heidelberg, Germany. If you have any concerns regarding our products, please contact ProductSafety@springernature.com

Printed and bound by CPI Group (UK) Ltd, Croydon, CR0 4YY
01/10/2026
02225922-0007